World Sustainability Series

Due to its scope and nature, sustainable development is a matter which is very interdisciplinary, and draws from knowledge and inputs from the social sciences and environmental sciences on the one hand, but also from physical sciences and arts on the other. As such, there is a perceived need to foster integrative approaches, whereby the combination of inputs from various fields may contribute to a better understanding of what sustainability is, and means to people. But despite the need for and the relevance of integrative approaches towards sustainable development, there is a paucity of literature which address matters related to sustainability in an integrated way.

Notes on the quality assurance and peer review of this publication

Prior to publication, the works published in this book are initially assessed and reviewed by an in-house editor. If suitable for publication, manuscripts are sent for further review, which includes a combined effort by the editorial board and appointed subject experts, who provide independent peer-review. The feedback obtained in this way was communicated to authors, and with manuscripts checked upon return before finally accepted. The peer-reviewed nature of the books in the "World Sustainability Series" means that contributions to them have, over many years, been officially accepted for tenure and promotion purposes.

Tony Wall · Laís Viera Trevisan ·
Walter Leal Filho · Adam Shore
Editors

Sustainability in Business Education, Research and Practices

Springer

Editors
Tony Wall
Liverpool Business School
Liverpool John Moores University
Liverpool, UK

Walter Leal Filho
HAW Hamburg
Hamburg, Germany

Laís Viera Trevisan
HAW Hamburg
Hamburg, Germany

Adam Shore
Liverpool Business School
Liverpool John Moores University
Liverpool, UK

ISSN 2199-7373 ISSN 2199-7381 (electronic)
World Sustainability Series
ISBN 978-3-031-55995-2 ISBN 978-3-031-55996-9 (eBook)
https://doi.org/10.1007/978-3-031-55996-9

This Springer imprint is published by the registered company Springer Nature Switzerland AG
The registered company address is: Gewerbestrasse 11, 6330 Cham, Switzerland

Paper in this product is recyclable.

Preface

Sustainability in business education focuses on integrating principles of sustainable development into teaching, research, and practice within business schools. It aims to prepare current and future leaders to understand and address the environmental, social, and economic challenges facing the world. This involves teaching about sustainable business practices, corporate social responsibility, environmental management, and ethical decision-making. It also often includes hands-on projects and partnerships with organisations, to apply sustainability concepts in real-world contexts. The goal is to create a new generation of business leaders who prioritise long-term well-being and responsible management over short-term gains.

This book presents a wide range of perspectives, showcasing how sustainability is being implemented in business education. Departing from a specialist workshop organised by the European School of Science and Research (ESSSR) and hosted by Liverpool Business School (part of Liverpool John Moores University) in the UK, the book compiles experiences from various experts and institutions, which showcase how the principles of sustainable development are being taken into account and implemented across business schools.

Thanks to its design and the contributions by experts from various areas, this book provides a welcome contribution to the literature on business education and sustainable development, and we hope it may inspire further works in this field. Indeed the book has already led to additional articles, symposia, and projects underpinned by international collaborations. We thank the authors for sharing their knowledge and their experience by means of their chapters, and those colleagues who have contributed to this book by assisting with peer-review.

Liverpool, UK — Tony Wall
Hamburg, Germany — Laís Viera Trevisan
Hamburg, Germany — Walter Leal Filho
Liverpool, UK — Adam Shore
Summer 2024

Contents

Social Value Impact in Business Education: Beyond Ethos, Towards Principles, Methods, and Technical Practice

Tony Wall, Adam Shore, Ann Hindley, Jan Brown, Konstantina Skritsovali, and Madeleine Stevens

Abstract Business education is increasingly integrating sustainability and moving towards the responsible management agenda, with the rise of the United Nations' Principles of Responsible Management Education initiative. This includes reference to the United Nations' Sustainable Development Goals, although its reach remains limited in terms of the operations and culture of business schools. Empirical evidence continues to highlight that economic and financial ideologies and considerations often usurp wider issues of sustainability including social impacts. This study draws from an international expert inquiry into the understanding and use of social value and social impact principles and methods in business education and involved stakeholders across 16 countries. The findings highlight that whilst there are practices aligned with the broad notion of social value and social impact, there needs to be a more nuanced understanding of social value impact methodology and its technical methods and tools. This paper therefore outlines eight key principles and associated methods of social value methodology as a basis to prompt new directions, and an urgent call for more holistic thinking across the university ecosystem.

T. Wall (✉) · A. Shore · A. Hindley · J. Brown · K. Skritsovali · M. Stevens
Liverpool Business School, Liverpool John Moores University, Redmonds Building, Brownlow Hill, Liverpool L3 5UG, UK
e-mail: T.Wall@ljmu.ac.uk

A. Shore
e-mail: A.P.Shore@ljmu.ac.uk

A. Hindley
e-mail: A.M.Hindley@ljmu.ac.uk

J. Brown
e-mail: J.Brown3@ljmu.ac.uk

K. Skritsovali
e-mail: K.Skritsovali@ljmu.ac.uk

M. Stevens
e-mail: M.Stevens@ljmu.ac.uk

T. Wall et al. (eds.), *Sustainability in Business Education, Research and Practices*, World Sustainability Series, https://doi.org/10.1007/978-3-031-55996-9_1

Keywords Social value · Principles of Responsible Management Education · Social impact · Impact co-creation

1 Introduction

Business education has been subject to critique for decades (Starkey and Madan 2001; Pfeffer and Fong 2002) on the grounds of questionable ethicality (Ghoshal 2005), questionable rigor (Tushman et al. 2007) questionable relevance (Knights 2008), questionable actionability (Bartunek and Egri 2012), questionable impact on society (Pettigrew and Starkey 2016), or its 'nonsensical' nature (Bamberger 2020). Even within the first few months of 2023, news reports claim "Survey portrays field as especially dangerous case of obsession with journal metrics, although chief accreditor promises aggressive bid on societal awareness" (Basken 2023: 1). Responding to these ongoing criticisms, business education has increasingly integrated the sustainability and the responsible management agendas (Durepos et al. 2020), especially with the rise of the United Nations' Principles of Responsible Management Education initiative (Greenberg and Hibbert 2022).

Such initiatives have promoted the use of the United Nations' Sustainable Development Goals as a pedagogic device to design educational approaches, methods, and interventions, although its reach remains limited in terms of impacting the operations and culture of business schools (see for example (Willness et al. 2022). Empirical evidence continues to highlight that economic and financial ideologies and considerations often usurp wider issues of sustainability including social impacts (Filho et al. 2019). This is not surprising given that the internal structures and cultures of business schools (as with organisations more generally) shape and influence the way its staff and students think and act (Akrivou and Bradbury-Huang 2015). It also highlights the mode of engagement with stakeholders which is intimately *relational* to ensure a direct and close link with changes that are valued by those stakeholders (Anderson et al. 2017; British Academy of Management 2021; MacIntosh et al. 2017).

Alongside this *relational* turn in business education, social value and social value impact management practice have become increasingly popular in the accounting and accountability practices of public services. Here, driven by various legislative changes across countries (such as the UK), public organisations were required to demonstrate an understanding of their decisions on their communities. Specifically, the adoption of social value was a response to contemporary decision making in organisations where "All too often key decisions about resources and policies are made using a limited economic concept of value, which fails to consider important effects on people and the environment" (Social Value International 2021) p1). In contrast, 'social value' is "the value that *stakeholders experience through changes in their lives*. Some, but not all of this value is captured in market prices." (ibid: p3, emphasis added). Such processes involve making more diverse forms of value visible because "it relates to outcomes experienced by people who have little or no power in decision making" (ibid: p3). Similarly,

> An account of social value is a story about the changes experienced by people. It includes qualitative, quantitative and comparative information, and also includes environmental changes in relation to how they affect people's lives. (ibid: p3).

As such, the management of social value involves identifying, using, and reporting social value, typically adopting existing methods from the fields of "social accounting and audit, sustainability reporting, cost benefit analysis, financial accounting, and evaluation practice" (ibid: p3). International best practice has synthesized key principles in professional and technical standards and guidance against eight key principles (see (Social Value International 2021) and Table 2 which summarises the key principles and methods). Empirical work, however, suggests that social value assessments or impact assessments have not yet been adopted in the context of business schools and universities more widely, with broad sustainability reporting generally still at an early stage (Alonso-Almeida et al. 2015; Gutiérrez-Goiria et al. 2021). However, this is not just the case for business schools or universities; the lack of agreed standards for nonfinancial reporting within and across industries remains a key barrier to widespread and rapid adoption (Amiano Bonatxea et al. 2022; Turzo et al. 2022).

This study therefore examined social value and social value impact management practice in business education. It draws from an international expert inquiry into the understanding and use of social value and social impact principles and methods in business education, and involved stakeholders across 16 countries. The findings highlight that whilst there are practices aligned with the broad notion of social value and social impact, there needs to be a more nuanced understanding of social value impact methodology and its technical methods and tools. This paper outlines eight key principles and associated methods of social value methodology as a basis to prompt new directions, and an urgent call for more holistic thinking across the university ecosystem.

This chapter is organised as follows. The next section outlines the methodological approach to this study, which involves expert inquiry sessions with an appreciative frame and participants around the world. The findings of implementing the method are then reported in the following section. This provides the platform for highlighting and showcasing social value and impact management practices to stimulate future practices. The final section is the conclusion, with a call to action.

2 Methods

This study draws from an international expert inquiry into the understanding and use of social value and social impact principles and methods in business education. In expert inquiry, people with a working knowledge of a situation or topic are invited together to explore issues, questions, and working solutions (Slocum-Bradley 2003). As an exploratory mode of research (rather than statistically representative), the aim was to explore the focal research question: *to what extent is social value impact management used in universities?*

The research team facilitated two expert inquiries (typically 4 h) with participation from 120 people including: multiple stakeholders groups (universities, social enterprises, private businesses, professional bodies), 16 countries (UK, Ireland, USA, Canada, China, Australia, New Zealand, India, Italy, France, Germany, Belgium, Taiwan, Japan, Russia, and South Africa), multiple disciplines (business, education, engineering, environmental sciences, health, arts), and at multiple levels (lecturers, professors, directors, vice chancellors). This closeness to practice not only underpins the strength of validity in exploratory studies, but also provides a platform for change and impact beyond the research activity (Bartunek and Egri 2012; Stokes and Wall 2017; MacIntosh et al. 2017). The participants, in effect, were people in practice who could institute and drive change beyond the expert inquiry sessions.

In addition to this exploratory mode of research, an appreciative frame was adopted which meant exploring the (a) current practices and resources which were relevant to the research question as well as (b) areas that could be positively developed (Cooperrider and Whitney 2011). Through an appreciate lens, these were analysed into two broad themes of *affirmative* aspects (those aspects or practice that were already present in practice and that positive enabled social value management) and *developmental* aspects (those aspects of practice that were already present in practice and that needed some development to enable social value management). These two areas were examined across (1) university practices and (2) business school practices (there was further time imagining new curricula using the principles of social value, but they are not the focal point for this study). Table 1 highlights the key prompts used in the expert inquiry sessions.

The expert inquiry events were recorded and fieldnotes were taken by facilitators in each group (consisting of 3–7 experts). Data were then analysed through an iterative thematic analysis procedure (Braun et al. 2022). Ethical review and approval were granted by Liverpool Business School, Liverpool John Moores University.

3 Results

Affirmative aspects

In terms of the affirmative aspects of social value impact management that are present in business education, findings highlight the early development of practice. It was recognised that business schools (and wider universities) are increasingly *positioning themselves in relation to the 'public good'*, and in relation to their stakeholders. Bodies and initiatives such as the United Nations' Principles of Responsible Management initiative were seen to have been largely instrumental in enabling this shift. As such, this was seen as a strategic, foundational platform for further social value impact management (to then be able to take a strategic approach to all of the social value principles).

Table 1 Prompts related to social value

University practices	
Aspect of social value	Examples of university practices
Identifying social value	Understanding what difference the university or its departments has made from the perspective of its stakeholders—surveys, consultations, co-creation. For example, students completing surveys that aim to capture their opinion on their experience and learning on social value and sustainability
Using social value to make decisions	Using social value assessments to inform decisions, such as the design of programmes and the investment in new capital builds. For example, capital projects building sports facilities and social spaces; design of programmes such as integrating sustainability, Carbon Literacy, digital skills; all benefitting individuals for employability but also in society
Reporting social value	Sharing its value assessments (and how it is using them) for accountability to other bodies or communities. This might include financial and nonfinancial or integrated reporting on an annual basis. For example, creating an open access index of assessment that could benefit other bodies or communities
Building capacity in all of the above	Investing in skills and processes to identify, use and report its social value and social value practices. For example, providing training to staff members to introduce them to social value and social value practices
Business school educational practices	
Aspect of social value	Examples of university practices
1. Involve stakeholders (relevant to the learning) to decide what gets valued and how	Working with a wide range of organisational stakeholders from the outset to ensure that multiple perspectives and definitions of value are accounted for
2. Understand changes (relevant to the learning) by articulating and then evidencing how change is created	Co-creating methods to evaluate changes with organisational partners from the outset. Evidence of changes shared between stakeholders and mapped over time
3. Value the things that matter to stakeholders	Creating forums in which stakeholders can clarify and communicate their definitions of value without judgement. Multiple forms of value acknowledged and celebrated
4. Only include what is material, i.e. those impacts that would make a difference to an organisational decision	Developing and using assessment methods that clearly define and evaluate the impacts that make a difference to an organisational decision

(continued)

Table 1 (continued)

University practices	
Aspect of social value	Examples of university practices
5. Do not over claim: ensuring assessments relate specifically to the activities being evaluated	Ensuring that effective assessment methods are created and used to ensure over claiming does not occur
6. Be transparent about how judgements of social value are made	Developing and communicating judgement of social value methods that are relatable and can be understood by all stakeholders
7. Verify their results	Checking and collaborating with stakeholders to ensure results are correct and agreed upon
8. Be responsive	Developing a sense of co-creation, collaboration and community with all stakeholders at all times

Source Adapted and extended from Social Value International (2021)

Similarly, business schools were also seen to have *extensive relationships with their network of stakeholders*. This was facilitated by a wide range of existing engagements including arranging external talks within the curriculum, volunteering opportunities for students, and work-based learning or work-integrated learning placements for students in local organisations. As such, this meant that business schools were often in a strong position to involve their stakeholders (principle 1) and to potentially be responsive to their needs (principle 8).

There was a view that business schools already provided *learning connected to social value* in a broad sense. In terms of implicit curriculum, this was often linked to curriculum in business ethics, corporate social responsibility, equality and diversity (and often the thematic areas of the 17 Sustainable Development Goals). Whilst this did not often provide the technical details and practices of social value impact management, participants did consider this to be relevant foundational learning and knowledge for involving stakeholders and their value perspectives, and understanding change (principles 1, 2, and 3).

Linked to the extensive networks of business schools, and learning connected to social value, participants also highlighted how business schools had already *established processes for involving stakeholders* in some parts of the design and delivery of curricula. This often took the form of engaging stakeholders in the design of new, or revamping of existing, courses and programmes as part of an employer engagement agenda. Within the delivery of curricula, models of work-based learning or work placement learning are long standing models which work to the value needs of organisations. Similarly, case studies involving stakeholders are typically adopted which create opportunities to compare and contrast value propositions.

Developmental aspects

Whilst there seemed to be the early foundational aspects of social value impact management in place in business schools, there were various areas that need visible

Table 2 Summary of social value principles and exemplar teaching practices (Wall et al. 2022)

Social value processes (Adapted from Social Value International 2021)	Example pedagogic practices and infrastructures	Further information
1. Involve stakeholders (relevant to the learning) to decide what gets valued and how	Time/space for students to ask local businesses what's important from their perspective: invited talks; student conferences; advisory boards; joint events; employment fairs to prompt dialogue; business clinics involving student participation	https://www.socialvalueint.org/principle-1-involve-stakeholders
2. Understand changes (relevant to the learning) by articulating and then evidencing how change is created	Student-led focus groups with organisations that map out **chains of impacts**, followed by the use of research methods to collect evidence of the **depth** and **duration** of changes Social value impact projects are optional/mandatory in programmes, with specific training in how to map and measure changes	https://www.socialvalueint.org/principle-2-understand-what-changes
3. Value the things that matter to stakeholders	Teaching '**weighted approaches**', '**cost-based**', '**revealed preference**', '**stated preference**', '**willingness to pay**', or '**wellbeing valuation**' techniques Exposure, mentoring and training in each of these (beyond economics teaching) Practice for organisational clients (especially social enterprises who will often be engaged in this work to acquire new governmental or public funding) Student involvement in creating project outcomes with stakeholders to understand what is defined as value from all stakeholders' perspectives from the start	https://www.socialvalueint.org/principle-3-value-the-things-that-matter

(continued)

Table 2 (continued)

Social value processes (Adapted from Social Value International 2021)	Example pedagogic practices and infrastructures	Further information
4. Only include what is material, i.e. those impacts that would make a difference to an organisational decision	See principle 1. Time/space for students to ask local businesses what's important from their perspective Ongoing relationships with key stakeholders help identify important decision areas (also see (MacIntosh et al. 2017)	https://www.socialvalueint.org/principle4-only-include-what-is-material
5. Not overclaim: ensuring assessments relate specifically to the activities being evaluated	Teaching how to find relevant **baselines, trends**, and **benchmarks** Linked to: understanding statistics, and the teaching and mindset development related to evidence-based management, research ethics, and decision making ethics and moral reasoning	https://www.socialvalueint.org/principle-5-do-not-overclaim
6. Be transparent about how judgements of social value are made	Teaching how to **record and report** decisions in ways that are accessible for audiences (e.g. use of plain language, formatting, use of executive summaries/foot-end notes/appendices) Creating the philosophy of co-creation and partnership so that trust is built between all stakeholders that allow judgements of social value to be expressed with confidence	https://www.socialvalueint.org/principle-6-be-transparent
7. Verify their results	Teaching how to **evaluate the trustworthiness** of reports Embed the validation or accreditation of skills related to principles 1–8	https://www.socialvalueint.org/principle-7-verify-the-result
8. Be responsive	Teaching/facilitating the development of a stakeholder-engagement **mindset**/timely response to organisational needs	https://www.socialvalueint.org/principle-8-be-responsive

Source Adapted and extended from Social Value International (2021) development. The first was related to *embedding social value and impact thinking throughout curricula*, at different levels and contexts of business education. This included, for example, full time undergraduate provision, as well as executive education, to reach students with varying levels of influence. This was to encapsulate not just *knowledge, but a 'stakeholder first' mindset* which highlights and inculcates an understanding of the importance of social value.

Teaching methods and opportunities were also seen to need greater direct and explicit linkage to social value and impact management knowledge and skills. This included opportunities within the curricula, but also extra-curricular (such as through student societies and clubs), and indeed, could use more co-creation approaches that have already been established. This could go further than the revamping of courses to extend across all processes from course conception, evaluation, academic assessment, academic feedback, and even teaching. Negotiated, vocational or work-based learning approaches are now well established and can be further embedded across full programmes as modes of study rather than units of study.

Business schools were also seen to need to *move from a primary focus on outputs, to include (or shift to) outcomes*. They were seen to be predominantly focused on measuring the activities (i.e. claiming outputs as impact) rather than the changes experienced by stakeholders (especially students) (i.e. claiming outcomes as impact). The latter of these were seen to be typically measuring student satisfaction of the activities rather than the changes as a result (e.g. upward mobility of students). For example, the National Student Survey in the UK in 2023, asks questions such as 'How good are teaching staff at explaining things', 'How often do teaching staff make the subject engaging?', and 'How often is the course intellectually stimulating?'. Similarly, there was a call for business schools not to overclaim or underclaim their impacts.

Connected with student satisfaction, was the perennial issue of student assessment and feedback (one of the most challenging areas related to the National Student Survey in the UK). Whilst participants recognised established systems for *student assessment and feedback* which are often required for international accreditation bodies (e.g., the Association to Advance Collegiate Schools of Business), these were not seen as particularly aligned with social value and impact management. Rather than focused on assessing the valued changes experienced by the student in ways that further promote change, it is often seen as superficial and pre-established by academics (i.e. those with the greatest power in the educational relationship).

It was also appreciated, however, that business schools needed to *balance needs across its stakeholders*, given the multiple challenging agendas they have to deliver (e.g. excellence in teaching, service, research, engagement, and impact). This however, needed business schools to have a clear understanding (and mapping) of their stakeholders to be able to make such decisions and prioritise their activities and investments. Business school leads for their United Nations' Principles of Responsible Management commitments were seen as potentially well placed to influence such understandings, should they have a place at the business school leadership team table.

4 Discussion: Moving Beyond Ethos, Toward Social Value Methods and Technical Practice

The results of this study are consistent with other studies which claim the early (and therefore limited) adoption of social value assessments or impact assessments in business schools and universities (Alonso-Almeida et al. 2015; Gutiérrez-Goiria et al. 2021; Amiano Bonatxea et al. 2022; Turzo et al. 2022). More specifically, however, this study finds that there are growing foundational, strategic intents and emerging pedagogic infrastructures (e.g. connection and engagement with stakeholder networks) that do align with social value impact management at the level of *ethos*, but not necessarily to the level of *technical practice*. Furthermore, whilst it was not clear how social value was identified, used, and reported in practice, participants could identify the potential adaptations to current practice to *develop* such technical practice in specific, actionable ways (as outlined in the Results section above).

Whilst these findings do reflect some of the ongoing criticisms of business education (Bartunek and Egri 2012; Pettigrew and Starkey 2016; Basken 2023), they also provide evidence of the ability and willingness of stakeholders internal and external to higher education institutions to develop technical practice. These insights extend other empirical work which highlights consistency in standards and ideological salience as barriers to adoption (Filho et al. 2019; Willness et al. 2022). Here, our results suggest that whilst there are significant gaps in technical practice, or the know-how to deliver social value impact management within and through business education, there is an appetite for change to develop it. Previous research has not identified this as a positive resource for driving change in business schools.

The ideas for the development of practice do reflect principles and strategies for engaging stakeholders in *research* such as involvement in the design of activity and its dissemination (Reed 2018). However, the participant's suggestions for developments were clearly framed and articulated in the context of business *education* rather than research or commercial activity. As such, they more closely reflected the intentions of socially-driven contributions often described in forms of *project-based learning* and *engaged scholarship*, where learning is designed for and in the service communities (Belwal et al. 2020; Koendjbiharie 2020; Lycko and Galanakis 2021).

Yet the extant literature has still to articulate teaching and learning in business schools which directly formulate routes to change, and measure the value generated (or destroyed) *from a stakeholder perspective*, i.e. social value.

In response to this gap, the development of technical practice in the field of social value impact management can be described through eight principles of international best practice (Social Value International 2021). These are highlighted in Table 2, with additional developmental guidance for technical practice. These provide a platform for further development of business education and business schools, in terms of understanding and auditing current practices, and identifying specific areas for professional and infrastructural development.

5 Conclusions

In conclusion, the advancement of embedding social value in business education has become increasingly important in recent years. As businesses and corporations play a significant role in shaping society and the environment, it is crucial for business leaders and future managers to have an understanding of the impact of their actions. By incorporating social value into the business curriculum, students are equipped with the knowledge and skills to make informed decisions that consider the well-being of stakeholders and society as a whole. This shift in business education not only benefits society but also contributes to the long-term success and sustainability of the business. It is therefore imperative that business schools continue to evolve their curricula to keep pace with these changing times.

As the *relational* turn in business school and business education policy and practice progresses (Anderson et al. 2017; Rigg et al. 2021; British Academy of Management 2021; MacIntosh et al. 2017), the lack of technical awareness around social value measurement and impact management is a direct barrier to its implementation in practice (Ahmadi and Vogel 2023). This technical practice is about the practical know-how of how *the changes experienced by stakeholders* are understood, measured, and used in decision making. Such technical expertise provides the underpinning capacity to understand how business schools and the educational programmes they provide *deliver* value in the communities they serve, and how to better deliver the types of value 'valued' in society. As such, social impact management is not just a statement of strategic intent, but is, rather, a realised organisational learning capability to better deliver value in society. This chapter calls for the development of practice and research in the field, with a view of fully embracing the relational turn, and countering some of the criticisms still pointing at business schools and business education.

References

Ahmadi A, Vogel B (2023) Knowing, but not enacting leadership: navigating the leadership knowing-doing gap in leveraging leadership development. Acad Manag Learn Educ 22(3):507–530. ISSN 1537-260X. https://doi.org/10.5465/amle.2020.0534

Akrivou K, Bradbury-Huang H (2015) Educating integrated catalysts: transforming business schools toward ethics and sustainability. Acad Manag Learn Educ 14:222–240

Alonso-Almeida M, del Mar F, Marimon FC, Rodriguez-Pomeda J (2015) Diffusion of sustainability reporting in universities: current situation and future perspectives. J Clean Prod 106:144–154

Bonatxea A, Iratxe J-G, Francisco M-D, Sianes A (2022) Is the global reporting initiative suitable to account for university social responsibility? Evidence from European Institutions. Int J Sustain High Educ 23:831–847

Anderson L, Ellwood P, Coleman C (2017) The impactful academic: relational management education as an intervention for impact. Br J Manag 28:14–28

Bamberger P (2020) Response to Dennis Tourish's "The Triumph of nonsense in management studies." Acad Manag Learn Educ 19:240–242

Bartunek JM, Egri CP (2012) Introduction: can academic research be managerially actionable? what are the requirements for determining this? Acad Manag Learn Educ 11:244–246

Basken, P (2023) Most business school research 'lacks real-world relevance. Times Higher Education. Accessed 1 Feb

Belwal R, Belwal S, Binti Sufian A, Al Badi A (2020) 'Project-based learning (PBL): outcomes of students' engagement in an external consultancy project in Oman. Education + Training 63: 336–59

Braun V, Clarke V, Hayfield N (2022) 'A starting point for your journey, not a map': Nikki Hayfield in conversation with Virginia Braun and Victoria Clarke about thematic analysis. Qual Res Psychol 19:424–445

British Academy of Management (2021) A shared vision of a business and management knowledge ecosystem. Accessed 9 Feb

Cooperrider D, Whitney D (2011) Appreciative inquiry: a positive revolution in change. San Francisco, CA, Berrett-Koehler

Durepos G, Maclean M, Alcadipani R, Cummings S (2020) Historical reflections at the intersection of past and future: celebrating 50 years of management learning. Manag Learn 51:3–16

Filho WL, Doni F, Vargas VR, Wall T, Hindley A, Rayman-Bacchus L, Emblen-Perry K, Boddy J, Avila LV (2019) The integration of social responsibility and sustainability in practice: exploring attitudes and practices in Higher Education Institutions. J Clean Prod 220:152–166

Ghoshal S (2005) Bad management theories are destroying good management practices. Acad Manag Learn Educ 4:75–91

Greenberg D, Hibbert P (2022) Beyond legitimacy: a bold agenda for MLE scholarship. Acad Manag Learn Educ 21:161–166

Gutiérrez-Goiria J, Amiano-Bonatxea I, Sianes A, Vázquez-De Francisco MJ (2021) Reporting the Social Value Generated by European Universities for Stakeholders: applicability of the global reporting initiative model. Front Psychol 12

Knights D (2008) Myopic Rhetorics: reflecting epistemologically and ethically on the demand for relevance in organizational and management research. Acad Manag Learn Educ 7:537–552

Koendjbiharie SR (2020) The distinct value of humanities students to employers in student consultancy projects. Ind High Educ 34:247–262

Lycko M, Galanakis K (2021) Student consultancy projects playbook: learning outcomes and a framework for teaching practice in an international entrepreneurial context. Int J Manag Educ 19:100285

MacIntosh R, Beech N, Bartunek J, Mason K, Cooke B, Denyer D (2017) Impact and management research: exploring relationships between temporality Dialogue, Reflexivity and Praxis'. Br Acad Manag 28:3–13

Pettigrew A, Starkey K (2016) From the guest editors: the legitimacy and impact of business schools—key issues and a research agenda. Acad Manag Learn Educ 15:649–664

Pfeffer J, Fong CT (2002) The end of business schools? Less Success than Meets the Eye. Acad Manag Learn Educ 1:78–95

Reed MS (2018) The research impact handbook. Huntly, Aberdeenshire, Fast Track Impact

Rigg C, Ellwood P, Anderson L (2021) Becoming a scholarly management practitioner—Entanglements between the worlds of practice and scholarship. Int J Manag Educ 19:100497

Slocum-Bradley N (2003) Participatory methods toolkit: a practitioner's manual. In: Brussels: ing Baudouin Foundation and Flemish Institute for Science and Technology Assessment

Social Value International (2021) The principles of social value. Accessed 14 Dec

Starkey K, Madan P (2001) Bridging the relevance gap: aligning stakeholders in the future of management research. Br J Manag 12:S3–S26

Stokes P, Wall T (2017) Research methods. London, Palgrave

Turzo T, Marzi G, Favino C, Terzani S (2022) Non-financial reporting research and practice: lessons from the last decade. J Clean Prod 345:131154

Tushman ML, O'Reilly C, Fenollosa A, Kleinbaum AM, McGrath D (2007) Relevance and rigor: executive education as a lever in shaping practice and research. Acad Manag Learn Educ 6:345–362

Wall T, Sachikonye C, Carline L, Shore AP, Stevens M, Brown J, Skritsovali K, Hindley A, Parasram I, Tyreman S, Westcott C, Harvey R, Carpenter B (2022) Social value at universities: policy and practice guidance. Liverpool

Willness C, Boakye-Danquah J, Nichols D (2022) How Arnstein's ladder of citizen participation can enhance community-engaged teaching and learning'. Acad Manag Learn Educ Forthcoming

From Surviving to Thriving—Leading the Way to Decent Hybrid Workplaces

Ute Franzen-Waschke

Abstract As we transition out of the pandemic, enjoying more flexible workplaces and worktimes, a bright future seems ahead for employees and employers. However, studies show that job types and industries determine choice, that remuneration policies often favour on-site over remote options, and that increasing the flexibility of working hours can affect the workforce's well-being. New skills and behaviours are needed in the use of technology and in leadership, and different aspects, compared to before the pandemic, may influence the career choices of the next-generation workforce. In these new circumstances, Business Schools are invited to adapt their curricula and companies might envision an update of their in-company continuous learning and development programmes to allow for these new workplaces and learning scenarios to become more sustainable. This study adopts an abductive approach to literature review which entwines literature and first-hand experience to identify lessons for business schools and beyond. The findings of this piece call for a more serious prioritisation of Decent Work (SDG8) for equal opportunities, equal pay, and well-being.

Keywords Hybrid work · Remote work · Decent work · Wellbeing · Leadership

1 Introduction

In September 2015, the United Nations General Assembly adopted an agenda for sustainable development with the goal to end poverty, reduce inequality and injustice, and to tackle climate change by 2030. However, by 2020, the pandemic brought shifts to many aspects of our lives. Those shifts were favourable for our climate, as commuting to and from work, transportation and logistics of goods and services,

U. Franzen-Waschke (✉)
Business English and Culture, Frackenstieg 15, 31188 Holle, Germany
e-mail: ute@discover-your-choices.de

Chester Business School, University of Chester, Chester, UK

T. Wall et al. (eds.), *Sustainability in Business Education, Research and Practices*, World Sustainability Series, https://doi.org/10.1007/978-3-031-55996-9_2

as well as traveling for business and leisure, were reduced to zero during lockdowns. Those shifts affected education and organisations, because suddenly, students, teachers, professors, and workforces from around the globe were working from home. The phenomenon created by the pandemic was described as "*the biggest experiment of our time*" (Bernstein et al. 2020). Especially for companies and employees it was a huge shift to have most of the workforce work from home. Before the pandemic hit in 2020, there were many industries, sectors, and job types, in which employees had never had the privilege to work from home (Desilver 2020). Overall, this had a short-term positive impact on the reduction of CO_2 emissions (Wołowiec et al. 2022).

These shifts and the changes offered opportunities and challenges at the same time. In the first half of 2020 both workers as well as companies were in survival mode. Working from home allowed both companies and staff to survive economically and mentally giving a focus on work as something that was left from everyone's 'normality' in those unprecedented times of the pandemic (Franzen-Waschke 2021; Singer-Velush et al. 2020). Only in the second half of 2020 and in 2021 it surfaced that workforces were more challenged than one would have thought. Well-being and mental health suffered due to loss of connection, unsustainable workload, and a general sense of loss of control over life and work situations. A 89% decline in workplace well-being was recorded among 1,500 respondents to an online survey during the pandemic in a study conducted by Moss et al. and published by Campbell and Gavett (2021). The fact that these new ways of working are here to stay are undisputed (Spataro 2022). Consequently, there is a need to learn and to adapt new skills and behaviours to navigate and thrive in those new settings long term (Hopkins and Figaro 2021). According to SDG#8 the focus of Decent Work and Economic Growth is to "*Promote sustained, inclusive and sustainable economic growth, full and productive employment and decent work for all.*" (see https://sdgs.un.org/goals/goal8). Post-pandemic, Business Schools and employers around the globe therefore have new opportunities to shape and co-create learning experiences for students and work experiences for workforces, which can help steer towards fulfilling what SDG#8 has been calling for.

This chapter is organized as follows: The introduction is followed by a definition of terminology and its use in the context of this work as well as a brief overview of the Method used. In the main section, the phenomena and paradoxes observed in remote and hybrid workplace settings are explored and juxtaposed with themes identified in the literature reviewed. The crucial role of technology, its availability and accessibility coupled with the complexities around varying human needs towards decent hybrid workplace settings are highlighted before the chapter concludes with recommendations for Business Schools and organisations and their part in helping to make hybrid workplaces more decent in a sustainable manner.

2 Terminology

According to Eurofound (2018) the term *'decent work'* was originally coined by the International Labour Organisation (ILO) in 1999 with the goal to bring more awareness to the fact that work is so much more than being economically safe. Whereas the idea of work providing purpose and fulfilment (Litchfield et al. 2016; Seligman 2018) used to be a 'nice-to-have' enhancement until research emphasised its importance for employee health and well-being. Yet a healthy balance of work quantity and quality must be kept, also with jobs that give us fulfilment and purpose besides a steady income allowing us to feel economically safe. Finally, for people to thrive in the workplace and to experience feelings of satisfaction, also value and appreciation are important to receive (Eurofound 2018). These multitudes of factors have led to an ongoing refinement of the original definition from 1999.

Two decades later, Gardner et al. say that a "*good job*" is a job that is engaging, excellent and ethical (Svicher and Di Fabio 2021). However, experiences and expectations of what makes individuals feel engaged in a job or task are unique and hence vary considerably. Similarly, the definition of what is considered ethical can differ depending on stakeholders and contexts. Blustein et al. (2019) have pointed out five job characteristics that could lead to decent work conditions. These five job characteristics can be summarised as two main domains (1) work fulfilment and (2) well-being (Blustein et al. 2019). Later sections of this chapter will outline the extent to which these new hybrid workplace scenarios can be fulfilling and can support or hinder professional and personal well-being, as well as the level of which economical safety can or cannot be offered in a world that has again become smaller and more competitive.

A *hybrid workplace* in this chapter refers to a workplace in which employees can do their job tasks either (1) together on-site, (2) together-remotely, (3) alone on-site, and (4) alone-remotely (Bayntun-Lees and Cross 2022). Therefore, hybrid work can be a combination of on-site and off-site work. Similarly, remote work can happen from home but also from co-working spaces, and from "anywhere". Consequently, in a hybrid work environment, technology and its use play a more important role than in an exclusively on-site work environment. Here, the role of technology is to help employees stay connected around work and non-work topics. Hence, technology plays a crucial role as an enabler as well as a stressor in these new work environments, and as such, can contribute to a more or a less decent work environment when technology is available and accessible on the one side, and easy or difficult to work with for users on the other side. Section five of this chapter will speak more to that aspect.

3 Method

The method adopted in this chapter is an abductive literature review. An abductive approach was chosen allowing the author to link phenomena observed and first-hand experiences made in her work as an executive and team coach in global corporate work environments to help identify lessons for Business Schools and organisations alike. The abductive approach is intuitive, starting with observations which are underpinned or refuted by theoretical knowledge (Dubois and Gadde 2002; Halecker 2016). The literature reviewed was sourced from online searches in the University of Chester Library, from Google Scholar Alerts, and lists of recommended articles based on the author's profile as a researcher on, e.g. Google Scholar. The World Wide Web was also used as literature in this field is emerging. Searches focused on the years 2020–2023 but were not limited to those years. The years 2020–2023 were particularly relevant for the topic of hybrid work, and at the same time, it was also necessary to consider literature addressing correlated topics in pre-pandemic years.

Search criteria were determined by phenomena observed in the workplace and investigated in this chapter, for example, well-being/wellbeing in hybrid and remote work environments, challenges in hybrid/remote work environments, remote and hybrid leadership, technology, techno stress. Inclusion criteria were peer-reviewed and scholarly articles in English language. Exclusion criteria were set for non-corporate and business sectors, such as health care, medicine, etc. References were examined according to their relevance towards the concepts of decent work and hybrid work. Reviewing literature asserted the many, and at times contradictory viewpoints of how hybrid work is experienced in contemporary workplace settings among the various stakeholders, and thus led to the identification of themes and paradoxes in hybrid work environments which are further elaborated on this chapter.

4 Paradoxes of Hybrid Work Environments

Theme 1: Onsite Versus Remote—Connection Versus Autonomy—Preferences Versus Practicalities

According to a survey conducted by Jennifer Moss, Michael Leiter, Christina Maslach, and David Whiteside among 1,500 HBR readers from 46 different countries in fall 2020, the second half of the first pandemic year had a negative impact on respondents' general and workplace well-being (Campbell and Gavett 2021). This decline was due to the lack of separation between home and work, increased work demands, and disconnection from co-workers (Campbell and Gavett 2021). In 2021, when the pandemic was more under control, and companies demanded their workforce back to the office, this resulted in what Klotz called and is known as the "Great Resignation" in the US (Hempel 2021; Hopkins and Figaro 2021; Kane 2021) and what lead within Europe in some areas to a return to old habits and pre-pandemic mindsets and thus to less equality and choice for possibilities for telework (Milasi et al. 2020). In 2022 organisations and workforces continued to experiment with the 'new normal' to make these new workplace scenarios work for all stakeholders. In 2023 the struggles for employers and employees were still manifold. The temptations on both sides for cherry-picking what is best practice, pursuing primarily unilateral needs and benefits, will not help to find sustainable solutions which would instead require a strategic approach to what is best for the employees' well-being, job satisfaction, engagement and productivity, and companies' economic bottom lines (Delany 2021; Ferreira et al. 2021). Such a strategic approach would require conversations and negotiations between employers and employees to co-create ways to allow for flexibility and autonomy that does not harm either side (Delany 2021; Hopkins and Figaro 2021).

There are proponents and opponents for either concept, on-site or remote/hybrid, and the reasons for the divide are multifaceted; some employees are worried about blurring boundaries and working from home making a negative impact on their work-life balance, as Campbell's and Gavett's (2021) work confirmed for the second half of 2020. Consequently, if work continues to move in at home on a permanent basis more sustainable solutions need to be found to support employees and help them to develop the skills to draw the necessary boundaries and stand up for their needs (Shirmohammadi et al. 2022). Organisations would need leaders who could offer spaces for such conversations from a neutral ground, looking out for solutions that serve both stakeholders (Bayntun-Lees and Cross 2022; Shirmohammadi et al. 2022).

Some opponents are also missing the connections to their co-workers, and again others have better work conditions (equipment, internet connection, etc.) in the workplace than at home (Lundqvist et al. 2022). Apart from the aforementioned personal preferences, cultural norms might also play a role because some prefer to work from home and others would like to return to the workplace on site (Sanhokwe 2022). According to Sanhokwe (2022) the word 'home' is in some cultures more than just the place where we live. It is a place where family gathers and considered private territory, thus calling for a stricter separation between the place we go to for work and the place we gather with our families. Those with a long commute to work and a well-equipped home office with good internet connection; those who are more introverted and can better focus and concentrate in the privacy of their homes, or those who work more independently from the type of work they do, might have stronger opinions on why they would like to continue to work from home and not return to the offices as a standard (Ferreira et al. 2021; Lundqvist et al. 2022). Disabilities and sexual orientation (Dowling et al. 2022) as well as race (Subramanian and Washington 2022) and gender (Tamin et al. 2021) appear to be factors that employees raise in favour of working-from-home arrangements. Many employers have negotiated agreements that allow their workforces to work a few days per week from home and a few days per week from the office (Vogel and Breitenbroich 2020). Evidence suggests that these arrangements are good for those employees, whose jobs allow that choice (Hopkins and Figaro 2021), however, this choice is not unilaterally available to all employees as we will see in the next section.

Theme 2—Choice, Job Crafting, Employability, and Remuneration

Industries and job types among other factors can influence the choice of the workforce and are one example when personal preferences can collide with job or sector practicalities (Milasi et al. 2020). A process engineer in a manufacturing location, for example, can work from home from time to time but not as a default. A resident engineer, who is there to support customers on their sites cannot work from home on a permanent basis either. Not to mention the many front-line, trade, tourism, and healthcare professionals, who do not work in offices, and thus do not have the same choices as workers in other sectors might have. Pre-pandemic, privilege of choice was often reserved for managers, executives, and knowledge workers (Milasi et al. 2020) and therefore was hierarchy and job type dependent. What types of jobs would fall into the category of a knowledge worker is not always clearly defined (Maruta 2012). Nowadays, however, the term of the knowledge worker is often used in the context of hybrid work to differentiate those who used to be referred to as white-collar or indirect workers working in an office from those known as blue-collar or direct workers producing a product. The controversy in the term knowledge worker is not subject of this chapter, but in the context of decent work conditions it highlights that inequality surfaces in new forms in hybrid work environments.

These new forms of inequality in companies are challenging existing paradigms and are bringing tensions to the workplace. Industry and sector as well as job type determine how flexibly one can choose workplace and worktime (Delany 2021). Consequently, the lack of such choices increases the sense of inequality in existing workplace structures within hybrid work environments and thus mitigates positive effects of job crafting. According to Svicher and Di Fabrio (2021), the possibility of job crafting has the potential to make work more decent for the workforce as it caters for greater meaning and self-determination, and it also offers the prospect to align job demands and resources with personal needs and preferences (Bakker and Demerouti 2017; Svicher and Di Fabio 2021). The privilege of choice or the possibility to craft one's job seems like a relevant topic of consideration not only for vulnerable workers, but also for the next-generation workforce when selecting their professions, and therefore is relevant for Business Schools and for organisations when it comes to retention and recruitment of talent.

Adequate remuneration, regardless from where a job is done, combined with the question if there are equal opportunities for promotion along with access to quality job tasks for either group of staff—remote and onsite—are new topics for employer-employee conversations not only in the hiring process (Hopkins and Figaro 2021; Delany 2021; Doyle 2021). Evidence has shown that remuneration and proximity bias bear risks for inequality in hybrid workplaces (Bernstein et al. 2020; Delany 2021; Future Forum 2022), as hybrid and remote work settings have opened doors for companies to threaten their remote workforce with pay cuts (Miller 2021). Here, companies are taking advantage more intentionally of the benefit that staff can be hired anywhere (Doyle 2021). Consequently, remote workers in high-cost countries could face disadvantages compared with remote workers in low-cost countries on a global scale when job positions are to be filled (Miller 2021) or even when work is allocated company internal. The latter has already led to more in-company competition in 2021 and employers taking advantage of savings on payroll costs, simultaneously re-igniting the debate of location-based salaries vs job-based salaries (Doyle 2021; Miller 2021). Remuneration and equal pay therefore continue to be hot topics also in this new work era and call for fair HR policies in organisations, and excellent leadership and communication skills to navigate such difficult conversations.

Theme 3—Skills Development for Leaders and Workforces

Leading at a distance requires not only the acquisition of new skills but also a shift of mindset (Bayntun-Lees and Cross 2022) and an awareness of those biases that might influence a leader's decision. Proximity bias is one example and is commonly known as the assumption that those who we do not see when they work, such as remote workers, are less productive and engaged than those we can see while they are at work (Tsipursky 2022). Proximity bias has been raised in many conversations with employers or leaders when discussing new workplace scenarios," (Future Forum 2022). Lower promotion rates among remote workers despite higher levels of productivity, and the opinion that remote workers can be replaced easier than on-site workers, are factors, according to Tsipursky (2022), for which companies need to raise awareness among their leaders. If leaders notice when proximity bias becomes an obstacle and how to mitigate it when it is unsubstantiated, this will have positive effects on hybrid work culture (Tsipursky 2022).

Additionally, organisations, leaders, as well as remote and hybrid workers are invited to become better at supporting each other (Shirmohammadi et al. 2022), ensuring that everyone can be heard and seen in hybrid meetings by adapting meeting structures and using tools to best suit this new environment and setup (Howspace 2022). Including the entire workforce in, for example, decision-making processes is vital and can be challenging already in traditional settings. Furthermore, leaders will benefit if they learn and further enhance skills and demonstrate behaviours which will make the workforce feel included, heard, and seen, regardless if they are on-site or remote (Howspace 2022; Tsipursky 2022).

Irrespective of where people work, it is beneficial if leaders are competent at creating a space for connections and relationships to be established and maintained (Hopkins and Figaro 2021). Research shows that leaders have always played an important role and have promoted workplace well-being and employee engagement even before the pandemic (Truss et al. 2014). Lundqvist et al. (2022) found out in a more recent study they conducted in Sweden in the public sector that regardless from where people work—at home or from the office—'*supportive leadership*' provided by the closest manager is what positively influences well-being in the workplace. These findings are backed up by Spataro's (2022) five trends that hybrid leaders must *"get right"* in order to be successful, and which include among four others how social capital is built differently in a hybrid working world (Spataro 2022). Delany (2021) even goes as far to say that line managers need to develop *"a sixth sense"* for when workers are struggling or when they are in flow. Conversations, that the author has had, with leaders in corporate environments around the globe underline a certain level of insecurity about how to deal with these new needs their workforces have and with which they are faced on an every-day basis. In the past leaders' primary tasks were to manage and lead their workforces on work topics. Since the pandemic, this mandate has more overtly and permanently been expanded. Not only have the requirements on skills been expanded for organisations and leaders (Delany 2021; Hopkins and Figaro 2021), also the workforce needs to further develop certain skills to make an active contribution to shape more decent hybrid work environments for themselves. Skills that are higher in demand for remote workers are decisiveness, pro-activity, and self-organisation (Delany 2021). Ferreira et al. also highlight the field of technology and the use of technology (Ferreira et al. 2021).

5 The Role of Technology and Its Potential to Promote Decent Workplaces in Hybrid Settings

Ferreira et al. (2021) conducted qualitative research by interviewing 129 professionals between March and September 2020, who already had experience in working remotely before the pandemic. Their research supports findings that certain types of jobs can be easier performed in a remote work environment than others, as well as that younger generations have fewer problems with technology. Technology as the means to communicate and collaborate in a remote or hybrid working world is key to employee productivity, engagement, a feeling of inclusion as well as higher levels of professional well-being (Ferreira et al. 2021; Howspace 2022). In their research Ferreira et al. (2021) also found out that technology was not always used in the best possible manner even by experienced workers. Sometimes the tools, suitable and available on the market, were banned from use in certain corporate settings, or the experience and skills of those expected to work with the tools, were missing. A lack of connectivity to the internet also challenged some workers in their home office leading to misunderstandings or breakdowns in communication as well as to inequality and exclusion from workplace communication (Ferreira et al. 2021).

Shirmohammadi et al. (2022) discovered in their research five stressors in the work domain, one of which was technostress. The literature that was reviewed for their study focusing on work-life balance during the pandemic also emphasised the importance of Information and Communication Technology (ICT) during the pandemic and while working from home, as they built the foundation for workers to be able to do their jobs. Technostress, i.e. factors that cause stress in the workplace when using ICT, had already negative impacts on job satisfaction and employee commitment before 2020 (Bondanini et al. 2020; Ragu-Nathan et al. 2008). With the sudden increase and dependence on ICT to get one's job done, the invasion of technology into both worlds, the workplace and the private sphere of workers when working from home, were and are easier to cope with and compensate for by some and harder by others. Here, incompatibility, inadequate tools, and limited or unstable internet connections contributed to stress and a decrease in well-being during the pandemic (Shirmohammadi et al. 2022).

Singer-Velush et al. (2020) confirm the steep learning curve of those who worked from home during the pandemic when it comes to the use of ICT, but the Work Trend Index published by Microsoft shows that an *'always on'* mentality is already on the decline (Spataro 2022). Here, it appears that the hybrid workforces are becoming more intentional about their work-life balance by not scheduling meetings during lunch time or outside normal business hours (Spataro 2022). Individuals with issues around boundary settings and with supervisors that are less understanding are challenged more than those who can draw those lines easier and where clear company policies are in place. With time spent in online meetings up by 252% (ibid), it has become so much more important to develop skills around boundary settings, to take advantage of technical support by inserting blockers in online calendars, and by ensuring there are breaks between online meetings so the body and brain can relax and recharge in between those online meeting marathons. For example, a 10-min break between online meetings can reduce stress levels as brain scans have shown (Mircosoft 2021).

6 Conclusions and Recommendations

With the complexities identified in hybrid work environments, the opportunities and challenges arising from those; the different needs that exist within organisations and among employees, because of personal preferences, and continuously changing situations at home and in the workplace, there will not be a single solution that will work permanently for the various cases and scenarios that exist. If hybrid work environments are truly meant to become a sustainable new way of working for most of the workforces around the globe, more focus must be put on what exactly makes hybrid work sustainable in a particular workplace setting for a particular individual or group of individuals. Standardised solutions will not cater for the plurality of interests that exists; understanding the needs and preferences of individuals and of organisations and their customers, crafting jobs so they can be decent for those who do them, are options to be explored and revisited again and again to ensure the necessary adaptations are made for thriving rather than just surviving in these new workplace arrangements.

Organisations are well-advised to take stock of how sustainable and decent their current work environments are from their own and their workforces points of view. Company internal conversations will help to identify relevant areas which need adapting and improving. Business Schools, researchers, coaches and consultants can support these processes by bringing new perspectives, knowledge, and research data to help organisations make informed decisions about which policies they would like to put in place, and how to adapt their in-company learning and development programmes to upskill their leaders and multigenerational workforce. Business Schools might want to reflect on the ways in which their curricula could be enhanced, so students and the next-generation workforces are better equipped right from the start with leadership, communication, and technology skills that are essential in these hybrid work environments.

Every challenge mentioned in this chapter can be turned into an opportunity to make this new working world more decent. Hybrid workplaces, which have the potential to be more than just a fad after the pandemic, will not disappear because they are not sustainable from a health or economic perspective. They are a joint effort of all stakeholders. Hybrid workplaces and the new ways of working and living, the lessons learned during and after the pandemic are another opportunity for humanity to demonstrate our ability to learn from crisis and experiences. This chapter identifies areas to focus on, it addresses the various stakeholders, and it gives recommendations about what could be done to help hybrid work environments become more sustainable and decent for current and future generations in the global workplace.

References

Bakker AB, Demerouti E (2017) Job demands–resources theory: taking stock and looking forward. J Occup Health Psychol 22(3):273–285. https://doi.org/10.1037/ocp0000056

Bayntun-Lees D, Cross A (2022) Rethinking Leadership for the hybrid world of work. What are the critical skills and tools leaders need? H. E. C. Education. https://www.hultef.com/en/insights/research-thought-leadership/rethinking-leadership-hyrid-work/

Bernstein E, Blunden H, Brodsky A, Sohn W, Waber B (2020) The implications of working without an office. Harv Bus Rev

Blustein DL, Kenny ME, Autin K, Duffy R (2019) The psychology of working in practice: a theory of change for a new era. Career Dev Q 67(3):236–254. https://doi.org/10.1002/cdq.12193

Bondanini G, Giorgi G, Ariza-Montes A, Vega-Muñoz A, Andreucci-Annunziata P (2020) Technostress dark side of technology in the workplace: a scientometric analysis. Int J Environ Res Public Health (17):1–23. https://doi.org/10.3390/ijerph17218013

Campbell M, Gavett G (2021) What Covid-19 has done to our well-being, in 12 charts. Harv Bus Rev. https://hbr.org/2021/02/what-covid-19-has-done-to-our-well-being-in-12-charts. Accessed 15 April 2023

Delany K (2021) What challenges will organisations face transitioning for the first time to the new normal of remote working? Hum Resour Dev Int 1–9. https://doi.org/10.1080/13678868.2021.2017391

Desilver D (2020) Before the coronavirus, telework was an optional benefit, mostly for the affluent few. https://www.pewresearch.org/fact-tank/2020/03/20/before-the-coronavirus-telework-was-an-optional-benefit-mostly-for-the-affluent-few/

Dowling B, Goldstein D, Park M, Price H (2022). Hybrid work: Making it fit with your diversity, equity, and inclusion strategy. https://www.mckinsey.com/business-functions/people-and-organizational-performance/our-insights/hybrid-work-making-it-fit-with-your-diversity-equity-and-inclusion-strategy. Accessed 13 July 2022

Doyle A (2021) Types of remote pay: location-based vs job-based—should your salary depend on location?

Dubois A, Gadde L-E (2002) Systematic combining: an abductive approach to case research. J Bus Res 55(7), 553–560. https://doi.org/10.1016/S0148-2963(00)00195-8

Eurofound (2018) Decent work. https://www.eurofound.europa.eu/observatories/eurwork/industrial-relations-dictionary/decent-work. Accessed 15 April 2023

Ferreira R, Pereira R, Bianchi IS, da Silva MM (2021) Decision factors for remote work adoption: advantages, disadvantages, driving forces and challenges. J Open Innov 7(1):70. https://doi.org/10.3390/joitmc7010070
Future Forum (2022) Leveling the playing field in the hybrid workplace. https://futureforum.com/research/leveling-the-playing-field-in-the-hybrid-workplace/
Franzen-Waschke U (2021) Working from home in 2020—Lessons learned to leverage these learnings going forward as emerging leaders and a remote office workforce. GiLE4Youth International Conference, The Development of Competencies for Employability. https://www.researchgate.net/publication/353794088_Working_from_home_in_2020_Lessons_learned_to_leverage_these_learnings_going_forward_as_emerging_leaders_and_a_remote_office_workforce
Halecker B (2016) Action case study—a research strategy based on abduction for relevant and rigorous management research. In: New perspective and insights on business model innovation using systems thinking and action case studies. Cuvillier Verlag
Howspace (2022) The State of Collaboration Report. https://howspace.com/guides/state-of-collaboration-report/
Hempel J (2021) Work-life balance In: The great re-norming. https://hbr.org/podcast/2021/07/the-great-re-norming
Hopkins JCD, Figaro KAD (2021) the great resignation: an argument for hybrid leadership. 9(4). https://ijbmr.forexjournal.co.in/archive/volume-9/ijbmr-090402.html. Accessed 11 Jan 2023
Kane P (2021) The great resignation is here, and it's real. Inc.Com. https://www.inc.com/phillip-kane/the-great-resignation-is-here-its-real.html
Litchfield P, Cooper C, Hancock C, Watt P (2016) Work and Wellbeing in the 21st Century. Int J Environ Res Public Health 13(11):1065. https://doi.org/10.3390/ijerph13111065
Lundqvist D, Reineholm C, Ståhl C, Wallo A (2022) The impact of leadership on employee well-being: on-site compared to working from home. BMC Public Health 22(1):2154–2154. https://doi.org/10.1186/s12889-022-14612-9
Maruta, R. (2012). Transforming knowledge workers into innovation workers to improve corporate productivity. Knowl-Based Syst 30:35–47. https://doi.org/10.1016/j.knosys.2011.06.017
Milasi S, González-Vázquez I, Fernández-Macías E (2020) Telework in the EU before and after the COVID-19: where we were, where we head to. J. R. C. The European Commission's Science and Knowledge Service. https://ec.europa.eu/jrc/sites/default/files/jrc120945_policy_brief_-_covid_and_telework_final.pdf
Miller S (2021) Google's salary cuts for remote workers renew location-based pay debate. https://www.shrm.org/resourcesandtools/hr-topics/compensation/pages/google-renews-location-based-pay-debate.aspx
Mircosoft (2021) Research proves your brain needs breaks. https://www.microsoft.com/en-us/microsoft-365/blog/2021/04/20/research-shows-your-brain-needs-breaks-outlook-and-microsoft-teams-can-help/
Ragu-Nathan TS, Tarafdar M, Ragu-Nathan BS, Tu Q (2008) The consequences of technostress for end users in organizations: conceptual development and empirical validation. Inf Syst Res 19(4):417–433. https://doi.org/10.1287/isre.1070.0165
Sanhokwe H (2022) The influence of nonwork resources, nonwork demands and external support on work engagement and productivity: a moderated mediation model. SA J Ind Psychol 48. https://doi.org/10.4102/sajip.v48i0.1957
Seligman M (2018) PERMA and the building blocks of well-being. J Posit Psychol. https://doi.org/10.1080/17439760.2018.1437466(RouteledgeTaylor&FrancisGroup)
Shirmohammadi M, Chan Au W, Beigi M (2022) Antecedents and outcomes of work-life balance while working from home: a review of the research conducted during the COVID-19 pandemic. Hum Resour Dev Rev 0/0. https://doi.org/10.1177/15344843221125834
Singer-Velush N, Kevin S, Erik A (2020) Microsoft Analyzed Data on its newly remote workforce. Harv Bus Rev
Spataro J (2022) 5 key trends leaders need to understand to get hybrid right. Harv Bus Rev

Subramanian S, Washington EFP (2022) Why flexible work is essential to your DEI strategy. Harv Bus Rev. https://hbr.org/2022/02/why-flexible-work-is-essential-to-your-dei-strategy. Accessed 13 July 2022

Svicher A, Di Fabio A (2021) Job crafting: a challenge to promote decent work for vulnerable workers. Front Psychol 12:681022–681022. https://doi.org/10.3389/fpsyg.2021.681022

Tamin J, Samuel O, Suraya A, Ebuenyi ID, Naicker N, Rajput-Ray M (2021) Vulnerable workers and COVID-19: insights from a survey of members of the international commission for occupational health. Int J Environ Res Public Health 18(1). https://doi.org/10.3390/ijerph18010346

Truss C, Delbridge R, Alfes K, Shantz A, Soane E (2014) Employee engagement in theory and practice. Routledge. https://doi.org/10.4324/9780203076965

Tsipursky G (2022) What is proximity bias and how can managers prevent it? https://hbr.org/2022/10/what-is-proximity-bias-and-how-can-managers-prevent-it. Accessed 03 Jan 2023

Vogel S, Breitenbroich M (2020) Industrial relations and social dialogue. Germany, Working life in the COVID-19 pandemic 2020. https://www.eurofound.europa.eu/de/publications/other/2021/working-life-in-the-covid-19-pandemic-2020

Wołowiec T, Myroshnychenko I, Vakulenko I, Bogacki S, Wiśniewska AM, Kolosok S, Yunger V (2022) International impact of COVID-19 on energy economics and environmental pollution: a scoping review. Energies (basel) 15(22):8407. https://doi.org/10.3390/en15228407

Measuring Sustainability Embeddedness in Business Education Curriculum

Rosa M. Fernandez Martin

Abstract Education for Sustainable Development (ESD) has become a prominent area in academic literature particularly thanks to the dedicated period promoted by the UN on the topic (2005–2014), and soon afterwards, with its inclusion in the UN Sustainable Development Agenda for 2030. Different proposals for embeddedness of ESD within curriculums at different levels of education have been appearing over time across the world, and of course, within European countries. The evaluation of said embeddedness is the object of this contribution. There is an increasing tendency to measure the performance of educational institutions through the creation of rankings that use a series of variables as indicators of success. Rankings exist at national and international levels, with education institutions using them as benchmarks, but also as competitive signals in a commodified education environment, particularly when it comes to Higher Education. This contribution analyses the use of rankings as indicators of embeddedness of sustainability in Business Schools, critically querying the adequacy of the indicators, the distinction between undergraduate and graduate studies for the classification in rankings and to which extent sustainability embeddedness can be considered achieved if it is applied to Business studies and not to the rest of curriculum within the same institution.

Keywords Business schools · Education for Sustainable Development (ESD) · Sustainability rankings

1 Introduction

The role of Higher Education Institutions (HEI) in the transition towards a more sustainable society, has been traditionally identified with the educational side, and the development of curricula conducive to Education for Sustainable Development (ESD) (Holdsworth and Thomas 2015). The UNESCO Decade on Education for

---

R. M. Fernandez Martin (✉)
Keele Business School, Denise Coates Foundation Building, Keele University, Staffordshire ST5 5AA, England
e-mail: r.fernandez.martin@keele.ac.uk

T. Wall et al. (eds.), *Sustainability in Business Education, Research and Practices*, World Sustainability Series, https://doi.org/10.1007/978-3-031-55996-9_3

Sustainable Development, together with previous initiatives, contributed to increase the academic research undertaken in this area, but leaving more work to be done by HEIs (Leal Filho et al. 2015). It makes sense in this context to extend the reach of the term 'education' to encompass teaching, research and engagement activities (Slager et al. 2020).

How do business schools fit into the efforts of HEIs to contribute to the transition to a more sustainable society? This is not an easy question to answer for a number of reasons: First of all, the legitimacy of business schools has been under scrutiny particularly in the aftermath of the international economic crisis of 2007–2008, since greediness and malpractice at the top of the leadership and management chains in big corporations were at the core of the issues identified as main cause for said crisis. Business schools were blamed for their role in training those professionals who caused the crisis (Miotto et al. 2020). Secondly and more related to the purpose of this paper, business schools tend to work separately from the rest of the HEIs they are part of. This is more the case in some countries than others, and also dependent on their status as public or private institutions, but the statement prevails if we realise that some business schools exist on their own (for instance ESADE in Spain) and they become the entire HEI under scrutiny. Changes being experienced by business schools across the globe give a blunt picture over their fragility, with reports of mergers and acquisitions in the UK, France or the US, amidst concerns for survival due to their financial struggles (Schlegelmilch 2020). This also makes business schools appear more like businesses than part of education institutions and speaks loudly about the commodification of higher education in the last decades.

Embedding sustainability in business schools' curricula is part of their reaction and their need to change and respond to new societal demands, particularly to play a role in solving global challenges (Rayment and Smith 2013). To which extent sustainability embeddedness has taken place is not easy to assess, and current academic literature tends to focus on the progress of whole institutions, addressing not only curriculum but estates and investment decisions, for instance, or individual programmes of study. Being part of rankings has become a common ground for HEIs as a signal used for competitive advantage by those at the top of the lists, and performance on specific indicators of certain rankings is commonly used when HEIs decide on their policies and strategies (Burman et al. 2021). Sustainability rankings for universities have started appearing in the last decades, and despite the criticism over their limitations (Ragazzi and Ghidini 2017), universities tend to publicise their place in rankings in their attempt to attract students.

This paper explores the use of sustainability rankings in business schools and their limitations to assess sustainability embeddedness in curriculum (research and practice) with the purpose of answering two main questions: Are these rankings measuring the right thing the right way? (ie. Are sustainability rankings reliable?) and, To which extent are business schools contributing to sustainability targets if they work in isolation from the rest of the HEI they are part of? To address these questions, an exploratory case study approach of business schools in the UK is used.

The paper is organised as follows: Sect. 2 summarises the policy and practice framework applicable to the embeddedness of education for sustainable development in curricula; Sect. 3 outlines the methodology used to undertake the analysis and provides relevant information about the sustainability rankings used; Sect. 4 presents and discusses the results for HEIs in the UK and particularly business schools, and Sect. 5 finishes the paper with concluding remarks.

2 Education for Sustainable Development Policy and Practice Framework

The UN Sustainable Agenda for 2030 (UN 2015) set a number of targets for the 17 Sustainable Development Goals, with indicators to measure progression against the targets for each goal. SDG4, Quality Education, includes target 4.7 '*By 2030, ensure that all learners acquire the knowledge and skills needed to promote sustainable development, including, among others, through education for sustainable development and sustainable lifestyles, […], global citizenship and appreciation of cultural diversity and of culture's contribution to sustainable development*' (UN 2017). There is only one indicator used to measure progress against this target, the *Extent to which (i) global citizenship education and (ii) education for sustainable development are mainstreamed in (a) national education policies; (b) curricula; (c) teacher education and (d) student assessment* (UN 2017). This paper focuses on the curricula aspect of this indicator to assess if and how it is being measured with regard to business schools. An important reason to focus on curricula is that information on teacher education and student assessment is almost non-existent. What is more, data on this indicator do not appear in the SDGs annual progress reports and in any case progression on SDG4 measures mostly primary and secondary education, with higher education aspects not featuring explicitly in main indicators and being hardly mentioned. Even so, UNESCO (Sabzalieva 2022) clarifies that SDG4 applies to higher education too, but it focuses on the area of HEIs governance and the predominance of self-regulation, which would include their decision to be part of rankings that measure their performance. In any case, the UN approved a new global framework on ESD at the end of 2019 (UN 2020), where higher education is only mentioned once, as a 'powerful driver for change', in the context of recovery from conflicts. The framework is accompanied by a detailed roadmap for 2030 (UNESCO 2020), where HEIs appear more explicitly as key actors that require a whole-institution approach for the transformation of learning and training environments, but also where research activities on ESD are highlighted as one of their key contributions. This may cause a deviation from ESD inclusion in curriculum in favour of treating it as a subject of research.

In the case of the UK higher education sector, the QAA (Quality Assurance Agency), with collaboration from the Higher Education Academy—now Advance HE, published the first guidelines on ESD for HE providers in 2014 (QAA 2014). The guidelines have been updated in 2021, and include numerous resources and links to the UK policy framework documents (QAA and Advance HE 2021), noting that there are separated framework documents for Wales and Scotland. The new guidelines build on the 2030 Agenda for Sustainable Development and particularly target 4.7, together with the UNESCO skills and competencies for ESD. It is worth pointing that as guidelines they are not compulsory, something that also applies to the subject benchmark statements that inform expectations about different university programmes of study. In that regard, the Business and Management Subject Benchmark Statement available to the institutions included in this study (QAA 2019) did not incorporate the initial 2014 guidelines, with sustainability only mentioned once, as part of the knowledge and understanding that students are expected to gain on 'management and sustainability of organisations'. With this lack of engagement at subject benchmark level, it becomes apparent that UK HEIs have been incorporating ESD in their business schools' curricula on an individual basis and not as part of an integrated country effort. Only very recently (March 2023) new Subject Benchmark Statements have been published for Business and Management courses incorporating for the first time the 2021 Guidelines. As a difference with most other disciplines, separated benchmark statements are published for undergraduate (QAA 2023a) and master programmes (QAA 2023b), but the ESD section of both documents is literally the same.

Within the UN framework, business actions towards sustainability are informed by the UN Global Compact initiative, where business can be signatories on a voluntary basis. The link with higher education is established through the joint initiative between UN and the Global Compact on the Principles for Responsible Management Education (PRME), of which business/management schools worldwide can be signatories, again on a voluntary basis. What began in 2007 with less than a hundred participating institutions, includes now more than 800 business schools in their membership, but it cannot be forgotten that they recognise there are more than 16,000 business and management programmes (not schools) worldwide (UNPRME.org, n.d.), which poses questions about representation and the reach of their actions. They also recognise that the existing rankings measuring the performance of business schools are not adequate to assess the way in which they contribute to 'equip business leaders with the skills needed to run a productive, inclusive, and sustainable economy for the twenty-first century' (Pitt-Watson and Quigley 2019). The exception they mention is the ranking created by *Corporate Knights*, which measures sustainability credentials of 40 MBAs across the world. The UK has the highest number of MBAs in this ranking in their 2022 list -12, followed by Canada with 8 (Lewington 2022). Hence the decision to use the UK as case study and the Corporate Knights ranking as one of those under analysis.

The results of the Corporate Knights ranking, specific to business schools, will be assessed against those of other sustainability rankings that measure whole universities' performance, to evaluate their effectiveness and the importance they give to the embeddedness of sustainability in HEIs curriculum. The paper will explore to which extent those rankings help measure ESD embeddedness. In doing so it will contribute to the body of literature that aims to define the role of business schools' education in the transition to a sustainable society.

3 Methodology: Sustainability Rankings in Higher Education

The analysis has taken place in two stages. First, four sustainability rankings have been compared to establish their appropriateness to measure ESD embeddedness and as such their usefulness to assess progress with regard to SDG4.7. Said rankings are the UI GreenMetric, created by Universitas Indonesia in 2010 and that currently includes around 1000 universities; The Times Higher Education Impact Rankings, which covers 1410 universities; the People and Planet University League, which evaluates universities in the UK only; and the aforementioned Corporate Knights ranking. Performance indicators for each ranking have been scrutinised against ESD guidelines and policy frameworks. A second step consisted on the exploration of business courses in the universities featured in each ranking, both undergraduate and graduate, focusing on those universities that scored highly in education-focused indicators. Information has been extracted from the publicly available data in the respective universities' websites. The purpose of this step is to assess to which extent business schools' courses are reflective of or contribute to (or not) the sustainability scores of whole institutions, in line with UNESCO's guidelines on ESD being part of whole institutions approach. These two stages allow addressing the two research questions posed by this paper.

The following paragraphs summarise the main characteristics of the four sustainability rankings, with specific information related to each of the relevant HEIs being displayed in the findings section. Data have been extracted from the websites of each of the rankings, including their specific methodology sites or documents, and refer to the latest ranking available, that of the year 2022 (or academic year 2022–2023).

We begin with the Corporate Knights ranking, which is based on 7 key performance indicators (KPIs) that are weighted. The highest individual weight (30%) is given to the content of core courses in terms their coverage of relevant sustainable development themes (Corporate Nights 2022). However, research aspects end up adding 50% of the score, if we add together faculty research intensity (20%), number of citations intensity (10%) and sustainability journal intensity (20%). Interestingly, nothing is mentioned about the quality (or ranking) of the journals, and it is only looked if they are peer-reviewed (Corporate Nights 2022). The other KPIs are the existence of research institutes and centres within business schools dedicated (fully or substantially) to areas of sustainable development (again research related); faculty diversity (gender and racial); and what is denominated as 'alumni impact' (5%), which includes the percentage of recent MBA graduates working for 'impact organisations', where they create their own definition including their global 100 and clean 200 companies, and all non-profit organisations (Corporate Nights 2022). The institutions that can be part of the ranking are international, but interestingly, the more recent FT100 MBA ranking participants (which do not cover sustainability), the top 40 MBAs from the previous year, and all current PRME 'champions' (membership of which requires the payment of a fee) are automatically included. It is also worth noting that this ranking only assesses MBAs, so nothing can be said about the undergraduate provision of the participant business schools.

The UI GreenMetric, in contrast, offers more information and evaluates full institutions, not just business schools, using six categories of indicators based on the three pillars of sustainability (economic, social, environmental), also weighted: Setting and infrastructure (15%), energy and climate change (21%), waste (18%), water (10%), transportation (18%) and education, including research (18%). Interestingly, only six UK universities participate in this ranking: Nottingham Trent (number 2), University of Nottingham (number 3), University of Warwick (36), Glasgow Caledonian University (120), Loughborough (226) and Teesside University (334) (UI GreenMetric 2023a), and if the education category is looked at in isolation, the UK universities get lower scores, with Nottingham Trent again as the first classified but as number 21 (UI GreenMetric 2023b). Surprisingly though, being a ranking that evaluates HEIs, the education category was not included when the ranking began, and it was first used in 2012. The indicators used for this category are interesting though, including the number of student organisations related to sustainability, number of sustainability community services projects organised and/or involving students, the ratio of sustainability research funding towards total research funding, number of sustainability-related startups, ratio of sustainability courses towards the total number of courses/subjects, and similarly to the Corporate Knights one, number of scholarly publications on sustainability. But it also includes number of events related to sustainability, number of cultural activities on campus, university-run sustainability website, sustainability report, and the number of university programmes to cope with the Covid19 pandemic (UI GreenMetric 2023c). In this case the curriculum aspect only measures the number of courses, but not their content or quality, and it is clear that education per se is not a significant proportion of the measurement.

The Times Higher Education (THE) Impact ranking evaluates the performance of HEIs against the 17 SDGs, grouping their categories of assessment in four areas: research, stewardship, outreach and teaching. The score is calculated by combining each university weighted score in SDG17 (22%) with its top three scores out of the remaining 16 SDGs (26% each). It is universities who have to provide evidence to support their claims, particularly with regard to policies and initiatives, and the information on publications to build the research metrics is provided by Elsevier, looking into those relevant for each SDG. This is combined with additional publications identified using artificial intelligence (THE Impact Rankings 2022). Out of 1410 entries, 53 universities from the UK appear in this ranking, and it is possible to isolate those who have a score for SDG4 (Quality education). Interestingly, at this level the number of participant UK HEIs gets reduced to 37 (out of 1180), and those universities listed are not the same as those featured in the general list, with the order of the ranking for those that appear also changing (THE Impact Rankings 2023). A key point to note is that the methodology to score SDG4 explicitly recognises that this element focuses on early years education and lifelong learning, which are not the main activities of HEIs. As such their methodology asks for these scores not to be taken as a measure of universities' quality of teaching (THE Impact Rankings 2022).

The People and Planet ranking is the only one that includes teachers' training as part of its indicators, making it more in tune with SDG4.7 target and related indicators. It uses a similar approach to that of THE Impact Rankings and UI GreenMetrics insofar it aligns with the SDG overarching goal of leaving no one behind, and indicators cover not just teaching and research aspects but also institutional efforts to reduce and manage carbon, water usage, boost staff and student engagement, ethical investment and banking, careers and recruitment, and others (People and Planet 2023a). The ranking is made of fourteen indicators, weighted, with ESD adding 9% of the total score, even if research is also embedded in other areas. However the biggest proportion of the ranking is made up by efforts to reduce carbon (15%) to which another 7% is added under the 'manage carbon' category. This is followed by 'environmental auditing and management systems', which make 10% of the score (People and Planet 2023a). This is an indication that the educational aspect is in any case given a limited importance within the ranking, in fact the lowest of all the rankings considered. There is a chance to find the classification of HEIs by individual indicators as well as the overall ranking, which provides the opportunity to look into the best performers in the ESD category. Similarly to the other rankings the order changes when it comes to the comparison between the overall ranking and the ESD ranking, but here all the institutions remain accounted for (People and Planet 2023b). Their data are obtained from the official university websites and the information published by the Higher Education Statistics Agency Estates Management Record (People and Planet 2023a).

The information provided by these four rankings has been used to select a number of institutions to look in more detail into their business school courses. The following section provides the relevant findings of this analysis.

4 Data Analysis and Findings

Table 1 provides the list of UK universities whose MBAs are part of the Corporate Knights top 40 ranking. This list has been used as baseline to compare with the presence of the related institutions in the other three rankings. Looking into the characteristics of these institutions and their adherence or not to wider sustainability frameworks helps with providing a wider picture of UK business schools. In that light it is worth noting that as of March 30th 2023, there were 97 UK based business schools signatories of the PRME (UNPRME.org 2023a), and nine of them are PRME champions (UNPRME.org 2023b). Table 1 indicates that of the twelve UK institutions featured in the Corporate Knights ranking, not all of them are signatories of the UN PRME, and only three of them would have been automatically included in the ranking due to their status as PRME champions. It could be argued that ranking position in this case has been secured on own merits.

Comparing the Corporate Knights ranking with the other three shows that only one institution appears in all the rankings: Glasgow Caledonian University, with the caveat that it does not appear in the education category of THE Impact Factor, which

Table 1 UK business schools in the Corporate Knights ranking 2022

Ranking order	Business School	PRME signatory	PRME champion
2	Warwick Business School	No (used to be)	No
7	University of Bath-School of Management	Yes	No
12	University of Edinburgh Business School	Yes	No
14	University of Exeter Business School	Yes	No
17	University of Winchester	Yes	No
18	Glasgow Caledonian University: School of Business and Society	Yes	Yes
23	Imperial College Business School	No	No
25	Durham University Business School	Yes	No
26	University of Strathclyde-Strathclyde Business School	Yes	No
29	Nottingham University Business School	Yes	Yes
33	Newcastle Business School	Yes	Yes
35	Alliance Manchester Business School	Yes	No

Source Corporate Knights (2022) and UNPRME (2023a, b)

means that the educational aspect is not among the areas they report and does not make the top three categories that compose said ranking. By looking at the scores in the different rankings, six institutions have been chosen to look more closely at their business schools' courses. They appear in Table 2, with their positions in each ranking, when they are part of them.

It can be observed that in general, when there is an education specific category, universities tend to perform worse than in the general ranking, with two exceptions: In the People and Planet University League the selected universities perform better in the education category than the general one, with the exception of Glasgow Caledonian; and in the THE Impact Rankings, the University of Winchester performs better in the education category than in the general score.

Going into each of the six selected universities presents an interesting picture about the embeddedness of sustainability or ESD in business schools' courses. Beginning with Glasgow Caledonian, at undergraduate level its BA Business Management programme has been looked at, showing that at the forefront of the information it appears the commitment towards the triple bottom line (people, planet, profit), as something financially good for businesses, recognising the demand to demonstrate corporate social responsibility (CSR), and for graduates to have leadership skills to bring responsible and ethical management (Glasgow Caledonian University n.d.a). The programme specification includes the mention of contribution to 'sustainable value for both business and society', and the adherence to QAA benchmark statement guidelines, together with PRME and the UN SDGs (Glasgow Caledonian University 2022). However from the titles of the modules it is difficult to ascertain to which

Table 2 Universities selected for business courses' analysis: Position on different rankings

University	Corporate Knights ranking	UI GreenMetric		THE impact rankings		People and planet university league	
		Overall ranking	Education ranking	Overall ranking	Education ranking	Overall ranking	Education ranking
Glasgow Caledonian	18	120	473	70	//	51	63
Exeter	14	//	//	42	301–400	6	1*
Manchester	35	//	//	9	301–400	56	1*
Newcastle	33	//	//	8	301–400	21	13*
Strathclyde	26	//	//	101–200	201–300	98	1*
Winchester	17	//	//	301–400	101–200	33	13*

Source Corporate Knights (2022), UI GreenMetric (2023a), THE Impact Rankings (2023), People and Planet (2023b). *In the People and Planet education ranking universities that appear to have the same score are given the same position in the ranking, so there are twelve universities in first place, followed by fourteen universities in 13th place, followed by nine universities in 27th place and so on and so forth

extent the guidelines and principles have been followed, with mention of ethics only in one Year 2 module and social aspects limited to optional modules (Glasgow Caledonian University 2022). At graduate level its MBA has been looked at, since it is the course included in the Corporate Knights ranking. Again, from the word go there is a focus on the UN PRME and Common Good principles, responsible leadership, sustainability and the UN's SDGs' as underpinning all the modules (Glasgow Caledonian University n.d.b). Only the module names appear as part of the information available on the website, but this time it is observable that the sustainability related aspects are more evident in module names, both in core and optional modules (Glasgow Caledonian University n.d.b).

The University of Exeter offers different courses at its Streatham (Exeter) and Cornwall campuses so to help comparison the BSc Business and Management course has been chosen from the main campus. It is worth noting no specific mention of ESD or sustainability in the general information about the course, no mention of adherence or focus on PRME or SDGs (University of Exeter n.d.a), nor specific module titles that could indicate coverage of sustainability related topics beyond a 'Business and society' module and a series of optional modules in Management for the final year (University of Exeter n.d.b). At Master level things change, with the MBA site specifically mentioning its position in the Corporate Knights ranking, a focus on responsible leadership and the 'potential of business to profitably solve the problems of people and planet, without doing harm' (University of Exeter n.d.c). The approach to the structure also changes, with blocks of content addressing global challenges and module titles specifically linked to sustainability related topics (University of Exeter n.d.d).

The University of Manchester makes no mention of ESD or sustainability for its BSc Management programme, though it includes very early on the possibility of lower entry requirements for those in care or with refugee status (Alliance Business School n.d.a). Instead it publicises the position some of its courses in general international university rankings (Alliance Business School n.d.a). The information about modules is very generic, with topics instead of module titles provided, though for years two and three a list of optional modules appears including sustainability related content and what seems to be the possibility of taking some of the modules indistinctively in both years (Alliance Business School n.d.a). At master level its MBA include several options with special versions of a Global Executive MBA, which can be part-time, online, and for selected executives. For the purpose of the analysis we have looked at the full-time MBA, which does not make any reference to ESD or sustainability, at least explicitly, and with only one optional module dedicated to collated aspects, being 'Not-for-profit Consultancy Project' (Alliance Business School n.d.b).

Newcastle University presents similar characteristics to previous programmes, with no mention about sustainability, ESD, SDGs or PRME in the information related to its BA Business Management course (Newcastle University n.d.a), and only an inclusion of a core CSR and ethics module in Year 3, with a few options that potentially could cover sustainability related topics (Newcastle University 2023). The approach of its MBA is very similar, again no ESD initial mention, but a module on CSR and ethical business available. Some of the content block have the potential to address sustainability challenges, for instance under 'transitions in the global economy' or 'leading organisational change', but it is clear that the focus is on leadership in the traditional sense of the term (Newcastle University n.d.b).

The organisation of courses for Strathclyde Business School differs from the traditional business/management classification of other universities. For comparison purposes the BA Business Enterprise programme has been selected, though it does not completely fit the characteristics of a typical business management course. Similarly to the rest of courses already explored, there is no mention of ESD or sustainability, and the only topics connected to sustainability challenges appear in a 'management development programme' block that is included as part of the course, with year 2 and 3 topics including working in the third sector, and a pathway to internship/charities (Strathclyde n.d.a). At MBA level several options appear, one of them with specialisation in sustainable energy futures, but we have looked at the full-time traditional MBA for comparison purposes. Though again no explicit mention of ESD appears in the frontline information, this MBA includes three pillars, of which one of them is 'responsible leadership' with specific sustainability related modules. The rest follows traditional business strategy and management content (Strathclyde n.d.b).

Last but not least we have the University of Winchester. The information about its business school includes the adherence to the PRME and the alignment with the institutional strategy in support of the UN SDGs (University of Winchester n.d.a). Its BA in Business Management mentions the position of the Business School as PRME Champion, and includes core sustainability related modules in years 2 and 3 and a few options also linked to sustainability topics (University of Winchester n.d.b). The information on its MBA also includes the mention of PRME Champion for the Business School, and aims to attract students who want to have careers in 'organisations with a sustainable outlook'. There are no specific modules listed though the themes covered include: leadership, stakeholder management, business sustainability and digital insight management (University of Winchester n.d.c).

It cannot be said that looking into six universities only and a specific type of course presents the full picture, but it can be argued that some distinctive features emerge. Glasgow Caledonian and Winchester present a more evident commitment with the embeddedness of ESD in their business curriculum, through their explicit adherence to PRME guidelines. In the case of Winchester its overall position in the sustainability rankings may well be helped by the commitment of the business school to this area, though further research beyond the scope of this paper would need to be undertaken. Conclusions about rankings and business schools will now be presented in the final section of this paper.

5 Concluding Remarks

The analysis of sustainability rankings indicators and their application to business schools' curricula in the UK offers some valuable insights on the role of said rankings and the commitment of business schools to the sustainable development agenda. In answering the first research question posed, it can be argued that the validity of existing rankings to assess ESD embeddedness is extremely limited, with a focus more on quantitative indicators than a proper quality assessment. It also becomes apparent that most business schools, unless in our selected sample, have not incorporated ESD elements as distinctive of their offer, which may be related to the delay in the inclusion of said elements in the corresponding subject benchmark statements. This calls for a rethink on the design of existing HEI sustainability rankings in favour of specific ESD embeddedness indicators.

With regard to the second research question, the institutions that appear as outliers have done so following international initiatives at individual level, and it is also observable that different approaches are used for undergraduate and graduate degrees. All in all, it cannot be said that the whole-institution approach advocated by UNESCO is widely followed, nor that business schools contribute in a significant way to HEIs sustainability efforts, so it can be easily concluded that UK HEIs in general and business schools in particular have a long way to go to provide a genuine engagement with the sustainable development agenda.

References

Alliance Business School (n.d.a) BSc (Hons) Management. https://www.alliancembs.manchester.ac.uk/study/undergraduate/bsc-hons-management/course-details/#course-profile. Accessed 30 March 2023

Alliance Business School (n.d.b) The Manchester Full-time MBA Degree. https://www.alliancembs.manchester.ac.uk/study/mba/full-time/. Accessed 30 March 2023
Burmann C, Garcia F, Guijarro F, Oliver J (2021) Ranking the performance of universities: The role of sustainability. Sustainability 13:13286. https://doi.org/10.3390/su132313286
Corporate Knights (2022) Better World MBA ranking methodology. Corporateknights.com
Glasgow Caledonian University (n.d.a) BA (Hons) Business management. https://www.gcu.ac.uk/study/courses/undergraduate-business-management-glasgow. Accessed 29 March 2023
Glasgow Caledonian University (n.d.b) Global Master of Business Administration (MBA). https://www.gcu.ac.uk/study/courses/postgraduate-global-master-of-business-administration-mba-glasgow. Accessed 29 March 2023
Glasgow Caledonian University (2022) Appendix A: Programme Specification Pro-forma PSP) BA (Hons) Business Management
Holdsworth S, Thomas I (2015) Framework for introducing education for sustainable development into university curriculum. J Educ Sustain Dev 9:137–159. https://doi.org/10.1177/0973408215588246
Leal Filho W, Manolas E, Pace P (2015) The future we want. Key issues on sustainable development in higher education after Rio and the UN decade of education for sustainable development. Int J Sustain High Educ 16:112–129. https://doi.org/10.1108/IJSHE-03-2014-0036
Lewington J (2022) Top 40 MBAs double down on commitments to sustainability. Corporate Knights. https://www.corporateknights.com/rankings/top-40-mba-rankings/2022-better-world-mba-rankings/top-40-mbas-double-down-on-commitments-to-sustainability/. Accessed 01 Feb 2023
Miotto G, Blanco-Gonzalez A, Diez-Martin F (2020) Top business schools legitimacy quest through the Sustainable Development Goals. Heliyon 6:e05395.https://doi.org/10.1016/j.heliyon.2020.e05395
Newcastle University (n.d.a) Business Management BA Honours. https://www.ncl.ac.uk/undergraduate/degrees/n200/. Accessed 30 March 2023
Newcastle University (n.d.b) Master of Business Administration MBA. https://www.ncl.ac.uk/postgraduate/degrees/5823f/?utm_campaign=postgrad&utm_source=google&utm_medium=cpc&utm_content=pg-subject-dsa&utm_term=&utm_campaign=&utm_source=adwords&utm_medium=ppc&hsa_acc=7175836991&hsa_cam=11760854817&hsa_grp=114199650637&hsa_ad=537520697478&hsa_src=g&hsa_tgt=dsa-1586850151356&hsa_kw=&hsa_mt=&hsa_net=adwords&hsa_ver=3&gclid=CjwKCAjw5pShBhB_EiwAvmnNV6R9xX6t2mXAlnJ6SwtRpYRFNkrhSP8yNZKm_bqjF9KwdVtcWf2KmxoCTccQAvD_BwE. Accessed 30 March 2023
Newcastle University (2023) Business Management BA Honours Course Summary
People and Planet (2023a) People and planet University league methodology. https://peopleandplanet.org/university-league-methodology. Accessed 22 Feb 2023
People and Planet (2023b) People and planet University League 2022/23. https://peopleandplanet.org/university-league. Accessed 20 Feb 2023
Pitt-Watson D, Quigley E (2019) Business school rankings for the 21st century. United Nations Global Compact
QAA (2014) Education for sustainable development: guidance for UK higher education providers. June 2014. QAA763. Gloucester, UK
QAA (2019) Subject benchmark statement—business and management. QAA2471. Gloucester, UK
QAA and Advance HE (2021) Education for sustainable development guidance. Gloucester and York, UK
QAA (2023a) Subject benchmark statement—business and management. Gloucester, UK
QAA (2023b) Subject benchmark statement—master's degrees in business and management. Gloucester, UK
Ragazzi M, Ghidini F (2017) Environmental sustainability of universities: critical analysis of a green ranking. Energy Procedia 117(2017):111–120. https://doi.org/10.1016/j.egypro.2017.07.054

Rayment J, Smith J (2013) The current and future role of business schools. Education + Training, 55:478–494. https://doi.org/10.1108/00400911311326081
Sabzalieva E (2022) Making the link for SDG4: Higher education governance and the quality and equity of higher education systems. UNESCO. https://www.iesalc.unesco.org/en/2022/07/13/making-the-link-for-sdg-4-higher-education-governance-and-the-quality-and-equity-of-higher-education-systems/. Accessed 20 Feb 2023
Schlegelmilch BB (2020) Why business schools need radical innovations: drivers and development trajectories. J Mark Educ 42:93–107. https://doi.org/10.1177/0273475320922285
Slager R, Pouryousefi S, Moon J, Schoolman ED (2020) Sustainability centres and fit: how centres work to integrate sustainability within business schools. J Bus Ethics 161(2020):375–391. https://doi.org/10.1007/s10551-018-3965-4
Stratclyde (n.d.a) BA Hons Business Enterprise. https://www.strath.ac.uk/courses/undergraduate/businessenterprise/. Accessed 30 March 2023
Stratclyde (n.d.b) MBA Master of Business Administration, Full-time, Glasgow. https://www.strath.ac.uk/courses/postgraduatetaught/mba/mba-fulltime/. Accessed 30 March 2023
THE Impact Rankings (2022) Impact Rankings Methodology 2022. Version 1.3. THE DataPoints
THE Impact Rankings (2023) Impact Rankings 2022. https://www.timeshighereducation.com/impactrankings#!/page/0/length/100/sort_by/scores_overall/sort_order/asc/cols/undefined. Accessed 21 Feb 2023
UI GreenMetric (2023a) Overall Ranking 2022. https://greenmetric.ui.ac.id/rankings/overall-rankings-2022. Accessed 20 Feb 2023
UI GreenMetric (2023b) Ranking by Category 2022–Education. https://greenmetric.ui.ac.id/rankings/ranking-by-category-2022/6. Accessed 20 Feb 2023
UI GreenMetric (2023c) Methodology. https://greenmetric.ui.ac.id/about/methodology. Accessed 20 Feb 2023
UN (2015) Transforming our world: the 2030 Agenda for Sustainable Development. Resolution adopted by the General Assembly on 25 September 2015. A/RES/70/1
UN (2017) Work of the Statistical Commission pertaining to the 2030 Agenda for Sustainable Development. Resolution adopted by the General Assembly on 6 July 2017. A/RES/71/313
UN (2020) Education for sustainable development in the framework of the 2030 Agenda for Sustainable Development. Resolution adopted by the General Assembly on 19 December 2019. A/RES/74/223
UNESCO (2020) Education for sustainable development. A roadmap. Education 2030. Paris, France
University of Exeter (n.d.a) BSc Business and Management 2023 entry. https://www.exeter.ac.uk/study/undergraduate/courses2023/business/businessman/#course-content. Accessed 30 Feb 2023
University of Exeter (n.d.b) Business School Undergraduate modules. http://business-school.exeter.ac.uk/study/undergraduate/modules/. Accessed 30 March 2023
University of Exeter (n.d.c) Exeter MBA. https://business-school.exeter.ac.uk/study/masters/mba/. Accessed 30 March 2023
University of Exeter (n.d.d) Full time MBA. https://business-school.exeter.ac.uk/study/masters/mba/about/fulltime/. Accessed 30 March 2023
University of Winchester (n.d.a) University of Winchester Business School. https://www.winchester.ac.uk/about-us/leadership-and-governance/our-faculties/bdt/university-of-winchester-business-school/. Accessed 30 March 2023
University of Winchester (n.d.b) BA (Hons) Business Management. https://www.winchester.ac.uk/about-us/leadership-and-governance/our-faculties/bdt/university-of-winchester-business-school/. Accessed 30 March 2023]
UNPRME.org (n.d.) Signatory members. https://www.unprme.org/signatory-members Accessed 20 Feb 2023

UNPRME.org (2023a) UNPRME UK members. https://www.unprme.org/search?title=&country=206454. Accessed 30 March 2023
UNPRME.org (2023b) PRME Champions. https://www.unprme.org/prme-champions. Accessed 30 March 2023

Using a Group Podcast and Policy Brief in an Economics Undergraduate Course to Enhance Sustainability Literacy, Communication and Teamwork Skills

Paulina Navrouzoglou, Lovleen Kushwah, and Geethanjali Selvaretnam

Abstract The continuing calls for Education for Sustainable Development require Higher Education Institutions to accelerate their curriculum transformation efforts. One way to boost curriculum transformation without additional cost and time is to strengthen student assessment approaches. The purpose of this paper is twofold. First, the development of a holistic assessment approach for promoting graduates' sustainability-oriented knowledge and skills through collaborative learning is articulated. Secondly, its application in an undergraduate Environmental Economics course is illustrated. Data were collected through a survey questionnaire to examine students' views about the assessment approach. The results from the analysis show that the proposed assessment approach can enhance students' engagement, knowledge and skills needed to promote sustainable development. Implications for practice and further research are discussed.

Keywords Education for sustainability · Sustainability literacy skills · Collaborative learning · Active learning · Student-centred · Authentic assessment · Scaffolding

1 Introduction

Accelerated climate change has made our planet highly vulnerable and perversely uncertain. Education needs to be reimagined to prepare learners of all walks of life to plan for, cope with, and find solutions for issues that are threats to our the planet's sustainability. This call for change has been mobilised by the United Nations'

P. Navrouzoglou (✉) · L. Kushwah · G. Selvaretnam
Adam Smith Business School, University of Glasgow, University Avenue, Glasgow G12 8QQ, UK
e-mail: Paulina.Navrouzoglou@glasgow.ac.uk

L. Kushwah
e-mail: Lovleen.Kushwah@glasgow.ac.uk

G. Selvaretnam
e-mail: Geethanjali.Selvaretnam@glasgow.ac.uk

T. Wall et al. (eds.), *Sustainability in Business Education, Research and Practices*, World Sustainability Series, https://doi.org/10.1007/978-3-031-55996-9_4

Decade of Education for Sustainable Development (ESD) initiatives (UNESCO 2014). Education is a key piece of the 2030 Agenda for Sustainable Development, so it is linked to the achievement of Sustainable Development Goals (SDGs) in the current United Nations ESD framework for 2020–2030 (UNESCO 2020).

Many countries and disciplines around the world are supporting these initiatives. Globally, the United Kingdom (UK) aspires to become a world-leading education sector in sustainability and climate change by 2030 (Department of Education (DfE), 2022). The 'Securing The Future' strategy in the UK is designed to deliver this mission emphasising that "*We need to make sustainability literacy a core competency for professional graduates*" (HM Govt 2005, p. 39). Similarly, the Scottish Curriculum of Excellence (2011) identifies learning for sustainability, enterprise and global citizenship as cross-curriculum priorities. ESD is also a key theme of the Quality Assurance Agency Subject Benchmark Statements for Economics (2023). It is stated that a principal purpose of Economics degrees should be to foster critical and creative capacities in students beyond the application of existing tools and models. Essentially, curriculum transformation is needed because current sustainable economic behaviours or solutions may not remain sustainable in the future.

These widely recognised initiatives provide Higher Education Institutions (HEI) with a relevant framework to integrate sustainability into the curriculum. Leal Filho and Vasconcelos (2022) document initiatives and best practices from various disciplines and institutions around the world to disseminate them. Despite the wide range of HEI's initiatives to transform education and learning, the Sustainability Skills Survey 2021–2022 shows only 37% of respondents understand how human activity is affecting nature during their time in HEI. At the same time, the survey indicates that 60% of students are keen to be engaged and learn more about sustainable development and 67% report a positive influence on their own attitudes and behaviours. Researchers have pointed out that there may be systemic obstacles in attaining ESD such as curricular constraints, administrative barriers and lack of training and time (Moore 2005; Doh and Tashman 2014; Leal Filho et al. 2019). Further, learning and teaching approaches may lack theoretical underpinnings making systematic and cross-discipline transferability difficult (Neus et al. 2017). Therefore, there is a need for better-informed learning and teaching ESD practices (Holdsworth and Sandri 2021).

Curriculum transformation for ESD infusion requires going beyond subject knowledge and vocational expertise to help students develop skills for sustainability literacy. This involves developing students' thinking through complexity, learning through dialogue and communication, engaging in deep reflection, developing worldview, valuing sensitivity and assessing activities that support or detract from achieving sustainability (Leal Filho et al. 2019). It also includes skills for identifying problems, solving problems, systems thinking and interdependence, economic restructuring, nudging policy makers effectively, decision-making; and building mindsets such as stewardship of natural resources/environment, respectful limits, social justice and fair distribution, intergenerational perspective, global citizenship and local champions (Nolet 2009). In sum, ESD requires a holistic learning and teaching approach—a golden thread linking knowledge, skills and mindsets that learners need when engaging with sustainability issues.

This paper presents a holistic assessment approach that was designed to promote students' sustainability-oriented knowledge and skills. The design was inspired by the theory of transformative learning (Mezirow 2000; O'Sullivan 2001; Sterling 2005; Wals 2007, 2010; Thomas 2009). Its key feature is that learning should be conducted in groups of individuals with different skills, ideas, characteristics and sustainability objectives so that learners can collaborate by discussing, debating and figuring out solutions. This form of learning focuses on building the capacity of critical thinkers in the present that can increase the likelihood of a citizenry that examines new challenges and acts wisely in the future. 'Transformative' thus refers to "*a shift or a switch to a new way of being and seeing*" (Wals 2011, p. 181). Such a transformative change creates conditions for future growth and development, both intellectually and morally (Dewey 1938). Therefore, transformative learning is a moral, resilient and forward-looking vision of ESD.

A range of activities can create conditions for transformative learning such as open debates, peer involvement, role playing, problem-based learning, simulations, games, case studies, expert engagement, and reflection, multimedia (Deyer and Bongero 1981; Woolfolk 2020; Ornstein and Lasley 2000; Howe 2003; Steinemann 2003; Truscheit and Otte 2004). We build on these insights and unveil new ones in the context of sustainability by incorporating pedagogical principles such as authentic, scaffolding, student-centred, and active.

This paper presents an assessment approach that was designed for an undergraduate environmental economics course. Students were required to work in groups to produce an interview podcast to highlight a real-world environmental problem convincingly to the attention of the relevant authorities, followed by a policy brief suggesting solutions to overcome the problem. In the final piece of assessment, each student was asked to reflect on their learning experience, which is not analysed for the purpose of this paper.

Students' learning can be influenced by their satisfaction with their learning experience (Lizzio et al. 2002) so it is important to consider the students' views about assessment tasks that are new to them. We used a survey questionnaire to understand students' learning experience of the first two assessments—podcast and policy brief—and what skills they developed. The data were analysed using quantitative and qualitative analysis showing that this assessment approach can effectively enhance students' engagement, knowledge and skills needed to promote sustainable development.

The paper is structured in four sections. First, the assessment design is articulated. Second, the methodology is described. Third, the results of the student survey are reported and implications for practice are discussed. Finally, conclusions are drawn and future research lines are discussed.

2 The Assessment Design: Principles and Description

The Aims and Intended Learning Outcomes (University of Glasgow 2022) of the elective undergraduate course, Environmental Economics, are to embed sustainability into the Economics programme curriculum. In previous years, the course was assessed by traditional exams and coursework (during the COVID-19 lockdown period these were online exams). Currently a fresh assessment approach has been introduced. The reason for this change was that ESD requires a more complex range of knowledge, skills, values, and ways of thinking. COVID-19 also demonstrated that future education should be flexible and resilient to different contexts and needs of changing societies. These realisations probed a deeper question—what learning and teaching approach would allow students to cope with the dynamic and complex nature of sustainability challenges and uncertainties? We propose that it can be achieved by creating a transformative learning environment in which students are encouraged to be more open-minded and think about a common good, rather than solely on an individual good. We propose that it can be achieved by creating a transformative learning environment in which students are encouraged to be more open-minded and think about a collective good, rather than individual interests. In this context, transformative learning enables alternative, immersive and novel thinking in finding solutions for sustainability issues that our students can "*co-create [and] co-own [by more reflexive citizens], living in a more reflexive and resilient society*" (Wals 2011, p. 42). The benefits of collaborative learning have been widely recognised in the pedagogical literature (Michaelsen et al. 2004; Koles et al. 2010; Fatmi et al. 2013).

The assessment approach design aligns with the following five principles suggested by Meyers and Nulty (2009, p. 4):

- authentic; real-world and relevant;
- constructive; sequential and inter-linked;
- require students to use and engage with progressively higher-order cognitive processes;
- are all aligned with each other and the desired learning outcomes; and
- provide challenge, interest and motivation to learn.

In that paper the authors explain that these principles can adjust the learning environment in ways that empowers students to adopt an active, student-centred learning approach which, in turn, maximises the quality of their learning outcome and experience.

The assessment schedule used for the analysis of this paper is summarised in Table 1. Students were randomly allocated to groups of four. Group size should be such that new ideas are brought up, discussed, resolved and effectively acted upon without members being obscure or free-riders, for which a group size of three to five would be not too small or large (Davies 2009; Race 2015; Francis et al. 2022). The random allocation of groups aimed to enable the pluralism of ideas for transformative disruptions to emerge (Wals 2011). Moreover, the idea behind letting students work together for a long period and assigning group assessment a high share in the final course grade (80%) was to understand whether these features can improve group dynamics and social cohesion among the diverse group of students for better listening and empathy (Wals 2011; Cheng and Selvaretnam 2022).

Students were allocated to groups in week 2 of the semester and the submission deadlines of the podcast and policy brief were in weeks 6 and 10, respectively. This period gave students ample time to perform both tasks. In-class discussion activities (e.g. review of rubrics and exemplars, think-pair-share) were used throughout this period to support students and encourage a collaborative learning environment and attitude.

The group assessment tasks were chosen to give students opportunities to express their knowledge, opinions and creativity using different communication mediums, and co-build new knowledge and skills. An imaginary real-life work scenario was provided to the students which required them to demonstrate skills and competencies that are needed to resolve environmental problems, with the expectation that this learning experience will encourage them to be active citizens and empowered professionals in future.

The first group assessment task was to create an interview podcast to identify an urgent environmental problem of a country and convince the 'Minister-in-charge' about the urgency of tackling it. To meet the marking criteria, students had to clearly describe the problem using evidence and well-persuasive argument, produce a podcast using content and style which is original and appropriate and use effective communication, production, and editing. Creating a podcast was a new assessment task for students thus training and recording facilities were offered to ensure nobody was excluded.

Students selected noteworthy urgent environmental problems from all over the world that are associated with the SDGs. Examples are:

- floods in Pakistan,
- desertification in China,
- air pollution in Delhi,
- air pollution in Nigeria,
- water scarcity in Egypt,
- glacial melting in Kyrgyzstan,
- drought in Madagascar,
- air pollution in India,
- rising sea level in Maldives,
- floods in the Philippines,

Table 1 The assessment schedule

	Assessment	% of Total grade	Duration/ length
Interview podcast	Group	30%	10 min
Policy brief	Group	50%	1000 words
Reflection	Individual	20%	500 words

- deforestation in Australia,
- deforestation in the Brazilian Amazon.

This type of assessment task involves a meaningful demonstration of knowledge and skills in real life contexts so it can be considered as an authentic assessment task (Swaffield 2011). By doing so, students promote higher order thinking/cognition and self-regulated learning that goes beyond the knowledge of issues (Newmann 1996).

Podcasting is an authentic learning and assessment experience on its own (Thompson 2007). Several other benefits are highlighted in the literature such as it promotes reflective thinking (Fisher and Baird 2006), provides a platform for collaborative learning and student knowledge creation (Lee et al. 2008), meets a range of learning styles and develops key skills including critical thinking (Dale 2007).

For the second group assessment task, students had to write a policy brief to propose to the 'Minister-in-charge' solutions for tackling the environmental problem presented in their podcast. Students had to showcase the ability to collate, integrate, synthesise and analyse research findings as well as communicate in a way that grabs the attention of someone who is busy and a non-subject specialist, which is a key employability skill for economics graduates (Jenkins and Lane 2019). Training on how to write a policy brief was offered since this assessment task was new to students too.

To perform this task, students needed to have a deep understanding of the problem at hand and why the current policy approach does not work. This sequential and interlinked design of assessment promotes higher-order thinking as experiences gained in the first assessment could be built upon the second assessment (Meyers and Nulty 2009). This represents itself as a scaffolding that enables a learning pathway for enhancing student understanding and development (Ge et al. 2000). Another critical characteristic of the scaffolding process is that it is shifting the learning responsibility to students over time (Van de Pol et al. 2010).

After completing each group assessment, students had to complete an evaluation form to acknowledge everyone's (including their own) contribution. These forms were used to avoid unfairness in marking resulted from unequal contributions. Finally, students submitted an individual reflection about their course learning experience, the deadline of which was in week 12. This assessment is not analysed for the purpose of this paper.

3 Methods

The data were collected from students in the course Environmental Economics, which is a popular elective course in the Economics undergraduate programme at the University of Glasgow. Among the 129 students who have currently taken the course, 86 (67%) consented to be part of the research. The participants consisted of 48 males (56%) and 38 females (44%), 35 domestic (40%) and 51 (60%) international students. At the University of Glasgow, all research involving human data is subject to formal ethical review. To comply with this regulation, we obtained an ethics approval from the University College of Social Sciences Ethics Committee for Non-Clinical Research Involving Human Subjects of the College where this research has taken place.

Firstly, the grade distribution of both assessments is presented to show how the students performed and compare their performances between the two assessments. However, the main objective of this paper is to understand the students' perceptions of their learning experience with the assessments and what skills they have developed. To elicit this information, a survey questionnaire was developed asking students to indicate if they agreed or disagreed with a series of statements about the expected skills and learning experiences of each assessment on a Likert scale (Strongly Agree, Agree, Neutral, Disagree, Strongly Disagree). It is noteworthy that the skills are not limited to the context of sustainability but go beyond; they are 21st Century Skills (Joynes et al. 2019). Table 2 presents the statements for each assessment.

The data obtained from the Likert-scale survey questions were analysed using visual and primarily descriptive analysis. Responses were also examined by gender and nationality (domestic/international) to identify possible differences.

At the end of the survey, there was an open-ended question asking: 'If you would like to share anything related to the podcast and/or policy brief assessment, then please add your comments below'. Out of the 86 students, 31 responded to it, but only 29 open comments provided data with justification that were used for analysis.

Table 2 The likert-scale survey questions

	Podcast	Policy Brief
Skills		
Developed social skills	•	•
Developed digital skills	•	
Developed research skills		•
Developed skills transferable to the world of work		•
Developed the ability to communicate research findings to a non-specialist audience		•
Developed professional communication skills	•	
Developed skills for identifying environmental problems	•	
Actively engaged in resolving environmental problems	•	
Actively engaged in working toward resolution of environmental problems		•
Learning experience		
Should continue to be used in future	•	•
Promoted independent learning	•	•
Internalised the course content in an interactive manner	•	•
Made learning more interesting	•	•
Was enjoyable	•	

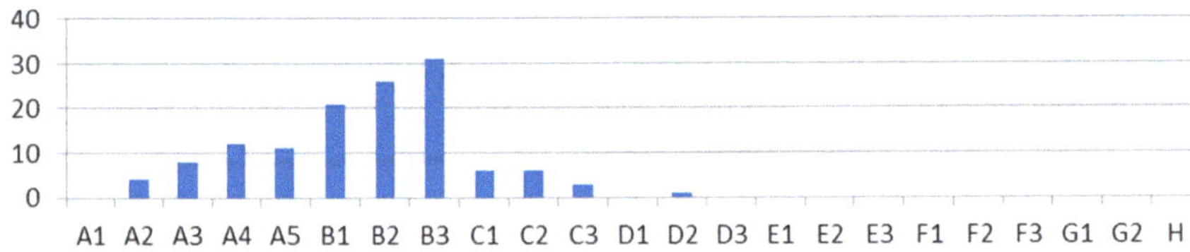

Fig. 1 Podcast grades

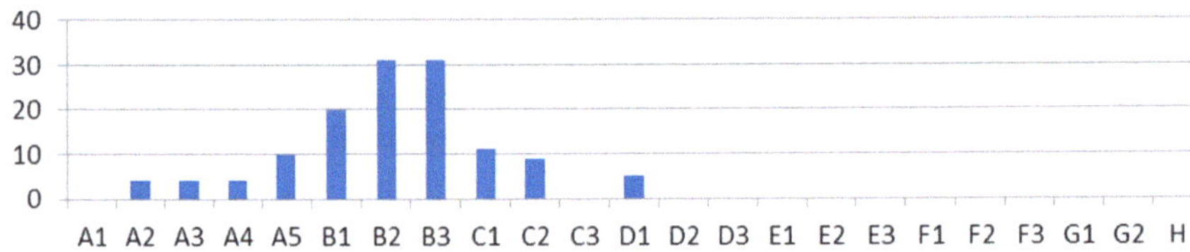

Fig. 2 Policy brief grades

After a reflective and thoughtful reading, the authors of this paper split these open comments into individual statements and categorised them (Braun and Clarke 2019; Waltzer et al. 2023). It's worth noting that these open comments provide limited data for any further analysis.

The instructor of the course is one of the authors of this paper whose expertise and drive to embed sustainability into the Economics programme resulted in this new form of assessments being introduced. So, we have also included the instructor's reflection on her experience and views.

4 Results

4.1 Students' Performance

Students' performance was of a high standard, with a mean grade of B2 which is indicative of a very good performance and achievement of the desired learning outcomes.[1] Figures 1 and 2 show the grade distribution of podcast and policy brief, respectively. It can be regarded that there is not much difference in grade distributions across the two assessments, despite that these were different assessment tasks; and the second one was designed in a sequence that required students to involve themselves in higher-order cognitive processes. These attained outcomes show that the assessment tasks inspired students to both stay engaged with their learning and demonstrate relevant skills and competencies.

[1] University of Glasgow grades are on a 22-point scale from A1 to G. More information can be found here.

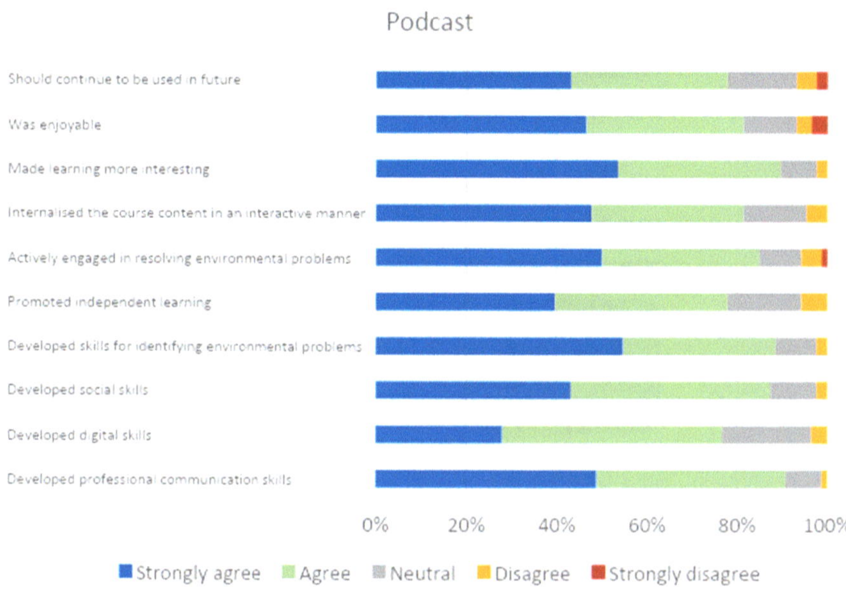

Fig. 3 Frequency of student perceptions about podcast

Fig. 4 Frequency of student perceptions about podcast: male and female

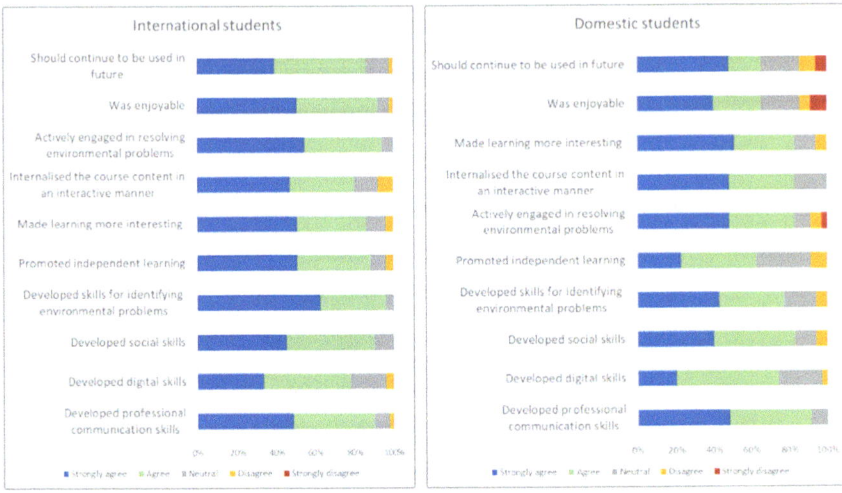

Fig. 5 Frequency of student perceptions about podcast: international and domestic

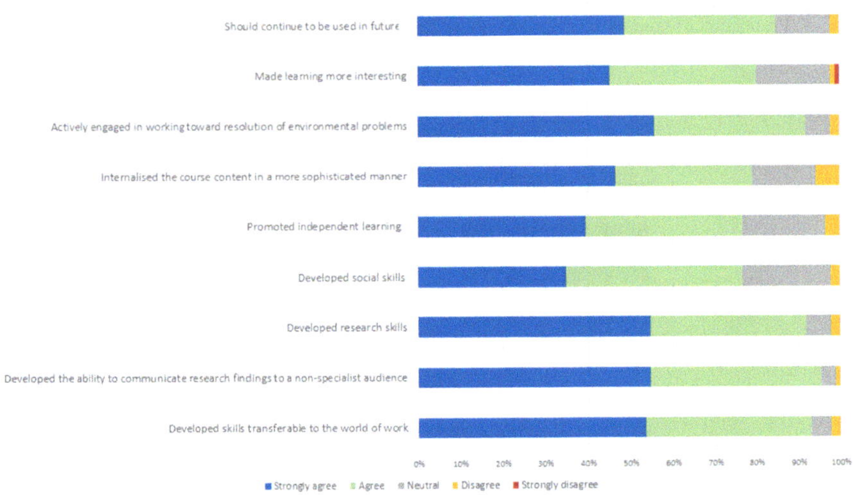

Fig. 6 Frequency of student perceptions about policy brief

4.2 Students' Perceptions

The findings of the Likert-scale survey questions enabled us to get some clear insights into whether the students' perceptions concerning the design of the two assessments are compatible with our assumptions and expectations that these assessments enhance students' sustainability-oriented skills and learning experience. The results are presented in Figs. 3, 4, 5, 6, 7 and 8.

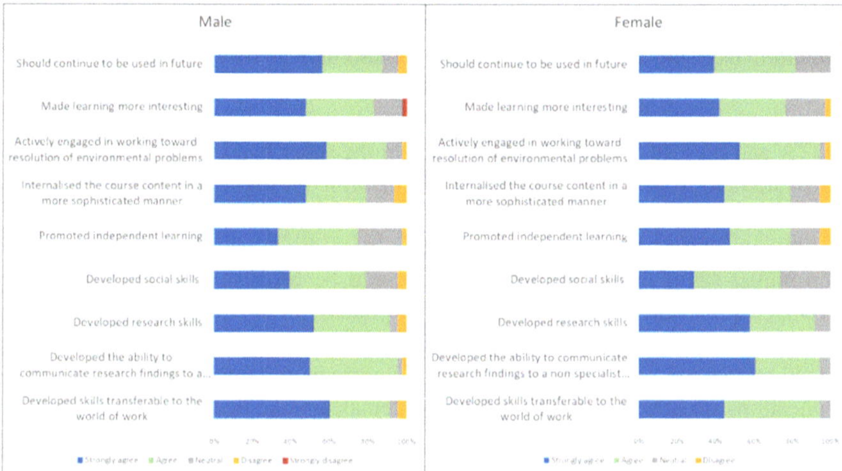

Fig. 7 Frequency of student perceptions about policy brief: male and females

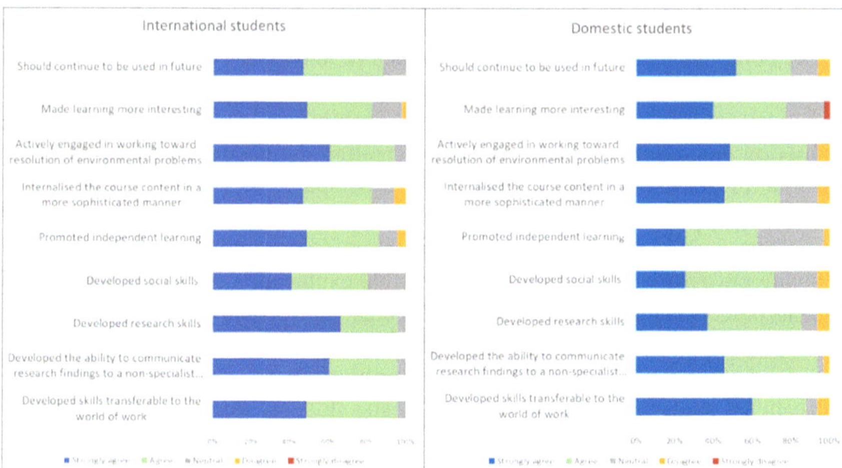

Fig. 8 Frequency of student perceptions about policy brief: domestic and international

Generally, the results are very encouraging. Students' perceptions of creating the podcast and writing the policy brief were overwhelmingly positive for both genders (females and males) and nationalities (international and domestic). There are however some noteworthy differences in the level of agreement with the statements between these two sub-groups that are highlighted and discussed.

Podcast

Figures 3, 4 and 5 present the results of the podcast. In terms of skills, more than 75% of students agreed and strongly agreed that working on the podcast helped them develop the skills outlined in Table 2. At the same time, the podcast helped students to learn independently, with more benefits indicated by international students (88%) compared to domestic students (63%). More than 90% of males benefitted in developing social skills (92%) and communication skills (96%), in comparison to 82% and 84% of females respectively. On the other hand, 79% of females said they benefitted from digital skill development, which is higher compared to 75% of males.

Although the agreeability rate is also high for digital skills, why this was not even higher (as it is for the other skills) could be because students had already developed these skills, or they didn't perceive podcast production as an opportunity to develop their digital skills. Furthermore, the observed gender differences can portray the old stereotype that 'technology' is a male-oriented domain, which might have implications on how differently males and females make decisions when it comes to technology adoption and use (Morris et al. 2005). On the other hand, studies show that female students have a more positive attitude toward communication skills learning than male students. This may be because female students perceive that their communication skills need improvement (Cleland et al. 2005) or because they perform better in the dimensions of empathy, structure, verbal expression and non-verbal expression than male students (Graf et al. 2017). In light of this observation, a more comprehensive investigation is needed to understand gender differences in digital and communication skills.

Another important result was that 89% of students said they 'developed skills in identifying environmental problems' after creating the podcast. It is worth noting that 77% of domestic students were positive compared to a much higher of 96% international students and 83% of males compared to 95% females. Around 85% of students gave a positive response that the podcast 'actively engaged them in resolving environmental problems'. A possible explanation of these results is that students perceived that having experience in creating a persuasive podcast improved their capability for action or developed their activism skills for raising public awareness about environmental problems. A further break down suggests that 94% of the international students and 84% of the domestic students were positive. Moreover, 90% of international students affirmed that the podcast developed their social skills compared to 83% of domestic students.

In terms of learning experience, students agreed that working on the podcast was enjoyable (81%) and interesting (90%), and that they want it to continue in the future (78%). The difference between domestic/international students can be observed once again. International students indicated a preference for the podcast assessment to continue (86%) which is much higher than that indicated by domestic students (66%). Further, international students found working on podcast was far more enjoyable (92%) than domestic students (66%). It should be emphasised that only 7% (or 6 out of 86) of students disagreed or strongly disagreed that the podcast was enjoyable and didn't want such an assessment to continue in the future.

In sum, the responses indicate that international students are more agreeable with the survey statements about the podcast in comparison with domestic students. There may be several reasons for this difference that could be better explored using longitudinal studies. An intuitive explanation is that international students are more positive because they perceived this assessment as an opportunity to improve their skills, especially English, social and communication skills. Studying online over the last three years didn't allow international students to develop these skills so they might have been keener to be engaged with the assessment to practice the English language and develop social and communication skills. On the contrary, domestic students might have perceived that these skills were already acquired, and hence they didn't perceive that these skills were further developed because of the assessment, especially if they overestimated themselves (Kruger and Dunning 1999) or perceived themselves to be more competent in such skills in comparison with the other students of their group. Moreover, students who think their communication skills need no improvement might have more negative attitudes towards communication skills (Rees et al. 2002).

Policy Brief

Figures 6, 7 and 8 present the results of the policy brief. The results are not that different from the results of the podcast, in fact, there are even more positive responses. More than 92% of students agreed and strongly agreed that the policy brief assessment helped them develop the skills, except for social skills (77%) which is not surprising given that this skill already evolved while working on the podcast (87%).

Similar to the podcast, international students showed a higher agreeability rate than the domestic students across all skills (research at 96% for international vs 86% for domestic students, communication to non-specialist at 96% vs 95%, transferrable skills at 96% vs 89%, and social skills at 80% vs 72%). Further, 80% of males compared to 74% of female students were positive about developing social skills, as in the podcast. Moreover, even though the task was to be completed in groups, the policy brief helped students to learn independently (77%), more so for international students (86%) than domestic students (63%). Most importantly, students felt that they were able to actively engage in working toward the resolution of environmental problems. With 92% of agreeability among students, consistent across all demographics, it confirms that the policy brief has been successful in further enhancing students' sustainability-oriented knowledge and skills.

A high agreement rate was also observed when students were asked about their learning experience. Working on the policy brief was interesting (80%) and they wanted it to continue in the future (85%), with both agreeability percentages being slightly higher for male and international students.

It is important to note that not more than 2.3% (or 1–2 out of 86 students) disagreed or strongly disagreed with the survey statements about the policy brief, except for the statement about whether the policy brief 'internalise course content in a more sophisticated manner' for which the percentage of disagreement was 7% (or 5 out of 86 students). This discrepancy will be further explored in the next section.

In sum, the results show again that international students benefitted more across all variables.

4.3 Reflection on Students' Open Comments

Students' open comments (29 students) were carefully categorised to shed further light on their perspectives about the two assessments. Three categories (themes) of statements emerged from the analysis: 'Benefits of', 'Drawbacks of' and 'Going forward'. These categories were split into subcategories to evaluate the assumptions and theoretical underpinnings of the proposed assessment approach. Table 3 presents the categories and subcategories (frequencies are given within brackets).

Students' open comments stated more benefits than drawbacks about the assessments. They highlighted the value students placed on their learning experience and skills acquisition through the assessments and referred to collaborative learning and sustainability-oriented knowledge. In terms of drawbacks, some students found it difficult to connect the course content with the assessment tasks, while others found that working as a group was challenging or that the weight assigned to the group assessment tasks was too high. Hence, the open comments are discussed below from the context of ways to reinforce benefits, address drawbacks, and respond to students' suggestions for the future.

In line with the results of the analysis in the previous sub-section, students' open comments clearly show their excitement for new positive learning experience and transferable skills development that motivated them to stay engaged. Students also acknowledged the various benefits of collaborative learning showing evidence of emotional intelligence. This is consistent with research that shows that collaborative work promotes academic achievement and collaborative abilities (Johnson and Johnson 2004; Gillies and Boyle 2010). Students also recognised the opportunities to expand sustainability-oriented knowledge. It is also worth mentioning that students had been more reflective in terms of sustainability-oriented knowledge, skills and mindsets as well as collaborative learning in their reflection assessment.

A few students commented that the course content and the assessments were not aligned, however, it should be noted that four of them (out of 5) worked in the same group. Therefore, their experience might be related to the nature of the problem they worked on together. It should be noted that these students disagreed with the survey statement that the policy brief 'internalise course content in a more sophisticated manner'. To better facilitate autonomous and self-regulated learning in future, some case studies can be discussed in the class rather than provided as additional reading to help students to understand how the course content fits into the 'bigger picture' and why the assessment tasks do not require reproduction of course content, but demonstration of deeper engagement with their learning.

Students referred to the common group work challenge of uneven student contribution. It is noted that these students were also negative with the survey statements related to 'enjoyment' and 'should be continued'. The presence of conflicts seems to hamper the learning and experiences of these students (Hammar 2014).

A solution to this problem might be to reduce the weight of the group work below 80% of the final course grade, as two students pointed out in their open comments. However, a deeper investigation is required since the students who reported uneven

Table 3 Description of the emerging categories from students' open comments

Category	Description/Subcategory	Examples quotes
Benefits of (17)	Learning experience (8)	"Enjoyable experience and an interesting yet captivating change to assessment"
		"I think both assignments make the course more dynamic. I enjoyed it a lot"
		"Great assessments which were different to traditional forms used in Economics. Class time to get to know groups was useful"
		"Fun and engaging items of work already recommended the module to junior honours students and younger"
	Skills (5)	"Taught me a lot more skills than an individual exam would"
		"Really awesome set of assessments which are directly transferable into work outside academia"
	Collaborative learning (2)	Although it was difficult in places to organise things and make sure everyone was working equally while still trying to achieve good grades I think the skills learnt were more transferable to work than other assignments I have partaken in" "
		I have enjoyed the atmosphere of working together as a group and hope to continue this form of work in the future. With a lot of searching for information and discussing with each other, it was a real experience of what it is like to play a professional and to practice, which is more important than just reading the content"
		"Improve diversification of work for students"
	Sustainability-oriented knowledge (2)	"While I enjoyed engaging with existing literature related to a specific environmental issue"
		"Writing a policy brief was a really entertaining activity as it gave me an idea of how it would be to be a policy maker in favour of the environment."
Drawbacks of (10)	Difficulties in alignment with the course content (5)	"I found that there was a disconnect between the course material and the assignments"
	Assessment weight (2)	"I do not think that 80% of my grade should depend on work of multiple people"

(continued)

Table 3 (continued)

Category	Description/Subcategory	Examples quotes
	Unequal contributions (3)	"I feel that my above answers have been informed by a really unequal distribution of labour, I felt the burden of work fell mainly on my shoulders and that of one other group member and as a result found the activities quite stressful. Were it not for this I feel I would have enjoyed it much more. However, given the random nature of assigned groups I feel having such a large portion of the assessment based on group work was unfair and emotionally taxing"
		"The problem of group projects, especially if you cannot choose, you will always end up with poor dynamics. When it came to the podcast a couple team members struggled speaking English"
Going forward (8)	Suggestions for the future cohort (2)	"When doing these two, consider it as a real problem that happened in real life, instead of an academic work. Tackling the problem and give suggestions based on the evidence."
		"For podcast and policy brief, two group tasks, communication efficiency among team members is crucial. Try contacting your team members and talking to each other before the podcast and policy brief even begin. This is very conducive to the development of the group work. Because everyone is already familiar with each other and more willing to express their own views and accept the views of the team members before carrying out the project."
	Suggestions for the instructor (6)	"Assessments more closely related to the lecture material might be more beneficial. podcast in which real-world examples of abatement costs are discussed might help to consolidate knowledge of pollution controls"
		"The work of the outstanding group can be published with the permission of the students, and they can be invited to present the experience"

student contribution in their peer evaluation form typically mentioned English language barriers. There might have been cultural differences in explanations, styles, attitudes, beliefs and even aspirations. To overcome such issues, anonymous evaluations while the project is under way can be used in future.

Wals (2010) points out that although there is more learning between those who are different than between those who are similar, it is important for there to be 'chemistry' between the group members without which the differences will prevent learning. Hammar (2014) found that a lack of time to know each other can influence students' ability to learn and experience with group work. It seems that the interaction over a long period and discussion activities in class helped students to develop social cohesion, as it is evident from other students' open comments. Further work is however needed to understand students' perceptions of social cohesion in diverse groups. For example, allowing groups to choose an alternative assessment format, instead of taking a type that might disadvantage them, is the main approach used in the inclusive assessment literature (Tai et al. 2021).

Finally, students proposed ways to move forward. Some comments focused on the next class cohort, and they will be shared in class as 'letter from your peers'. Another student suggested the best podcasts are disseminated, and indeed there is such a plan. By doing that, students will also perceive that their collective voice is heard which might increase their activism action in future.

Instructor's Reflection

The learning objectives of the assessments are to provide students with opportunities to demonstrate progress when applying their knowledge and skills in increasingly demanding and unfamiliar contexts. They are assessments *for* learning (Scottish Curriculum of Excellence 2011). In the context of sustainability however, knowledge and skills of a single student against the range of challenges of environmental problems may not be sufficient to foster new knowledge and skills, inspire behaviours for future collective action and cause transformative learning changes. Thus, their present learning experience may not act as seeds of positive future changes. To achieve that through learning and education, we (educators) should all work together on what constitutes ESD. My co-authors and I have been inspired to start exploring that through this paper.

Students had to visualise the scenario of being a team of environmental consultants working for the Department of Environment Management of a country of their choice which is undergoing new changes to tackling the country's most important environmental challenges. This scenario should have inspired students to behave as global professional citizens working towards a common good. By doing so, students became aware of the environmental problems of numerous countries, and through dialogue and compromise, they selected a critical threat to sustainability. For the podcast, the students collected evidence to prepare an advocative and convincible interview podcast that calls for immediate policy action. Building upon this new knowledge, awareness and skills, they wrote a policy brief to propose a co-created solution in a way that a non-specialist (e.g. public) can follow.

I will use Arran Stibbe's (2009, p. 11) definition of *sustainability literacy* to express my belief (and hope) in the student's learning journey:

> As people gain sustainability literacy skills, they become empowered to read society critically, discovering insights into the unsustainable trajectory that the society is on and the social structures that underpin this trajectory. But more than this, they become empowered to engage with those social structures and contribute to the re-writing of self and society along more sustainable lines.

5 Conclusion

Education is a key driver to overcome sustainability challenges posed by the anthropogenic climate change. It can seed a 'good' Anthropocene. It has been argued that transformative learning can prepare learners better to respond and contribute to the current and future sustainability challenges. We considered how education toward transformative learning can be applied though a careful design of an assessment approach that combines pedagogical principles to offer a more holistic ESD practice.

Following the specification of the principles of assessment design for both sustainability literacy skills development and student-centred learning, this paper illustrated the design, application and effectiveness of a scaffolded group assessment approach that combines an interview podcast and policy brief for advocative highlighting and addressing a real and urgent environmental problem around the world. We utilised a questionnaire to gather students' views on the assessment tasks of an undergraduate elective course on Environmental Economics at the University of Glasgow. Following data collection, we employed qualitative and quantitative analysis methods to examine the responses.

The results show that these assessment tasks can encourage students to work productively and enthusiastically with their group members to pursue research about sustainability problems and solutions as well as to build sustainability-oriented knowledge and skills by sharing similar interests but different skills, ideas, characteristics, and experiences. Students agreed that the assessment tasks facilitated the development of various skills such as research, independent thinking, digital, social and communication skills. They also recognised that these skills are transferable and related to the identification and resolution of environmental problems. Further, students were affirmative that the assessment tasks were interesting, and enjoyable and should be continued in future.

Our findings reiterate the importance of opportunities to work in groups, support and encouragement that students need for collaboration and working as a group, and the joy of building knowledge and transferable skills independently. It is not surprising that there were a few disgruntled comments about group work being unfair and queries as to why the course content was not directly assessed. Since these types of assessment tasks are new to the students, we can do more about raising students' understanding of what these assessments are setting out to achieve—the development of their sustainability literacy skills that will empower them to engage and contribute to a common good, "*the re-writing of self and society along more sustainable lines*" (Stibbe 2009, p. 11).

Our principles of assessment design can be applied to other assessment tasks such as vodcasts, posters and panel discussions taking into consideration the benefits of having a variety of assessment types across a course or programme. In addition, the assessment task can be about proposing solutions to any SDG at a global, country, or local level (e.g. cities, universities, industries). The students' work can be disseminated or presented to the relevant stakeholders so that their voices can bring out real change.

There is potential for further valuable research to be carried out in this strand. It would be beneficial to do a follow-up survey after some period, especially once the students start seeking employment or higher studies to understand whether the course learning experience has any effect on their desire for action competence for sustainability. We could also do further data analysis of the responses given by students in the same group to understand whether there were any differences and why. This is particularly important for the groups with international and domestic students as the results indicate different perceptions in terms of skills development and learning experience as well as potential language barriers for building social cohesion across diverse groups. Moreover, the results highlight gender differences when it comes to communication and digital skills so it could be useful to explore the inclusivity and accessibility of the assessments. Finally, to contribute further to this literature, we plan to perform a qualitative analysis of the final reflection assessment, which is not included in this paper, to shed further light on our ESD practice as well as on the values of reflection practice in the context of ESD practice.

References

Braun V, Clarke V (2019) Reflecting on reflexive thematic analysis. Qual Res Sport Exerc Health 11(4):589–597. https://doi.org/10.1080/2159676X.2019.1628806

Cheng W, Selvaretnam G (2022) Effects of mixed groups on multicultural interaction and student experience. Learn Teach 15(2):1–28. https://doi.org/10.3167/latiss.2022.150202

Cleland J, Foster K, Moffat M (2005) Undergraduate students' attitudes to communication skills learning differ depending on year of study and gender. Med Teach 27:246–251. https://doi.org/10.1080/01421590400029541

Curriculum for Excellence Factfile—Overview of Key Terms and Features (2011). Education Scotland. https://education.gov.scot/media/0g2cthxv/cfefactfileoverview.pdf

Dale C (2007) Strategies for using podcasting to support the learning of tourism students. J Hosp Leis Sport Tour Educ 6(1):49–57. https://doi.org/10.3794/johlste.61.155

Davies WM (2009) Groupwork as a form of assessment: common problems and recommended solutions. High Educ 58(4):563–584. https://doi.org/10.1007/s10734-009-9216-y

Department for Education (DfE) (2022) Sustainability and climate change: a strategy for the education and children's services systems. https://www.gov.uk/government/publications/sustainability-and-climate-change-strategy/sustainability-and-climate-change-a-strategy-for-the-education-and-childrens-services-systems

Dewey J (1938) Experience and education. First Collier Books ed. 1963. Collier Books, New York

Deyer CA, Bongero A (1981) Trigger films for teaching in the affective domain. Am J Med Technol 47(4):255–258. PMID: 7223773. https://pubmed.ncbi.nlm.nih.gov/7223773/

Doh JP, Tashman P (2014) Half a world away: the integration and assimilation of corporate social responsibility, sustainability, and sustainable development in business school curricula. Corp Soc Responsib Environ Manag 21:131–142. https://doi.org/10.1002/csr.1315

Fatmi M, Hartling L, Hillier T, Campbell S, Oswald AE (2013) The effectiveness of team-based learning on learning outcomes in health professions education: BEME Guide No. 30. Med Teach 35:1608–1624. https://doi.org/10.3109/0142159X.2013.849802

Fisher M, Baird DE (2006) Making mLEarning work: Gen Y, learning and mobile technologies. J Educ Technol Syst 35(1):3–30. https://doi.org/10.2190/4T10-RX04-113N-8858
Francis N, Allen M, Thomas J (2022) Using group work for assessment: An academic's perspective. National Teaching Repository Presentation. https://doi.org/10.25416/NTR.21270621.v1
Ge X, Yamashiro KA, Lee J (2000) Pre-class planning to scaffold students for online collaborative learning activities. Educ Technol Soc 3(3):159–168. https://www.jstor.org/stable/jeductechsoci.3.3.159
Gillies RM, Boyle M (2010) Teachers' reflections on cooperative learning: issues of implementation. Teach Teach Educ 26:933–940. https://doi.org/10.1016/j.tate.2009.10.034
Graf J, Smolka R, Simoes E, Simoes E, Zipfel S, Junne F, Holderried E, Wosnik A, Doherty AM, Menzelet K, Herrmann-Werner A (2017) Communication skills of medical students during the OSCE: gender-specific differences in a longitudinal trend study. BMC Med Educ 17:75. https://doi.org/10.1186/s12909-017-0913-4
Hammar CE (2014) Group work as an incentive for learning—students' experiences of group work. Front Psychol 5:558. https://doi.org/10.3389/fpsyg.2014.00558
HM Government (2005) Securing the future: delivering UK Sustainable Development Strategy (publishing.service.gov.uk)
Holdsworth S, Sandri O (2021) Investigating undergraduate student learning experiences using the good practice learning and teaching for sustainability education (GPLTSE) framework. J Clean Prod. https://doi.org/10.1016/j.jclepro.2021.127532
Howe A (2003) Twelve tips for developing professional attitudes in training. Med Teach 25(5):485–487. https://doi.org/10.1080/0142159031000160562 4
Jenkins C, Lane S (2019) Employability Skills in UK Economics Degrees: a report for the Economics Network. https://www.economicsnetwork.ac.uk/research/employability
Johnson DW, Johnson RT (2004) Assessing students in groups: promoting group responsibility and individual accountability. Corwin Press. ISBN-13: 978-0761939474
Joynes C, Rossignoli S, Fenyiwa Amonoo-Kuofi E (2019) 21st Century Skills: Evidence of issues in definition, demand and delivery for development contexts (K4D Helpdesk Report). Institute of Development Studies, Brighton UK. https://opendocs.ids.ac.uk/opendocs/handle/20.500.12413/14674
Koles PG, Stolfi A, Borges NJ, Nelson S, Parmelee DX (2010) The impact of team-based learning on medical students' academic performance. Acad Med 85(11):1739–1745. https://doi.org/10.1097/ACM.0b013e3181f52bed
Kruger J, Dunning D (1999) Unskilled and unaware of it: how difficulties in recognizing one's own incompetence lead to inflated self-assessments. J Pers Soc Psychol 77(6):1121–1134. https://doi.org/10.1037/0022-3514.77.6.1121
Leal Filho W, Shiel C, Paço A, Mifsud M, Veiga Ávila L, Londero Brandli L, Molthan-Hill P, Pace P, Azeiteiro MU, Ruiz Vargas V, Caeiro S (2019) Sustainable development goals and sustainability teaching at universities: falling behind or getting ahead of the pack? J Clean Prod 232:285–294. https://doi.org/10.1016/j.jclepro.2019.05.309
Leal Filho W, Vasconcelos C (2022) Handbook of best practices in sustainable development at university level. Springer Nature. ISBN: 978-3-031-04763-3
Lee MJ, McLoughlin C, Chan A (2008) Talk the talk: learner-generated podcasts as catalysts for knowledge creation. Br J Edu Technol 39:501–521. https://doi.org/10.1111/j.1467-8535.2007.00746.x
Lizzio A, Wilson K, Simons R (2002) University students' perceptions of the learning environment and academic outcomes: implications for theory and practice. Stud High Educ 27(1):27–52. https://doi.org/10.1080/03075070120099359
Meyers NM, Nulty DD (2009) How to use (five) curriculum design principles to align authentic learning environments, assessment, students' approaches to thinking and learning outcomes. Assess Eval High Educ 34(5):565–577. https://doi.org/10.1080/02602930802226502
Mezirow J (2000) Learning as transformation: critical perspectives on a theory in progress. Jossey Bass, San Francisco, USA. ISBN-13: 978-0787948450

Michaelsen LK, Knight AB, Fink LD (2004) Team-based learning: a transformative use of small groups in college teaching. Centers for Teaching and Technology—Book Library 199. https://digitalcommons.georgiasouthern.edu/ct2-library/199

Moore J (2005) Barriers and pathways to creating sustainability education programs: policy, rhetoric and reality. Environ Educ Res 11(5):537–555. https://doi.org/10.1080/13504620500169692

Morris MG, Venkatesh V, Ackerman L (2005) Gender and age differences in employee decisions about new technology: an extension to the theory of planned behaviour. IEEE Trans Eng Manag 52(1):69–84. https://doi.org/10.1109/TEM.2004.839967

Neus E, Stevenson RB, Lasen M, Ferreira J, Davis JM (2017) Approaches to embedding sustainability in teacher education: a synthesis of the literature. Teach Teach Educ 63:405–417. https://doi.org/10.1016/j.tate.2017.01.013

Newmann FM (1996) Authentic assessment in social studies: standards and examples. In: Phye GD (ed) Handbook of classroom assessment. Academic Press, pp 359–380. https://doi.org/10.1016/B978-012554155-8/50014-4

Nolet V (2009) Preparing sustainability-literate teachers. Teach Coll Rec 111(2):409–442. https://doi.org/10.1177/016146810911100207

Ornstein AC, Lasley TJ (2000) Strategies for effective teaching, 3rd edn. McGraw-Hill, Boston, MA. ISBN-13: 978-0697298850

O'Sullivan E (2001) Transformative learning: educational vision for the 21st century. Zed Books, London, UK. ISBN-13: 978-1856496995

Quality Assurance Agency Subject Benchmark Statements for Economics (2023) https://www.qaa.ac.uk/the-quality-code/subject-benchmark-statements/subject-benchmark-statement-economics

Race P (2015) Making small group teaching work. Chapter 4 in The Lecturer's Toolkit. The Lecturer's Toolkit (4th edn). Routledge, Abingdon. https://doi.org/10.4324/9781315767277

Rees C, Sheard C, Davies S (2002) The development of a scale to measure medical students' attitudes towards communication skills learning: the Communication Skills Attitudes Scale (CSAS). Med Educ 36:141–147. https://doi.org/10.1046/j.1365-2923.2002.01072.x

Scottish Curriculum of Excellence (2011). Building the curriculum 5: a framework for assessment. https://education.gov.scot/Documents/btc5-framework.pdf

Stibbe A (2009) The handbook of Sustainability literacy skills for a changing world. Green Books Ltd, ISBN: 978 1 900322 60 7. https://sustainability.glos.ac.uk/wp-content/uploads/2017/07/Handbk-Sustainability-literacy-EC-16092020.pdf

Steinemann A (2003) Implementing sustainable development through problem-based learning: pedagogy and practice. J Prof Issues Eng Educ Pract 129(4):216–224. https://doi.org/10.1061/(ASCE)1052-3928(2003)129:4(216)

Sterling S (2005) Higher education, sustainability, and the role of systemic learning. In: Corcoran PB, Wals A (eds) Higher education and the challenge of sustainability: Problematics, promise, and practice. Kluwer Academic Press, Dordrecht. https://link.springer.com/chapter/https://doi.org/10.1007/0-306-48515-x_5

Sustainability and climate change: a strategy for the education and children's services systems- UK Department for Education. https://www.gov.uk/government/publications/sustainability-and-climate-change-strategy/sustainability-and-climate-change-a-strategy-for-the-education-and-childrens-services-systems

Sustainability skills survey 2021-22. Higher Education Report. https://www.sos-uk.org/post/sustainability-skills-2021-22-research-report-launched#:~:text=66%25%20say%20they'd%20like,from%2035%25%20the%20previous%20year

Swaffield S (2011) Getting to the heart of authentic assessment for learning. Assess Educ Princ Policy Pract 18(4):433–449. https://doi.org/10.1080/0969594X.2011.582838

Tai J, Ajjawi R, Umarova A (2021) How do students experience inclusive assessment? A CritAl Rev Contemp Lit Int J Incl Educ. https://doi.org/10.1080/13603116.2021.2011441

Thomas I (2009) Critical thinking, transformative learning, sustainable education, and problem-based learning in universities. J Transform Educ 7(3):245–274. https://doi.org/10.1177/1541344610385753

Thompson L (2007) Podcasting: the ultimate learning experience and authentic assessment, in: ICT: Providing Choices for Learners and Learning, 1019–1023, Proceedings ascilite Singapore. https://www.ascilite.org/conferences/singapore07/procs/thompson-poster.pdf

Truscheit A, Otte C (2004) Sustainable games people play: teaching sustainability skills with the aid of the role-play. Greener Manag. https://www.jstor.org/stable/greemanainte.48.51

UK Government Agenda 2030: Delivering the Global Goals. https://www.gov.uk/government/publications/agenda-2030-delivering-the-global-goals

UNESCO (2014) Shaping the Future We Want - UN Decade of Education for Sustainable Development. (Final report)

UNESCO (2020) Education for sustainable development: a roadmap. http://unesdoc.unesco.org/ark:/48223/pf0000374802.locale=en

Van de Pol J, Volman M, Beishuizen J (2010) Scaffolding in teacher–student interaction: a decade of research. Educ Psychol Rev 22(3):271–296. https://doi.org/10.1007/s10648-010-9127-6

Wals AEJ (2007) Think piece. learning in a changing world and changing in a learning world: reflexively fumbling towards sustainability. South Afr J Environ Educ 24(1):35–45. https://www.ajol.info/index.php/sajee/article/view/122738

Wals AEJ (2010) Mirroring, gestaltswitching and transformative social learning. Int J Sustain High Educ 11(4):380–390. https://doi.org/10.1108/14676371011077595

Wals AEJ (2011) Learning our way to sustainability. J Educ Sustain Dev 5(2):177–186. https://doi.org/10.1177/097340821100500208

Waltzer T, DeBernardi FC, Dahl (2023) A student and teacher views on cheating in high school: perceptions, evaluations, and decisions. J Res Adolesc 31(3):108–126. https://doi.org/10.1111/jora.12784

Woolfolk A (2020) Educational psychology. 14th Edition Pearson ISBN-13: 978-1292331522

Building a Student Community for Social Change: How Institutional Collaboration and Peer Learning Supports a Strategy for Embedding Sustainability Within Fashion Business Education

Naomi Braithwaite and Lisa Trencher

Abstract This chapter presents the rationale and initial results from an ongoing collaborative research project undertaken by two Fashion Business academics working within UK Higher Education Institutions. Drawing from our expertise as academics specializing in responsible business and sustainable consumption, we have developed innovative pedagogical approaches that support the development of skills and knowledge that will enable our students to become agents for sustainable change. The first phase of our project recognized the value of peer led learning by engaging our students in active learning focused on sustainable business and consumption. The proposed next phase will create an opportunity for fashion business students to collaborate across institutions, sharing knowledge and expertise. Our approach proposes a community of students committed to achieving social change, realized through the development of an online platform as a space for sharing industry insights, career opportunities and the showcasing of student research projects. This is supported with an annual student conference, fostering dialogue and knowledge exchange through business practice and consumer focused research. The focus on student collaboration across Higher Education Institutions will be discussed as an innovative model that can positively embed sustainability within fashion business education. Drawing from diverse student perspectives, our proposed community unites responsible business practice with sustainable consumption, equipping students with the knowledge skills and aspirations to become agents for social change.

N. Braithwaite (✉)
Nottingham School of Art and Design, Nottingham Trent University, 50 Shakespeare Street, Nottingham NG1 4FQ, UK
e-mail: naomi.braithwaite02@ntu.ac.uk

L. Trencher
Manchester Fashion Institute, Manchester Metropolitan University, Righton Building, Cavendish Street, Manchester M15 6BG, UK
e-mail: l.trencher@mmu.ac.uk

T. Wall et al. (eds.), *Sustainability in Business Education, Research and Practices*, World Sustainability Series, https://doi.org/10.1007/978-3-031-55996-9_5

Keywords Education for sustainable development · Fashion business education · Transformational learning · Green skills for employability · Student-centered curriculum

1 Introduction

The impact the fashion industry has had on sustainability stems from the negative environmental impacts of production (Geneva Environment Network 2023) and the overconsumption of garments (UNEP 2022), all of which are symptomatic of the traditional linear business model (Ellen MacArthur Foundation 2021) that has underpinned fashion. The addressing of these issues merits advanced consideration within the realms of fashion education, to ensure that future generations are equipped with the knowledge and skills to make a positive change. The problem the fashion industry presents to the environment through waste with over 92 million tonnes of garments going to landfill in the UK alone (Carpani 2020) and the negative elements of the fashion supply chain in terms of sustainability challenges are well documented (Moorhouse and Moorhouse 2019). This has led to fashion being identified as one of the most polluting industries in the world (Urbancic 2022). In contrast the fashion industry has made headway towards a more sustainable future from production to consumption with business models that are more circular and challenge the linear traditions, such as clothing rental (Braithwaite and Schlemann 2018) and recycling (Roberts-Islam 2021). Despite this shift there is still much that needs to be addressed to achieve a truly sustainable fashion industry. While these innovations from materials to business may speak for change, they have not yet lessened the industry's planetary impacts (Pucker 2022). As academics based in fashion business departments at two UK Higher Education Institutions Manchester Metropolitan University (MMU) and Nottingham Trent University (NTU)), we recognize the urgent need for education to support our students in developing their sustainable knowledge and skills. In response to personal growing concerns over fashion's unsustainability and our passion for ensuring our students have the skills to drive forward the industry of the future we have collaborated to develop pedagogical approaches that will embed sustainability in our respective fashion business courses and could become a model that can be translated across additional institutions.

Higher Education Institutions (HEIs) have an integral role in the successful implementation of the United Nations (UN) Goals for Sustainable Development (SDG) and can contribute towards students becoming sustainable citizens, enabling them to play a transformative role in society (Žalėnienė and Periera 2021). The increase in sustainable principles in the strategies of HEIs, particularly through the development of curricula sets the context for the research project presented in this chapter. This project recognizes the integral role that education has for the embedding of sustainable development within the fashion industry. A collaborative project established by two academics from the discipline of fashion business forms the basis for discussion, the intention of the work is to apply transformative pedagogical approaches which place students at the heart of their academic journey, including skills and knowledge development for their future employability. Through education the project encourages students to become proactive members of their academic communities through research and networking with industry to build the skills required to become agents of change in the industry. The project objective, through research and reflection, was to create an online platform focused on knowledge and skills development which will become a student community-based sustainability forum. The project methodology has been underpinned by pedagogical research and reflection which supports the ongoing conceptualization of this community platform that will be grounded in personal, practical experiences as lecturers, and through discussions and evaluations from colleagues, industry partners and students. The chapter will discuss the research project's development with an evaluation of the results that have informed the rationale for our respective pedagogical approaches that have led to the proposal of a unique student community platform.

2 Embedding Sustainability Within Business Education

The role of universities is no longer solely focused on vesting students with academic knowledge, but it is about ensuring each student is ready to work through the development of employability skills. The 17 Sustainable Development Goals adopted by the United Nations in 2015 have shifted focus from primary and secondary education to include higher education, which was previously absent from the development agenda (Chankseliani and McCowan 2020). Education for Sustainable Development (ESD) has encouraged a focus on how and what students learn in the context of sustainability, instilling them with sustainable behaviours and values (Egan 2004). The objectives for education for sustainable development are to build competencies, skills, attributes and values in relationship to specific subject knowledge and knowledge of sustainable development (QAA 2021). In this context universities have a transformative role for encouraging a more sustainable world. Through curriculums, research and employment opportunities, staff and students can actively collaborate to encourage sustainability (UNESCO 2020).

Previous studies (Mason et al. 2009; Pardo et al. 2020) have shown how higher education increases student skills, propensity to entrepreneurship and the path to higher level jobs. For universities to successfully implement ESD they must develop curriculums based on sustainable principles (Žalėnienė and Periera 2021). To support a green economy, Higher Education institutions should invest in developing the green skills that are needed for this (Ruwoko 2022). The COP27 summit held in Egypt in 2022 included an event themed as "Capacity building on green skills to enable local, regional and international climate action" (Ruwoko 2022). From universities to policy and civic society there is a pressing need to build skills relevant for a transition to a low carbon society is recognized. Green skills are the skills needed in a low-carbon economy and will be required in all sectors and at all levels in the workforce as emerging economic activities create new (or renewed) occupations (Vona et al. 2021).

The Global Green Skills report (2023) identifies the emergence of current trends in green jobs which includes 'sustainable fashion' as one of the fastest growing jobs. There is, however, a noted lack of candidates with the green skills required to fulfill these jobs. (LinkedIn Economic Graph 2022) While the fashion industry may be making strides in addressing the challenges of sustainability there is a sense that it seems to be 'stuck in pilot phase' (Cernansky 2022). These points are integral to supporting the project objectives to ensure our students have the appropriate skills for the fashion industry. More broadly, graduate outcomes are a key metric for universities, that are captured through a survey fifteen months post-graduation that reports on graduate destinations. As Dunbar-Morris and Lowe (2023) discuss, graduate outcomes are now at the top of the agenda for higher education institutions, with results reported in the three UK league table publications. Graduate Outcomes are considered a core element of a student's overall university experience. Whilst the demand for green talent at all levels of education has increased, the fastest growth is for candidates with a first degree or higher. This highlights the urgency for universities to address the skills gap (LinkedIn Economic Graph 2022) and consequently this would positively impact universities results in this area whilst providing a societal need.

Embedding sustainability within Higher Education business courses has traditionally been challenging, given the legacy of economic models of growth, driven by marketing, profit maximization and consumption (Fisher and Bonn 2011; Armstrong and LeHew 2013). While some studies have discussed the embedding of sustainability in marketing, management, and business education (Stubbs and Cocklin 2008; Benn and Dunphy 2009), there is limited research that demonstrates the effectiveness of student led pedagogical approaches for embedding sustainability in the curriculum (Radclyffe-Thomas et al. 2018). Applying an integrated approach in business courses to support the development of skills and knowledge for students has been shown to enhance graduate attributes (Barber et al. 2014). In 2020 AdvanceHE introduced the Sustainable Development Goals (SDG) challenge to encourage business school educators to embed sustainability into their teaching at the University of Glasgow (AdvanceHE 2021). The objective was to support staff with the integration of the SDGS into their teaching, including the development of an invaluable online space with resources.

Bennetta and Oeppen Hill's (2022) research on consumer perceptions of sustainability within the fashion industry acknowledged that educational interventions from an early age would be the most effective method of ensuring a sustainable future. Recognizing the value of education for instilling change, higher education has an integral role in supporting students from all disciplines in this quest for a sustainable future. While the role of universities in supporting the development of identity, world views and epistemology during undergraduate study has been recognized (Hay and Eagle 2020), there is a limited understanding of how sustainable knowledge has evolved for students during the period of study. A previous study by Kim et al. (2012) used quantitative research to analyze changes in undergraduate student knowledge of sustainability and fashion at a higher education institution in Midwestern USA. The authors found that an increase in knowledge derived from the curriculum did not translate into meaningful change in purchasing behaviour (Kim et al. 2012). This emphasizes an opportunity for adopting pedagogical approaches that can encourage students to become sustainable citizens, demonstrated through their behaviours and actions. This is further reinforced through a research study which found that embedding sustainability within higher education has positively impacted students' sustainable knowledge and behavior at the end of their studies (Hay and Eagle 2020). While it is key to foster knowledge and skills that will support their professional development, education can nurture sustainable values for each student on a personal level.

Sustainability professionals are agents for change towards sustainable development. They require the knowledge and skills to support them in the initiation change and universities have an integral role to play here. Heiskanen et al. (2016) have actioned a study that explored the results of students being engaged in a real-world project as part of a Strategic Environmental Development course. Their study is part of a growing body that examines the value of learning experiences in the education of sustainability professionals. Within higher education opportunities that bring students together with industry through work integrated learning programmes are invaluable for developing skills such as entrepreneurship, problem solving and self-management, enhancing their employability (Ng et al. 2021). A review of the existing literature in this field enabled the establishment of the project that will be discussed further through this chapter.

3 Project Methodology; Supporting an Institutional Approach Towards Sustainability

The starting point for this project was an independent review of how sustainable development has been approached within our respective curriculums and how this fitted with our institutional strategies for sustainability. We noted a tendency to teach sustainability from a tutor led approach through themed lectures and seminar activities. Informal conversations in the classrooms reinforced our students' appetite for further learning opportunities focused on sustainability as well as the need to develop employability skills that would support a sustainable fashion industry. This was further confirmed by the annual sustainability skills survey led by the National Union of Students which found that 79% of students in 2021/22 would like to see sustainable development actively incorporated through courses (SOS 2022). Reflecting on our traditional pedagogical approaches we realised a need to move away from taking a top-down method to sharing sustainable knowledge, and instead encourage an active student led approach. This informed the first phase of the research project where we focused on developing employability skills for sustainability at MMU and a research informed approach to understanding sustainable consumption at NTU. The project has applied a constructivist approach (Taber 2011), where students take charge of their learning as an impactful method for knowledge creation (Burk-Rafel et al. 2020). Drawing from this the research evaluated whether a pedagogical approach that actively engaged students in co-creating sustainable knowledge could be effective in supporting their development as agents of change within sustainable business and society.

Purcell (2021) emphasized that sustainable development should be the mission that frames the learning, teaching and research that universities are involved in. Within the education strategy at Manchester Metropolitan University, the mission is to '*Welcome our students into a transformational active learning community where we support them to develop, grow, and realize their potential in order that they positively shape the societies of the future.*' (MMU 2023). At NTU inspiring learners to be ambitious and succeed in their chosen industry is central to the university's learning strategy. Sustainability is at the heart of NTU through research, teaching and estate management. In 2022 NTU was awarded the second most sustainable university in the world through the 2022 UI GreenMetric World University Rankings. Our research project aligns with the respective universities' missions and values as students are given opportunities to engage in transformational learning experiences to help shape a more sustainable fashion industry. Embedding this knowledge and allowing students to understand the challenges and consider potential solutions provides motivation for change. The following sections will discuss the different pedagogical approaches applied to support the embedding of sustainability within fashion business courses. In line with the ESD agenda both MMU and NTU offer students opportunities through the curriculum and extra curricula activities to support their engagement with sustainability. Currently fashion students at MMU can engage with the 'Mentor Me' scheme, securing an industry mentor. There is an opportunity to develop this scheme further, identifying 'green mentors' and connecting with students who would like to develop specialist knowledge and explore a potential career in this area. Recognizing the need for students to be equipped with skills and knowledge NTU runs a Sustainability in Practice certificate course focused on exploring the SDGs from a personal, disciplinary and supradisciplinary perspective. Similarly, students at MMU have access to carbon literacy training. Both institutions have staff and student focused funding opportunities, events and research projects which underpin the implementation of the ESD agenda. While these sustainability strategies continue to evolve, we found that there was still a need to embed sustainability more seamlessly through our curriculums. This was even more pressing given the identified challenges of embedding sustainability in fashion business education, and the continued environmental impacts of the fashion industry.

3.1 MMU: A Constructivist Approach to Embedding Sustainable Knowledge Within the Curriculum

One of the pedagogical approaches developed at Manchester Fashion Institute is to establish a network of alumni whose careers are focused on making a positive impact within sustainable fashion. Alumni return to university as guest speakers for current students, outlining their journey, challenges, and experiences as both green entrepreneurs and industry professionals, tasked with embedding sustainability within fashion businesses. This is both powerful and inspirational. Alumni attend networking events with current students to discuss and advise, offering mentorship and the potential in some cases for work experience. Alumni are introduced not only to each other, but to the wider university to explore additional opportunities to develop their network and reach; for example, the Enterprise and Sustainability teams. The aim is to further develop this network to build a sustainable fashion community. This is supported by LinkedIn Economic Graph (2022) who acknowledge the need to develop and foster opportunities for green entrepreneurs. However, of equal importance is to ensure networks and connections to collaborate and thrive within a like-minded community. This will facilitate the development of both global green talent and green economies. (LinkedIn Economic Graph 2022).

At MMU the academic is employability lead for Manchester Fashion Institute, working closely with industry partners to create opportunities for embedding employability initiatives within the curriculum, develop work experience, internships, and graduate roles. These opportunities include live briefs and projects, including more focused studies such as dissertations exploring solutions for specific business challenges. These approaches are not new, however, as both awareness of sustainability and the pursuit for sustainable solutions gain momentum, the conversations with both students and industry partners become more frequent and meaningful.

To build on this opportunity the pedagogical approach included the RISE Programme, a co-curricular initiative supporting the development of student employability skills, unique to MMU. Students can undertake amongst other initiatives research projects. During the first phase of the research this project was conceived; 'Careers in sustainable fashion.' This allowed students to undertake research in this area, exploring job opportunities, gaining industry perspectives, collating relevant learning resources. This initial content was built upon and sits within 'Employability in the Fashion Industry' a shared resource for staff and students to enhance knowledge in this area. This partnership between students and the academic is key. As co-creators of the curriculum, students gain confidence and academics gain insight into student approaches, which informs teaching practice and resource development.

Grounded in a constructivist approach, The RISE research projects encouraged students to explore and develop resources, whilst gaining different perspectives and viewpoints on a complex area. Reflecting on undertaking this activity allowed them to understand their own knowledge and attitudes to sustainable fashion and responsible fashion business practice. This supports the development of employability skills in this area and results in a transformational learning experience. Simsek (2012) describes transformational learning as often leading to significant changes in thoughts and values, this is required to support the change needed to move the fashion industry forward in terms of sustainability.

3.2 NTU: A Research Led Approach to Embedding Sustainable Knowledge

The value of experiential and interactive approaches has become significant in education for sustainable development through peer learning particularly in context of transformational learning (Sterling 2011). At NTU the academic applied a peer led approach to gain deeper insights regarding sustainable consumption and to create data that informs the curriculum. The approach was informed by the academic's position as a material culture researcher who uses object-based research to examine the relationship between fashion and identity. The development of the pedagogical approach was influenced by the need to better understand the drivers for the consumption of fashion and how that relates to sustainability. To succeed in fashion business students need to understand the complexities of consumer behaviour, particularly in the context of sustainability. During creative teaching sessions for undergraduate and postgraduate fashion business courses, students were asked to focus on their personal consumption practices by reflecting on their own clothing choices. The objective was to elicit understanding of what drives fashion consumption through a personalized approach.

Two research-led approaches were applied at NTU. The first was 'fashioning change' which involved students participating in garment-based workshops where they brought a selection of personal items and shared discussions with each other about them. They were asked to discuss the consumption choices behind garments, and their individual feelings towards these items and any specific relationships they had with them. Rather than asking the students questions which explicitly addressed sustainability, these more personal conversations around clothing explored the individual's relationship to sustainability. Each student was asked to bring four garments to the workshop for discussion, something they had owned for a long time, an item that was worn frequently, and something rarely or never worn, and a recent purchase, with the focus on what drove the purchase decision and how that garment was now valued in the use phase. This interactive teaching approach fleshed out attitudes towards garment ownership and how these were linked to sustainable behaviours. The students were asked to document their findings from the discussions and in subsequent sessions they worked together to draw out key insights and use them to create a thematic map related to different values, motivations, and behaviours of fashion consumption.

The second approach 'Fashion Stories' was undertaken with postgraduate students. Here students were asked to select any item of clothing or accessory that was meaningful to them and photograph it. The image was then supported with a written narrative response to why this garment was meaningful to them and how wearing it made them feel. Students shared their posts with peers drawing out the deeper meanings around clothing and the personal values related to clothing ownership. Both NTU approaches applied a personalized approach by focusing learning through the students' own behaviours and attitudes towards clothing, and through this they were able to debate the implications of fashion consumption and sustainability.

4 Results and Discussion

This section presents an evaluation of the findings from the first phase of the project. A reflective evaluation of the impact of our approaches on students' learning and experiences has inspired the next phase project where we propose a model that will support a community of students invested in sustainable development within fashion.

4.1 Phase One Findings

Student perspectives from pilot projects at Manchester Fashion Institute highlighted the breadth of opportunities available in terms of jobs. In addition, as students explored the challenges surrounding embedding more sustainable practices, it became clear there was much to do. Industry perspectives allowed students to hear contextualized, firsthand experiences of these challenges and development of solutions. Whilst the projects provided a transformational learning experience what was clear to academics was that it merely scratched the surface. Whilst there were some excellent outcomes including the development of key resources for dissemination, there was potential for further development. Perhaps most importantly was the motivation of students to create change.

Over 400 students have participated in the Fashioning Change workshops and Fashion Stories project at NTU and the students' individual garment stories have revealed a multitude of different attitudes and behaviors towards consumption and sustainability. As the data from the workshops has been mapped thematically by students it has emphasized several factors that motivate fashion consumption including emotional attachment to garments, the quest for individuality, bargain purchases and the need to fit in. In the realms of fashion business education these findings are significant as they shed light on some of the key motivating forces for consumer behaviour which underpin the mechanisms of ensuring a more sustainable fashion business. Reflecting on our different approaches we recognize the value of a collaborative approach that considers sustainable consumption and employability. In the initial phase of the project, we worked independently with our students and through the evaluation of the approaches we have acknowledged the advantage of bringing our students together to engage collaboratively in these approaches and share knowledge across institutions. This process of knowledge exchange should have a significant impact on the student's learning experience, while fostering the creation of a community of students who are invested in sustainability, without being confined to the bounds of their home institution. The first phase of the project with reflections has established the foundations for the next phase of this student focused research project.

Post session and project evaluations with participating students have evidenced the effectiveness of these student led approaches for supporting education for sustainable development within our curriculums. Student feedback demonstrated that they had taken on board knowledge and skills that they could apply both academically and in a real-world context through future employability. In both of these approaches, students have become co-creators of knowledge, generating data related to green skills and sustainable consumption that has become an invaluable resource for future cohorts to use and further develop. Reflecting on our work to date demonstrates the positive impact of our pedagogical approaches in student learning and their contribution to generating commercial awareness. As students co-create the curriculum and understand the impact and empowerment of independent learning through a constructivist approach, it is important to build upon this in phase two of our research. In addition to academic and professional skills these student-led learning approaches are invaluable in developing soft skills including emotional intelligence. Soft skills including communication, critical thinking, problem solving, self-motivation and time management are integral to academic and career success (Majid et al. 2012) and are therefore important considerations within pedagogical practice.

As professionals concerned with improving the fashion industry and educating students who will be responsible for implementing positive change, we see a significant opportunity to collaborate further, wider, more collectively as we strive to achieve a common goal. This must be the priority for all fashion educators. The community and resources proposed provide the opportunity for knowledge exchange, which will drive change quicker and more effectively.

4.2 Phase Two Proposal

The overarching aim of the second phase of the project is to create a community network for fashion business students across local and global institutions. A place for knowledge exchange, skills development, networking with peers, researchers, and industry professionals. The objective being to support curriculum development through the co-creation of sustainable knowledge and skills. This will result in graduates who are not only 'industry ready' to take on green challenges within the fashion industry, but to be the change makers who define the fashion industry of the future. The pedagogical approach informing this proposal is network learning, which is based on technology to promote connections between learners, tutors and learning community and learning material (Gourlay et al. 2021). It is our intention that this community platform will become the go to place for students to showcase their work, network, share sustainable knowledge and skills, supporting their studies, personal development and personal growth as sustainable citizens.

The academic at MMU is an advocate of LinkedIn as a strategic networking tool. The use of LinkedIn is key for fashion industry professionals and the academics undertaking this study continue to develop significant networks in this area, including professional connections concerned with sustainable fashion. Embedded within the curriculum, within the second-year employability focused unit, students create a LinkedIn account. This allows them to begin to develop their network and start to showcase themselves as potential industry professionals. Students develop their confidence on this platform and become comfortable using it. It is also an excellent tool for job searches and allows students to follow specific brands, industry specialists etc. Internal university LinkedIn groups allow academics to share opportunities, including industry events, internships, graduate roles, and knowledge exchange. As fashion sustainability and responsible business practice continue to gain momentum, the opportunity to develop 'green networks' is key.

Initial community network testing provided rationale to further develop the concept. Industry feedback acknowledged the need for skills in this area, initial discussions focused on the potential for collaborative live briefs, knowledge exchange and opportunities for work experience. Academic support is key to developing the concept; when this network was approached, significant support and interest was received. The student response provided positive feedback, with students recognizing the value of developing employability skills, and as an opportunity to showcase their work and network with others.

5 Conclusions

The project findings have demonstrated the need for educators to provide innovative pedagogical approaches which place students at the centre of their learning. As educators who promote positive change, it is our role to facilitate this, to inspire students to develop the confidence through this learning to drive positive change. The project has highlighted the advantages and opportunities that a collaborative approach can provide, the importance of sharing practice and working together to realise ambitions for a more sustainable industry.

Building on both transformative and networked learning approaches, the next phase of the project whilst facilitated by academics will empower students and alumni to drive change. The aim is to create a community of students invested in sustainability and responsible business practice across HE institutions. In addition to support a local to global network that connects students with peers, academics, industry and supports them in taking the lead in developing the knowledge and skills needed to become future change makers for a sustainable fashion industry.

The following objectives define the next phase of the project:

- Development of an online platform that includes student forums, podcasts, job repository, and provides an opportunity to showcase student projects.
- Create a 'green' LinkedIn group engaging students, industry professionals, and academics; an online community of like-minded individuals focused on achieving positive impact in this area.
- Deliver an annual student conference with industry speakers, student showcase and masterclasses.
- Create a series of webinars led by alumni working in sustainable fashion.

The platform will initially be led by students at NTU and MMU with a view to expand through our networks across other national and international Higher Education institutions. Our intention is that this can become a platform where students can showcase their work and further their employability through access to industry professionals, and other resources. It will become a resource that informs the curriculum while also supporting students in the development of professional and inter-personal skills.

The current climate of fashion industry challenges and shortfall of green skills highlights both the urgency and opportunity for change. Our students will define how sustainable the fashion industry will be moving forward, as HE educators we have a responsibility to facilitate an environment where students can explore, discuss, and collaborate to develop the skills, knowledge, attitudes, and confidence to positively shape the fashion industry of the future.

References

Advance HE (2021) The SDG challenge: encouraging business school educators to embed sustainability into their teaching. AdvanceHE practice guide. https://www.advance-he.ac.uk/knowledge-hub/sdg-challenge-encouraging-business-school-educators-embed-sustainability-their. Accessed 12 Feb 2023

Armstrong CM, LeHew MLA (2013) A case study in sustainability and fashion education: adventures on the green. J Sustain Educ 4. https://www.researchgate.net/publication/298424813_A_Case_Study_in_Sustainability_and_Fashion_Education_Adventures_on_the_Green. Accessed 28 Feb 2023

Barber NA, Wison F, Venkatachalam V, Cleaves SM, Garnhma J (2014) Integrating Sustainability into Business Curricula: University of New Hampshire case study. Int J Sustain High Educ 15(4):473–93. https://doi.org/10.1177/2F2379298119844975. Accessed 23 March 2023

Benn S, Dunphy D (2009) Action research as an approach to integrating sustainability into MBA programs: an exploratory study. J Manag Educ 33(3):276–295. https://doi.org/10.1177/2F1052562908323189. Accessed 23 March 2023

Bennetta K, Oeppen Hill J (2022) Educating for change?: An investigation into consumers' perception of sustainability and the educational drivers needed to support sustainable consumption. Int J Fash Des Technol Educ 15(3):418–429. https://doi.org/10.1080/17543266.2022.2083694. Accessed 23 March 2023

Braithwaite N, Schlemann A (2018) Product service systems: a viable business model for fashion brands. In: Becker-Leifhold C, Heuer M (eds) Eco friendly and fair: fast fashion and consumer behaviour. Routledge, London, pp 132–143

Burk-Rafel J, Harris KB, Heath J, Milliron A, Savage DJ, Skochelak SE (2020) Students as catalysts for curricular innovation: A change management framework. Med Teach 42(5):572–577. https://doi.org/10.1080/0142159X.2020.1718070. Epub 2020 Feb 4. PMID: 32017861. https://pubmed.ncbi.nlm.nih.gov/32017861/. Accessed 23 March 2023

Carpani, J (2020) Britain's fast fashion poses a grave risk to the environment. The telegraph 7th April 2020. https://www.telegraph.co.uk/news/2020/04/07/britains-fast-fashion-poses-grave-risk-environment/?. Accessed 28 Feb 2023

Cernansky R (2022) Sustainable fashion is stuck in 'pilot phase' | Vogue Business. Accessed 23 March 2023

Chankseliani M, McCowan T (2020) Higher education and the sustainable development goals. High Educ 81:1–8. https://doi.org/10.1007/s10734-020-00652-w. Accessed 15 March 2023

Dunbar-Morris, H, Lowe, T (2023) Higher education policy institute: graduate employability takes top spot in the battle of the metrics—so it's time we understood it better. Graduate employability takes top spot in the battle of the metrics – so it's time we understood it better—HEPI. Accessed 30 March 2023

Egan J (2004) Skills for sustainable development. Office of the Deputy Prime Minister, London. Accessed 15 March 2023

Ellen MacArthur Foundation (2021) The trends and trailblazers. Creating a circular economy for fashion. https://ellenmacarthurfoundation.org/articles/the-trends-and-trailblazers-creating-a-circular-economy-for-fashion. Accessed 15 March 2023

Fisher J, Bonn I (2011) Business sustainability and undergraduate management education: an Australian study. High Educ 62:563–571. Accessed 15 March 2023

Geneva Environment Network (2023) Environmental sustainability in the fashion industry. https://www.genevaenvironmentnetwork.org/resources/updates/sustainable-fashion/. Accessed 15 March 2023

Gourlay L, Rodríguez-Illera JL et al (2021) Networked learning in 2021: a community definition. Postdigit Sci Educ 3:326–369. https://doi.org/10.1007/s42438-021-00222-y

Hay R, Eagle L (2020) Impact of integrated sustainability content into undergraduate business education. Int J Sustain High Educ 21(1):131–143. https://doi.org/10.1108/IJSHE-05-2019-0174. Accessed 15 March 2023

Kim Y, Hillier C, Kozar JM (2012) Sustainability knowledge and behaviors of apparel and textile undergraduates. Int J Sustain High Educ 13(4):1467–6370. https://doi.org/10.1108/14676371211262335. Accessed 30 March 2023

Heiskanen, E Åke Thidell, A Håkan, R (2016) Educating sustainability change agents: the importance of practical skills and experience. J Clean Prod 123:218–226. https://doi.org/10.1016/j.jclepro.2015.11.063, https://www.sciencedirect.com/science/article/pii/S0959652615017291. Accessed 30 March 2023

LinkedIn Economic Graph (2022) Global green skills report 2022. li-green-economy-report-2022-annex.pdf linkedin.com. Accessed 29 Jan 2023

Majid S, Liming Z, Tong S, Raihana S (2012) Importance of soft skills for education and career success. Int J Cross Discipl Subj Educ 2(2):1036–1042. Accessed 15 March 2023

Manchester Metropolitan University (2023) Sustainability Strategy. https://www.mmu.ac.uk/sustainability/strategy. Accessed 25 Jan 2023

Mason G, Williams G, Cranmer S (2009) Employability skills initiatives in higher education: what effects do they have on graduate labour market outcomes? Educ Econ 17:1–30. https://doi.org/10.1080/09645290802028315. Accessed 15 March 2023

Moorhouse D, Moorhouse D (2019) Creating a sustainable luxury fashion brand. Bloomsbury Publishing, London, UK. https://doi.org/10.5040/9781350989955.0009. Accessed 29 Jan 2023

Networked Learning Editorial Collective (NLEC) (2021) Networked learning: inviting redefinition. Postdigital Sci Educ 3:312–325. https://link.springer.com/article/https://doi.org/10.1007/s42438-020-00167-8#article-info. Accessed 30 March 2023

Ng PML, Chan JKY, Wut TM, Lo MF, Szeto I (2021) What makes better career opportunities for young graduates? Examining acquired employability skills in higher education institutions.

Education + Training 63(6):852–871. https://doi.org/10.1108/ET-08-2020-0231. Accessed 15 March 2023
Pardo-Garcia C, Barac M (2020) Promoting employability in higher education: a case study on boosting entrepreneurship skills. Sustainability 12:4004. https://doi.org/10.3390/su12104004. Accessed 15 March 2023
Pucker KP (2022) The myth of sustainable fashion. Harv Bus Rev. https://hbr.org/2022/01/the-myth-of-sustainable-fashion. Accessed 30 March 2023
Purcell WM (2021) Sustainability in higher education: an agenda for transformational change: a manifesto for change and "third-way" people'. presentation at the international sustainable campus network. https://international-sustainable-campus-network.org/conferences/iscn-cop26/iscn-2021-preliminary-programme/. Accessed 20 Nov 2021
QAA (2021) Education for Sustainable Development. https://www.qaa.ac.uk/quality-code/education-for-sustainable-development. Accessed 22 March 2023
Radclyffe-Thomas N, Varley R, Roncha A (2018) Balancing the books: creating a model of responsible fashion business education. Art Des Commun High Educ 17(1):89–106. https://doi.org/10.1386/adch.17.1.89_1. Accessed 10 March 2023
Roberts-Islam B (2021) fashion is overselling circularity and recycling—but there is hope. Forbes. https://www.forbes.com/sites/brookerobertsislam/2021/10/01/fashion-is-overselling-circularity-and-recycling-but-there-is-hope/?sh=3352a7e95d95. Accessed 15 March 2023
Ruwoko E (2022) No green economy without green skills: HE's critical role. University World News. https://www.universityworldnews.com/post.php?story=20221123123756933. Accessed 15 March 2023
Simsek A (2012) Transformational learning. In: Seel NM (eds) Encyclopedia of the sciences of learning. Springer, Boston, MA. https://link.springer.com/referenceworkentry/https://doi.org/10.1007/978-1-4419-1428-6_373. Accessed 30 March 2023
SOS (2022) Students organising for sustainability United Kingdom. Sustainability skills survey. https://www.sos-uk.org/research/sustainability-skills-survey. Accessed 10 March 2023
Sterling S (2011) The future fit framework: an introductory guide to learning and teaching for sustainability in HE. The Higher Education Academy, York
Stubbs W, Cocklin C (2008) Conceptualizing a "sustainability business model." Sage J Organ Environ 21(2):103–127. https://doi.org/10.1177/1086026608318042. Accessed 30 March 2023
Taber KS (2011) Constructivism as educational theory: contingency in learning and optimally guided instruction. In: Hassaskhah J (ed) Educational theory. Nova Publishers, New York
UN Environment Programme (UNEP) (2022) Shifting the fashion narrative: rethinking aspiration in a world of overconsumption. https://www.unep.org/events/online-event/shifting-fashion-narrative-rethinking-aspiration-world-overconsumption
UNESCO (2020) Inclusion and education. Global education monitoring report. https://www.unesco.org/gem-report/en/inclusion. Accessed 28 Feb 2023
United Nations (2022) The sustainable development goals report. https://unstats.un.org/sdgs/report/2022/. Accessed 10 Feb 2023
Urbancic N (2022) Fast fashion will be held accountable for clothing quality, toxic chemicals and waste says the EU. Euronews.com. https://www.euronews.com/green/2022/03/31/fast-fashion-will-be-held-accountable-for-clothing-quality-toxic-chemicals-and-waste-says. Accessed 03 Feb 2023
Vona F, Popp D, Marin G (2021) Who's ready for the green economy? Measuring skills for the low-carbon transition. The OECD forum network. who's ready for the green economy? measuring skills for the low-carbon transition. The OECD Forum Network (oecd-forum.org). Accessed 28 April 2023
Žalėnienė I, Periera P (2021) Higher education for sustainability: a global perspective. Geogr Sustain 2:99–106. https://www.researchgate.net/publication/351750122_Higher_Education_For_Sustainability_A_Global_Perspective/citation/download. Accessed 30 March 2023

Using OODA Loops to Build Sustainable Practice into Business Education

Mike Ryder

Abstract The concept of 'reflective practice' has many applications in business education, research and practice. However, the theories that underpin this concept are arguably dated and not suited to the uncertainties of the modern world. This chapter explores ways we might use John Boyd's OODA loop theory to rethink reflective practice and put sustainability at the heart of an inclusive, forward-thinking curriculum.

Keywords Management education · Reflective practice · OODA loops · Sustainability · Strategy

1 Introduction

The idea of 'reflective practice' is an important one in higher education and is taught in various forms across a range of disciplines (cf. Bruno and Dell'Aversana 2016; Peabody and Noves 2017; Treacy and Gaunt 2021; Ali 2022). Many universities run compulsory modules on reflective practice to help placement students think critically about their time in industry. Reflective practice is also often taught as part of modules that include some form of 'practical' element, where students are required to write personal reflections on their group work and/or individual projects.

But of course, reflective practice is not just limited to the classroom. It is also a key part of continuing professional development (CPD), especially in the fields of healthcare (HCPC 2022a, b; BPS 2022) and medicine (Grant et al. 2017; AOMRC et al. 2022). This helps to ensure that professional standards are met and that practitioners are engaged in a continual process of self-reflection and improvement (NHS 2022; Mantzourani et al. 2019).

However, there is a problem.

Many of the most widely used models of reflection were designed several decades ago for a world that is very different to the one we live in today. Indeed, it has

M. Ryder (✉)
Marketing Department, Lancaster University Management School, Lancaster LA1 4YX, UK
e-mail: m.ryder@lancaster.ac.uk

T. Wall et al. (eds.), *Sustainability in Business Education, Research and Practices*, World Sustainability Series, https://doi.org/10.1007/978-3-031-55996-9_6

been argued that many of our most well-established models 'derive more from the prevailing ideological and institutional context of adult education in the 1970s than from focused programs of inquiry about the connections between experience and reflection in learning' (Seaman 2008: 8). This comment aligns with more recent observations that the process of developing reflexivity and critical reasoning is 'far more subtle, complex, disorderly and emotional [...] than is currently implied' (Dyer and Hurd 2016: 288). This has led some scholars to argue that existing models of reflective practice are therefore outdated and unfit for purpose (Ryder and Downs 2022).

This line of argument has many implications for business schools and management education more broadly. There has been much written in recent years about the struggles of business schools to stay relevant and meet the needs of a changing marketplace (cf. Thomas and Ambrosini 2021). One particular issue is the fact that 'Pedagogies that discourage critical self-reflection may contribute to the inculcation of habits that favour hubris' (Sadler-Smith and Cojuharenco 2021: 274). There is also a growing body of work that suggests students need to be better prepared to embrace chaos and uncertainty (Bureau and Komporozos-Athanasiou 2017: 51). Indeed, Hibbert (2013) notes that students tend to favour familiar understandings and struggle to deal with radical alternatives to their unquestioned cultural norms (816); a point that supports Seaman's (2008) claim that these models derive from a particular ideological stance.

One of the major issues is that many of the most popular models taught in higher education—such as those described by Kolb (1984, 2015), Gibbs (1988) and Driscoll (1994, 2007)—do not place enough emphasis on context, and therefore do not put the practitioner at the heart of the cycle (Ryder and Downs 2022). As such, the practitioner (or indeed, student) becomes a passive observer as events unfold. There is also a tendency to favour short-term thinking as each 'event' is reflected on in abstract isolation.

Clearly, this issue has many implications, not least in terms of sustainability—the focus of this book. If students and practitioners are not suitably equipped to reflect in a way that situates themselves at the heart of the event that they are reflecting on, then they are less likely to envisage the potential long-term consequences and ramifications of their actions. They are also less likely to take positive action to operate in a more sustainable, more environmentally conscious way.

To address these problems, this chapter makes a case for adopting John Boyd's OODA loop as our new primary model for reflection. At its most simplistic level, OODA stands for Observe, Orient, Decide, Act. However, the concept is far more nuanced than its acronym suggests. At its core, the OODA loop represents a philosophy that embraces chance and puts a strong emphasis on the role of 'orientation' as a key component that shapes how we see the world and the decisions that we make. By adjusting the way we teach reflection, and by moving away from the static models of the past, so we can help shape a new generation of practitioners going forward—practitioners who are critical, agile and reflexive, and who take a holistic view to self-reflection and how their actions relate to the world around them.

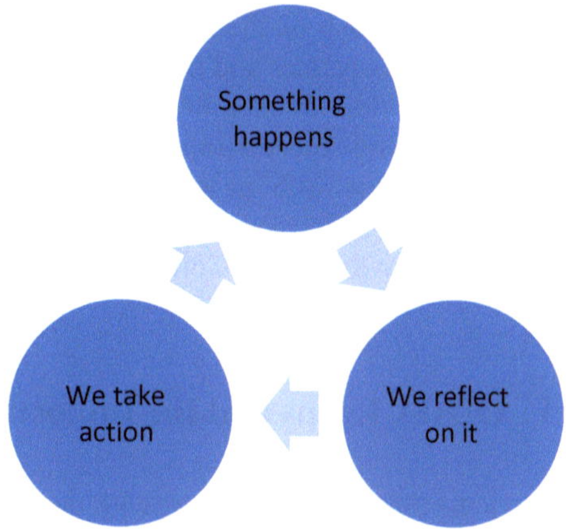

Fig. 1. A generalised view of reflective practice as popularised by Gibbs (1988), Driscoll (1994), Rolfe et al. (2011) and Jasper (2013)

The climate emergency is *the* critical issue of our age. Only by changing the way we *think*—changing the way we *think about thinking*—will we be able to respond to the crisis and bring about real and sustainable change to the way our world works.

2 The Problem with Reflective Practice

The concept of reflective practice has been in use for the best part of a century, starting with Dewey (1933) and his work on reflective thinking in education. However, it wasn't until some fifty years later with the works of Schon (1983) and Kolb (1984) that the concept really began to take a hold. Schon in particular was instrumental in framing reflection as a key part of professional development, laying the ground for many of the theories of reflection that were to follow. These include the likes of Gibbs (1988), Driscoll (1994), Rolfe et al. (2011), and Jasper (2013)—works that are all still actively taught in the classroom and promoted by professional training providers.[1]

While each of these models has its own particular nuance and approach to reflection, they can be generalised as follows: 1. Something happens, 2. We reflect on it (in whatever way the model suggests), and 3. We take some form of action [see Fig. 1].

What is notable about these models is that they each imply a certain linearity—that each event and action is related to the last. Crucially, they also fail to include the role of external factors in shaping what happens, how we reflect on it, and how we take action. Indeed, none of the established models—including Kolb (1984, 2015)—puts

[1] See for example Gibbs cited by NHS (2022) and Kolb cited in guidance for medical doctors and students (AORC et al. 2022). Private healthcare training providers such as HSCE (2022) also promote the cycles of Gibbs, Kolb, Rolfe et al. and Jasper (among other) as resources for professionals seeking to make use of reflective practice.

sufficient emphasis on the role of the person doing the reflecting as a central part of how the cycle operates and how each part of the cycle plays out.

This is a vital absence from established models as it means that as educators, we are teaching students (and therefore, future practitioners) to think in terms of isolated incidents without wider context. By doing this we risk oversimplifying the reflective process (Dyer and Hurd 2016), and thus inculcating habits that favour hubris (Sadler-Smith and Cojuharenco 2021).

3 OODA Loops and the Importance of Orientation

Given the long list of problems with our most popular models of reflective practice, this chapter examines the applications of a powerful alternative: John Boyd's OODA loop (Ryder and Downs 2022).

John Boyd (1927–1997) was an acclaimed pilot and military strategist who was at the heart of the US Air Force's Lightweight Fighter (LWF) programme in the 1960s. While he never published a full-length book, his theories and strategies exist as a series of slide presentations reproduced in digital format and preserved in online archives. Most important of these documents perhaps is 'The Essence of Winning and Losing', which Boyd delivered as part of a series of military briefings in 1995 and 1996. In it, Boyd gives a clear breakdown of the OODA concept, while also setting out the core of his philosophy around change and uncertainty.

At first glance, the OODA loop is a very simple concept. Its letters stand for: Observe, Orient, Decide, Act. However, it is far more nuanced than the acronym suggests. Figure 2 shows the OODA loop as depicted in Boyd's presentation. As the Figure shows, Orientation rests at the heart of the OODA loop, for it is Orientation that exerts implicit guidance and control over Observation and Action, with feedback loops built into the OODA loop at every stage.

In this way, the OODA loop represents a significant advance on the likes of Kolb, Gibbs, Driscoll and Jasper. This is because it is a loop that never ends, with user Orientation *shaped by*, and *shaping*, interactions with the environment. Quite simply: it is an adaptive model that changes as the world around it changes too.

This is a significant step beyond how existing models conceive of reflection as if it is a discrete 'thing' that reflective practitioners 'do'. Rather, the cycle should become a part of the individual such that the cycle and the practitioner co-exist as one, shaping each other and developing together in turn.

3.1 Embracing Change

Before moving on to the application of this theory to sustainability in education, it is worth pausing for a moment to consider the OODA loop in its wider context.

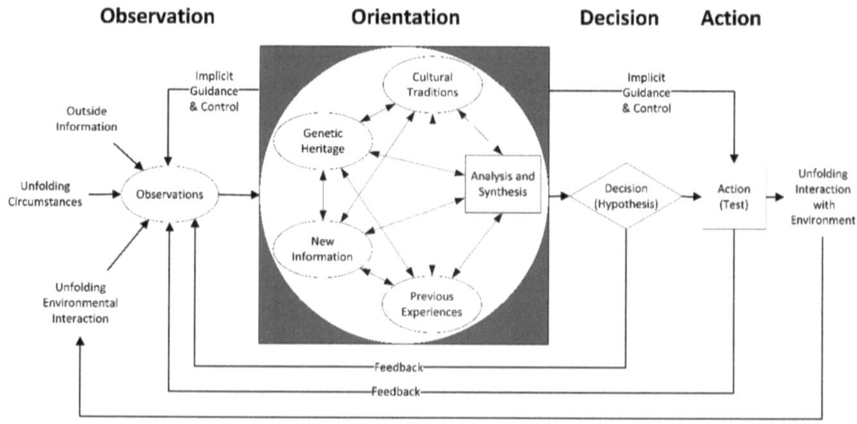

Fig. 2 John Boyd's OODA loop, adapted from 'The Essence of Winning and Losing' (Boyd 1995)

At its heart, the OODA loop is a form of mindset—a way of thinking about the world and our place in it. This is a point made by Boyd in 'The Essence of Winning and Losing' in which he prefaces his OODA loop design with a series of key statements that feed one into the other. For Boyd, it starts with all of the 'genetic heritage, cultural traditions, and previous experiences' that contribute to an 'implicit repertoire of psychophysical skills shaped by environments and changes that have been previously experienced' (1995: 1). These implicit skills and experiences then feed into how one goes about approaching a particular problem.

According to Boyd, it is critical to approach learning and reflection with an open mind, and not just to approach a problem from a standard or 'accepted' position. In many respects, counter-claims are just as useful as accepted norms as they challenge expectations; even in the process of rejecting them so further understanding can be gained. Indeed, Boyd argues that 'Without a many-sided, implicit cross-referencing process of projection, empathy, correlation, and rejection [...] we cannot even do analysis and synthesis' (Boyd 1995: 1). And it is here, under the umbrella of analysis and synthesis that we evolve means by which to deal with unfamiliar phenomena and unforeseen change.

What makes the OODA loop such a powerful concept is the way that it embraces change as a key driver of any system. While many of our most popular models of reflective practice struggle in the face of an ever-changing and unpredictable world, Boyd's theory is designed to respond to change and use it as a means for development and growth.

But before we can embrace change, first we need to disavow ourselves of long-held notions of static models to describe the world. For Boyd, no framework can ever encapsulate reality. This is because entropy leads all static systems to eventually stagnate and decay (Bousquet 2009: 221). It is therefore wrong to think that any single model can fully represent the way the world works at any given moment. This is certainly true of the models of reflective practice discussed previously. After all, no two individuals have the same subjective experience, and there are many factors affecting the way we perceive events and interact with the wider environment.

In response to the 'problem' of change, Boyd extols a theory of agility, which requires practitioners prepare themselves to adapt to changing circumstances. Change shouldn't therefore come as a surprise, but rather as an expected step. Boyd's theory on agility is described by his former associate Chet Richards, himself a Maths PhD who worked with Boyd from the 1970s. Writing in his book *Certain to win* (2004), Richards observes: 'The essence of agility [...] is to keep one's orientation well matched to the real world during times of ambiguity, confusion, and rapid change, when the natural tendency is to become disoriented' (2004: 30). To do so requires a level of strategic oversight as to the end goal, or rather, the 'big picture'; it is not enough to treat events in isolation, but rather to adapt and change to unfolding circumstances while keeping the end goal in sight. The 'means' may change, but the end remains the same.

However, it is important to note that our ability to keep our orientation matched to the real world is largely a function of how well we observe (Richards 2004: 63)—which itself is shaped by orientation. The inter-relatedness of the various aspects of the OODA loop may seem complex at first glance, however, it is only by understanding the loops in context and gaining a 'fingerspitzengefuhl', or 'finger-tip feeling' (Boyd 1987: 45) for our own individual loops—our own individual context—and those of our 'adversaries' (in whatever form they take) that we can make best use of OODA loops as an effective means of reflective practice and self-improvement.

As Boyd argues: 'Without OODA loops embracing all of the above [...] we will find it impossible to comprehend, shape, adapt to and in turn be shaped by an unfolding evolving reality that is uncertain, everchanging, and unpredictable' (1995: 1).

4 Putting Sustainability First

As we have seen, current models of reflective practice taught in universities and applied in a professional setting are outdated, inflexible and not suited to the modern world. They are essentially unfit for purpose (Ryder and Downs 2022). By embracing John Boyd's OODA loop theory, as educators, we can hope to situate our students (and therefore future practitioners) at the heart of reflection to take ownership of their reflection and their own personal context.

But how might this work in practice?

In Ryder and Downs (2022), my colleague and I propose sessions with students designed to encourage them to start thinking critically about their own particular context—about all the things that make them who they are—all the things that make them 'tick'. I have called this task 'Orientation-in-action', as the first step in any self-reflection should be to understand one's starting point before the reflection proper begins.

In my own teaching I often give a worked example based upon my own context and situation. I start by drawing a basic mind map with myself at the centre. I then add various elements of my background that effect my position as observer. These include (but are not limited to): my childhood; where I grew up; my education; my political views; my personality; my personal philosophy; my (lack of) religion etc. The important point to note here is that this process should not be seen as a formal exercise in listing every single category and concept possible. Rather, students should be reminded that the process of unpicking one's Orientation is a journey of self-discovery—a process that is ongoing and will change over time.

I should also note that there is no 'right' or 'wrong' answer; there are many diverse things that shape our perspective on the world, and our Orientation can include things such as: the football team we support; whether we are a 'cat person' or a 'dog person'; our preferred taste in music; our favourite books. All of these things shape us in some way, and reflect who we are as individuals. For example, as a certified 'dog person', I value loyalty and friendship above the detached sort of self-interestedness of cats, and this plays out in the relationships that I form and the way I view other people.

Mapping out a diagram of all the things that make us who we are can take a little time depending on how it is presented to a class of students. However, it is really worth the time and effort as it gives students a sense that Orientation will be different for each and every person in the room. It also has the added benefit of instilling some much-needed authenticity in the delivery of teaching—to show that as teachers, we are real people with real lived experiences and a connection to the real world (c.f. Rogers 1969; Cranton 2001, 2006a, 2006b; Kreber 2013).

Once this first example has been worked through, the next step is to get students to reflect on their own Orientations and compare it with others sitting around them. This is a useful way to get conversations working in the class and to encourage further critical self-reflection. It may also lead to discussions where students start to think about how they reflect in different ways. For example, someone who grew up in a rural location may discover they are more self-sufficient and willing to travel than someone who grew up in a city, or who went to boarding school. Equally, it may be that people with large, close-knit families find that they are by nature very loyal, and place great weight on inter-personal relationships and commitment to the group. This may differ from someone who is an only child who has grown up without those ties.

This session offers many possibilities for further development. For example, instructors may choose to spend some time offering up scenarios for students to consider, wherein they may have different perspectives on an event. This can be used as a way to encourage students to reflect back upon why they responded to the scenarios in a particular way—what from their personal orientation may have caused them to take the perspective that they did.

Once students have completed their initial reflection on their personal Orientation, tutors may choose to look at specific facets of Orientation in turn. For the purpose of this chapter, I will focus specifically on sustainability, as it is such a key facet of modern life, with a marked increase in sustainable lifestyles (Deloitte 2022).

In the first instance, a similar mind-mapping exercise might be started by placing sustainability at the centre of the map, and expanding the diagram with the various elements that students understand to relate to the subject. They should do this individually in the first instance. However, this is just the starting point. Students should *then* be encouraged to think about how these various elements of sustainability and/or sustainable practice impact on their own personal decision-making specifically. For example: Do they always recycle their coffee cups? Do they make efforts to turn off lights and other devices when they are not in use? How often do they drive a car on journeys of less than a mile?

The aim here is not to single out students who may behave in ways that defy established norms, but rather to facilitate conversation that leads to meaningful reflection on how and why they behave in the way that they do. This is the first step in bringing about change.

To give a worked example: in my own context, my grandparents were from the East End of London and were stereotypical Londoners of a certain vintage. Their 'blitz spirit' stuck with them throughout their lives—as did their attitude to waste, and their desire to 'make do and mend' rather than buy things anew. This background gives a context to my own views on waste. I don't like to buy new things unless I absolutely have to, and I am a very keen recycler; I don't like things going to landfill. This is reflected in my buying habits and my approach to shopping. It also impacts the way I look after my money, and the way I interact with others.

But while we can perhaps all agree that recycling is generally a good thing, it is not something we all view in the same way. Therefore, if we are to instil a sustainable mindset in our students, we need students to reflect on *why* they behave in certain ways as the first step to bringing about change.

In my own case, I sometimes persist in using worn-out items to the detriment of my personal comfort. My natural inclination to avoid spending money on myself also means I may not check where products are sourced when I purchase new clothes; I tend to prioritise function over anything else. This is something I recognise that I need to spend more time on when buying clothes, even though my natural inclination is to minimise the time I spend shopping wherever possible.

This simple example hopefully goes to show how reflecting on one's personal Orientation can start to unpick some of the many complex facets that feed into the ways that we behave in the ways that we do. Of course, there is no simple solution here—each person will work in a different way—however, it is only by first taking a step back to consider where we come from that we might then start to think about where we are now, and where we want to be. This is the power of John Boyd's OODA loop theory: as the system (i.e. the world) is constantly in flux, we therefore need to adopt a flexible, fluid mindset, and move away from the idea that we are fixed beings incapable of change. To do this, we first need to understand what it is that makes us who we are and see the world in the way that we do. Only then can we open doors to a more sustainability-focused way of living.

5 Adopting a Sustainability Mindset

There can be no doubting the world is in a global environmental crisis (UN Environment Programme 2022; United Nations 2022; World Health Organization 2022; UNICEF 2015). With this in mind, I propose re-framing Boyd's OODA loop concept to explicitly include its relationship with the wider world. This context is absolutely critical as each and every element in the loop is related to and relies upon the global environment, and there being a global environment left for us to prosper in.

While experienced advocates of Boyd's philosophy will already understand the role of feedback mechanisms and external factors in shaping Observation, Orientation, Decision and Action, the theory has at times been over-simplified and misinterpreted by some scholars who perhaps do not understand the nuance of Boyd's argument (cf. Rogers 2010). This can lead to the mistaken assumption that 'success' is merely a process of completing the loop faster (Osinga 2007: 5–6; cf. Papenfuhs 2012; Maor 2022). For this reason then, I propose that global context, or more specifically, *global environmental context* should be stated *explicitly* as the context within which all loops operate (see Fig. 3).

Starting on the left-hand side of the diagram, Observation should be considered in terms of its relationship with the global environment. What one perceives to be 'sustainable' or 'ethical' is of course shaped by Orientation; and yet one needs to be prepared to adjust observations based on environmental context. From this, we make a decision, which itself shapes observations and is tested through action—all of which we might then lead to unfolding interaction with the environment, which again, exists within the context of the global environmental context.

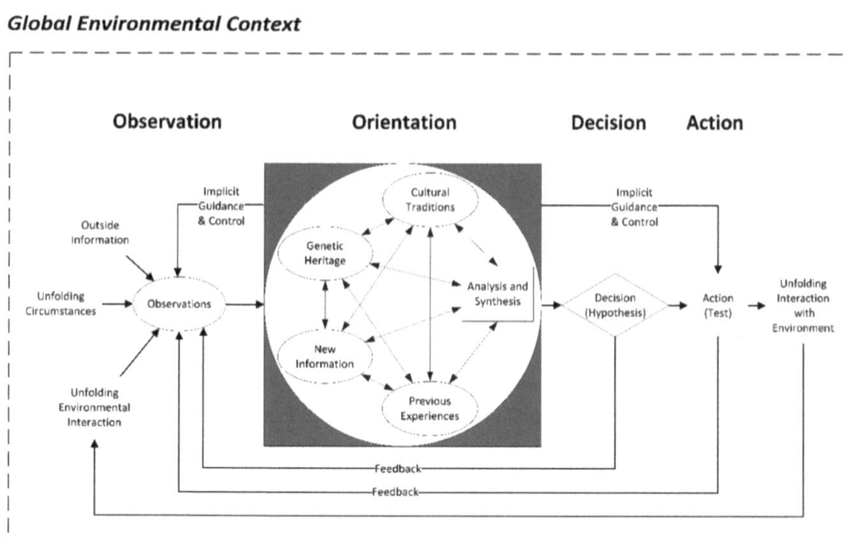

Fig. 3. OODA loop operating within the global environmental context. *Source* author's own

To give an example: over the period of Christmas 2022, UK Conservative MP and Environment Minister Rebecca Pow shared a video on Twitter in which she encouraged her followers to iron their used wrapping paper in order to re-use it for future gifting (Pow 2022). While clearly well-intentioned, the video focuses on only one aspect of sustainability to the exclusion of others. In this case, Pow's message promotes re-use—a positive message—while at the same time using a seemingly disproportionate amount of energy to heat an iron. It is perhaps no surprise then that Pow's video came in for much criticism, especially at a time when so many people in the UK were struggling to pay their energy bills (cf. Brazell 2022; Yadav 2022).

While it is hard to compare two different forms of consumption (buying new wrapping paper vs heating an iron), as critical reflective practitioners we need to consider all of the factors at play, while also appreciating how our own observations shape our reflections on the event being observed—not just for ourselves, but also for others. In this case, we need to factor in how our own political leanings may shape the way we instinctively respond to Pow's post. This includes the party we support, our class allegiances and our opinion of UK politicians. It also includes our understanding of the many complex issues involved in the sustainability (or otherwise) of wrapping paper such as its cost, where it is made, how easy it is to recycle, and so on.

Putting the political issues to one side, Pow's video reveals something of the tension at the heart of decision-making, especially when it comes to making choices around sustainability. This is because there is always a certain 'opportunity cost' associated with any decision as every action comes at the expense of acting in a different way. In this case, either we iron wrapping paper so as not to buy new wrapping paper, or we buy new wrapping paper so as not to iron old wrapping paper.

What this example goes to show is that not everyone has access to the information or insight to weigh up the environmental impact of two quite different activities (even though an Environment Minister really should know better!) Unfortunately, there is no clear and easy answer to this conundrum. While researchers such as Mike Berners-Lee (2020) have attempted to shed some light on picking more environmentally-friendly options, this information is not readily available to everyone, and some facets of the sustainability agenda may have passed some people by.

We might also reflect on the fact that even the best studies have their limitations. As Berners-Lee himself would admit, we cannot know the exact carbon cost of everything, as the world is just too complex a place. While cycling to work is typically better for the environment than say, taking the car, the impact of this activity is limited if the cyclist eats a lot of red meat and rides a bike imported from China. The alternative might be to get the bus, but if the bus is empty then that's a lot of excess fuel being burnt to transport a single passenger plus the driver. Another option may be to stay at home; however, working from home burns extra energy for things such as heating and lighting; plus, the carbon costs associated with new electronic equipment such as monitors, laptops (etc.). This makes the decision-making process very difficult indeed.

While there may be no clear 'right' or 'wrong' answer in this situation, as reflective practitioners we need to be able to embrace the complexity of the situation, and the fact that the situation could potentially change in the future. We also need to consider how our personal Orientation shapes how we perceive and respond to events. By first understanding ourselves and the way we see the world, we can make better, more environmentally-conscious decisions. We should also seek to pro-actively adapt our Orientation as we receive new information.

6 Conclusions

Clearly, reflective practice has an important role to play in education as we help to shape the leaders and decision-makers of tomorrow. However, the models that we teach currently are not well adapted to the modern world. A key reason for this is that they are rigid in structure and do not sufficiently consider context and the role of the individual in shaping how reflections are carried out in the first place. Crucially, they do not sit in relationship with a wider global environment—a factor that is ever important as we face an unprecedented climate emergency.

To respond to these challenges, we need to adapt our way of thinking. We need to think more creatively, more flexibly, more critically. The model I suggest for doing this is John Boyd's OODA loop (1995). While no single model can ever reflect the complexity of our modern world, Boyd's OODA presents a new way of thinking about the world and our relationship with it. While most other models of reflection are static in nature and rigid in their approach, the OODA loop is designed to adapt and change as the world changes about it. By considering Boyd's OODA loop within the global environmental context, I propose that educators, students and practitioners more broadly can better equip themselves to confront the challenges the climate emergency brings.

References

Cranton P (ed) (2006a) Authenticity in teaching: new directions for adult and continuing education, number 111. Jossey-Bass

Ali A (2022) Critically reflective practice in visual communication design teaching for higher education undergraduate program. Int J Technol Des Educ 32(2):1063–1078. https://doi.org/10.1007/s10798-020-09626-6

AOMRC, COPMeD, GMC, MSC (2022) The reflective practitioner: guidance for doctors and medical students. https://www.copmed.org.uk/images/docs/reflective_practice/The_reflective_practitioner_guidance.pdf

Berners-Lee M (2020) How bad are bananas?: the carbon footprint of everything. Revised edition. Profile Books

Bousquet A (2009) The scientific way of war: order and chaos on the battlefields of modernity. Hurst & Company

Boyd J (1987) The strategic game of ? and ? https://danford.net/boyd/strategic.pdf

Boyd J (1995) The essence of winning and losing. https://www.coljohnboyd.com/static/documents/1995-06-28__Boyd_John_R__The_Essence_of_Winning_and_Losing__PPT-PDF.pdf

BPS (2022) Best practice in CPD. The British psychological society. https://www.bps.org.uk/best-practice-cpd
Brazell E (2022) Tory minister tells families to iron old wrapping paper to save money. Metro, December 22. https://metro.co.uk/2022/12/22/cost-of-living-minister-says-iron-old-wrapping-paper-to-save-money-17981419/
Bruno A, Dell'Aversana G (2017) Reflective practice for psychology students: the use of reflective journal feedback in higher education. Psychol Learn Teach 16(2):248–260. https://doi.org/10.1177/1475725716686288
Bureau SP, Komporozos-Athanasiou A (2017) Learning subversion in the business school: an 'improbable' encounter. Manag Learn 48(1):39–56
Cranton P (2001) Becoming an authentic teacher in higher education. Krieger
Cranton P (2006) Integrating perspectives on authenticity. In: Cranton P (ed) Authenticity in teaching: new directions for adult and continuing education, number 111. Jossey-Bass, pp 83–89
Deloitte (2022) How consumers are embracing sustainability. https://www2.deloitte.com/uk/en/pages/consumer-business/articles/sustainable-consumer.html
Dewey J (1933) How we think: a restatement of the relation of reflective thinking to the educative process. Heath
Driscoll J (1994) Reflective practice for practise—a framework of structured reflection for clinical areas. Snr Nurse 14(1):47–50
Driscoll J (ed) (2007) Practising clinical supervision: a reflective approach for healthcare professionals. Ballière
Dyer S, Hurd F (2016) 'What's going on?' Developing reflexivity in the management classroom: from surface to deep learning and everything in between. Acad Manag Learn Educ 15(2):287–303
Enck R (2012) The OODA loop. Home Health Care Manag Pract 24(3):123–124
Gibbs G (1988) Learning by doing: a guide to teaching and learning methods. Oxford Polytechnic
Grant A, McKimm J, Murphy F (2017) Developing reflective practice: a guide for medical students, doctors and teachers. Wiley Blackwell
HCPC (2022a) Reflective practice. https://www.hcpc-uk.org/standards/meeting-our-standards/reflective-practice
HCPC (2022b) Reflection and meeting your standards. https://www.hcpc-uk.org/standards/meeting-our-standards/reflective-practice/reflection-and-meeting-your-standards
Hibbert P (2013) Approaching reflexivity through reflection: issues for critical management education. J Manag Educ 37(6):803–827. https://doi.org/10.1177/1052562912467757
HSCE (2022) Reflective practice. https://www.hsce.co.uk/lessons/reflective-practice
Jasper M (2013) Beginning reflective practice (2nd ed). Cengage
Kolb DA (1984) Experiential learning: experience as the source of learning and development. Prentice-Hall
Kolb DA (2015) Experiential learning: experience as the source of learning and development (2nd ed). Pearson Education
Kreber C (2013) Authenticity in and through teaching in higher education: the transformative potential of the scholarship of teaching. Routledge
Mantzourani E, Desselle S, Le J, Lonie JM, Lucas C (2019) The role of reflective practice in healthcare professions: next steps for pharmacy education and practice. Res Social Adm Pharm 15(12):1476–1479. https://doi.org/10.1016/j.sapharm.2019.03.011
Maor E (2022) Applying the OODA loop to cybersecurity and secure access service edge. Dark Reading, December 6. https://www.darkreading.com/attacks-breaches/applying-the-ooda-loop-to-cybersecurity-and-secure-access-service-edge
NHS (2022) The importance of reflection in good competency evidence. https://nshcs.hee.nhs.uk/services/train-the-trainer/stp-train-the-trainer/guide-to-producing-good-competency-evidence/the-importance-of-reflection-in-good-competency-evidence
Osinga F (2007) Science, strategy and war: the strategic theory of John Boyd. Routledge

Papenfuhs S (2012) The OODA loop, reaching time, and decision making. Police1, February 23. https://www.police1.com/use-of-force/articles/the-ooda-loop-reaction-time-and-decision-making-fEOcXtsXFutU07cY

Peabody MA, Noyes S (2017) Reflective boot camp: adapting LEGO® SERIOUS PLAY® in higher education. Reflective Pract 18(2):232–243. https://doi.org/10.1080/14623943.2016.1268117

Pow R (2022) My top Christmas #recycling tips Ironing old wrapping paper means you can re-use last year's. https://twitter.com/pow_rebecca/status/1605239119269199872

Richards C (2004) Certain to win: the strategy of John Boyd, applied to business. Xlibris

Rogers CR (1969) Freedom to learn. Merrill

Rogers LH (2010) Betrayed how the education establishment has betrayed America and what you can do about it. Rowman & Littlefield Education

Rolfe G, Jasper M, Freshwater D (2011) Critical reflection in practice: generating knowledge for care (2nd ed). Palgrave

Ryder M, Downs C (2022) Rethinking reflective practice: John Boyd's OODA loop as an alternative to Kolb. Int J Manag Educ 20(3):100703. https://doi.org/10.1016/j.ijme.2022.100703

Sadler-Smith E, Cojuharenco I (2021) Business schools and hubris: cause or cure? Acad Manag Learn Educ 20(2):270–289

Schon DA (1983) The reflective practitioner: how professionals think in action. Basic Books

Seaman J (2008) Experience, reflect, critique: the end of the 'learning cycles' era. J Exp Educ 31(1):3–18

Thomas L, Ambrosini V (2021) The future role of the business school: a value cocreation perspective. Acad Manag Learn Educ 20(2):249–269

Treacy D, Gaunt H (2021) Promoting interconnections between reflective practice and collective creativity in higher arts education: the potential of engaging with a reflective matrix. Reflective Pract 22(4):488–500. https://doi.org/10.1080/14623943.2021.1923471

UN Environment Programme (2022) Facts about the climate emergency. https://www.unep.org/facts-about-climate-emergency

UNICEF (2015) Unless we act now: the impact of climate change on children. https://sdgs.un.org/sites/default/files/publications/2161unicef.pdf

United Nations (2022) What Is Climate Change? https://www.un.org/en/climatechange/what-is-climate-change

World Health Organization (2022) Climate change: Impact. https://www.who.int/health-topics/climate-change#tab=tab_2

Yadav S (2022) Tory MP tells people to reuse old wrapping paper to save money. International Business Times, December 23. https://www.ibtimes.co.uk/tory-mp-tells-people-reuse-old-wrapping-paper-save-money-1710464

Sustainability: A Business Case for Adoption

Madhavi Venkatesan, Khadija Elsewedy, Shruti Vummadising, and Brandon Gunasti

Abstract A central focus of the Sustainable Development Goals (SDGs) is poverty elimination. However, given that poverty is both an input and outcome of our present economic system, there appears to be a contradiction in the expectation for businesses to operationalize the SDGs. The business focus on financial return, which defines fiduciary responsibility, has eliminated the inclusion of social and environmental parameters in decision-making, except when constrained by regulation. This discussion highlights the relationship between the United Nations Global Compact as an SDG integration tool for businesses and the use of corporate social responsibility reporting (CSR) and the environmental, social, and governance attribution (ESG), as business marketing channels to promote sustainability branding. The discussion surfaces the superficiality of present sustainability reporting with a review of the inclusion of the SDGs in the sustainability platforms of the top 10 companies listed in the Fortune 500. The findings are provided along with a suggested catalyst, sustainability marketing, to facilitate business sector transition and adoption of SDGs as parameters in business as usual.

Keywords SDGs · Corporate social responsibility · Brand premium · Fiduciary responsibility · Consumer perception

M. Venkatesan (✉) · K. Elsewedy · B. Gunasti
Department of Economics, Northeastern University, Boston, MA, USA
e-mail: m.venkatesan@northeastern.edu

K. Elsewedy
e-mail: elsewedy.k@northeastern.edu

B. Gunasti
e-mail: gunasti.b@northeastern.edu

S. Vummadising
Department of Economics, University of Texas, Austin, TX, USA
e-mail: shruti.vummadising@utexas.edu

T. Wall et al. (eds.), *Sustainability in Business Education, Research and Practices*, World Sustainability Series, https://doi.org/10.1007/978-3-031-55996-9_7

1 Introduction

The Sustainable Development Goals (SDGs) reflect 17 interconnected and ambitious goals that are a non-enforceable, voluntary, and nation-based, global call to action. The SDGs are linked to poverty eradication (SDG 1) by 2030 relative to a $1.90 a day consumption equivalent; further, SDG 1 is referenced as "an indispensable requirement for sustainable development" (United Nations (UN) n.d. para. 1). Additionally, the SDGs are focused on global environmental protection and climate change, as well as economic and political stability.

As a result, central to the goals is a foundation in equity, as peace, prosperity, and attainment of a suitable quality of life are implicitly viewed as universal human rights, echoing the Universal Declaration of Human Rights adopted by the United Nations (UN) in 1948.

> Sustainability is implicit within the SDGs, as well as explicit in the naming of the goals. The concept incorporates the intertemporal allocation of resources through a holistically assessed strategic utilization rate that includes environmental and social justice parameters and aligns with the recommendations of the Brundtland Report (UNESCO 2000).

Specific to sustainable development, this concept has been characterized as having three pillars (Burkart, 2009): Environment, Economic, and Social. The pillars are interconnected, so much so, that the actions in one area can reinforce the goals of another. This is readily observable by reviewing any one SDG. For example, in eliminating poverty, there is an acknowledgement of the disproportionate nature of women's poverty rates relative to male counterparts, highlighting the rationale for SDG 5, Gender Equality.

The SDGs provide clear guidelines and targets for all countries to adopt in accordance with their own priorities and the environmental challenges of the world at large. Though there is reporting oversight, there is no regulatory mandate or punitive process to ensure a stated goal or target is met or in process. This aspect of the SDGs promotes subjectivity in measurement and a limited prescriptive approach; while the lack of enforceability of SDG targets provides a basis for concern related to their attainment. However, the flexibility of SDG implementation within countries could also be looked upon from the perspective of cultural sensitivity as each country is able to develop targets that are consistent to their cultural and structural identity and capacity. One item of concern may be that though sustainability is embedded within the SDGs, the SDGs remain human-centric. Though they reference the environment, the focus of environmental protection is the quality of human life.

By not explicitly articulating and establishing the need for environmental protection independent of human perception of value, the SDGs arguably do not consider the risk related to what human knowledge and understanding does not know. When coupled with a lack of acknowledgement of the intrinsic value of the environment and its non-human inhabitants, this perspective does not directly promote a call for stewardship. Therefore, a significant component of sustainability, which seeks to establish intergenerational equity in resource allocation may not be addressed through the present SDGs. Arguably, not until the intrinsic value of the ecosystem is addressed will all aspects of sustainability evaluation be factored into sustainable development. An additional and related limitation relates to the measurement of SDG progress, which relies on quantitative metrics (GDP, income) related to the present market-based economy. Given the accepted perception among a growing number of academics across disciplines that a finite planet is inconsistent with infinite growth, an alternative to GDP that is focused on intergenerational quality of life would appear to be more aligned to sustainability. Therefore, in the measurement of SDG performance, an inherent inconsistency with sustainability is visible.

2 Corporations and SDGs

Though the focus of SDG deployment has been in large measure country-specific adoption, there are multiple reasons why their incorporation at the business-level would be more transformative. The most obvious reason may be that the largest corporations in the world exceed the financial capacity of many economies and have a foothold across many countries, effectively providing an opportunity for significant global operationalization of the SDGs related to business operations and through their political influence (Kim and Milner 2019).

In 2016, the World Economic Forum noted that Walmart, based on its annual revenue, ranked as the 10th largest economic entity, below 9th ranked Canada. Further, the report noted that the top ten largest corporations were within the 30 largest economic entities, see Table 1. Global Justice Now noted in 2018 that 157 of the 200 largest economic entities by revenue were corporations. The report, which was based on 2017 revenue data, highlighted that among the top 100 economic entities 69 were corporations, the remaining were countries.

Table 1 Top ten companies ranked by size (across all economic entities)

Company	Rank
Walmart	10
State grid	14
China national petroleum	15
Sinopec group	16
Royal dutch shell	18
Exxon mobil	21
Volkswagen	22
Toyota motor	23
Apple	25
BP	27

Source:World Economic Forum (2016)

Aligned to the economic and political status of corporations, another rationale for adoption of SDGs by businesses is that the SDGs were intended and therefore designed for business adoption.

> The new [SDGs] Global Goals result from a process that has been more inclusive than ever, with [g]overnments involving business, civil society, and citizens from the outset. We are all in agreement on where the world needs to go. Fulfilling these ambitions will take an unprecedented effort by all sectors in society—and business has to play a very important role in the process (UN Global Compact n.d.(a) para. 2).

Further, to advance sustainable development, starting with the Millennium Development Goals in 2000 through the 2015 adoption SDGs, the UN Global Compact initiated the Ten Principles of the UN Global Compact for business adoption; these also align to the 17 SDGs. As provided by the UN Global Compact:

> At the UN Global Compact, we aim to mobilize a global movement of sustainable companies and stakeholders to create the world we want. That's our vision.

To make this happen, the UN Global Compact supports companies to:

- Do business responsibly by aligning their strategies and operations with Ten Principles on human rights, labour, environment and anti-corruption; and
- Take strategic actions to advance broader societal goals, such as the UN Sustainable Development Goals, with an emphasis on collaboration and innovation (UN Global Compact n.d.(c) para. 1–4).

Corporations can become signatories to the Ten Principles, but the acknowledgement referenced as a "commitment" is non-binding and non-enforceable, unfortunately, another commonality with the SDGs. The benefit to corporations in signing a commitment to the Ten Principles is in branding given the increasing consumer awareness of the significance of sustainability as it relates to combatting global climate change. The consumer opportunity is highlighted in the findings of global strategy firms, Simon-Kucher & Partners. In their Global Sustainability Study 2021, which surveyed 10,000 people across 17 countries, sustainability was a significant variable in consumers' purchasing decisions. The study noted, "Globally, 85% of people indicate that they have shifted their purchase behavior towards being more sustainable in the past five years—however, we observe meaningful generational differences in attitude. When looking at both Baby Boomers and Generation X, 24% across each have significantly changed their behavior towards being more sustainable—this figure climbs to 32% for Millennials. Additionally, one third of Millennials will choose a sustainable alternative when available, whereas older generations are less likely to actively choose sustainable alternatives (24–29%)" (Jain et al. 2021). It is important to note that the issue with older cohorts may tie to fixed income and the perception and reality of the premium attributable sustainable products rather than pro-environmental bias (Venkatesan 2017).

As of July 2022, 6,666 companies (entities having 1,000 or more employees) were listed as "active", including small and medium sized enterprises the total increased to 14,500 with representation across 247 countries (UN Global Compact, n.d.(c)).

This compares with the stated 9,500 companies noted by the UN Global Compact, which were distributed across more than 160 countries in 2018 (UN Chronicle, n.d., para.16).

3 Significance of the Consumer

In 2020 the Boston Consulting Group conducted a survey of 3,000 consumers in eight countries: United States, United Kingdom, France, Indonesia, Brazil, South Africa, India, and China. The findings highlighted the significance of environment in their decision-making. 70% noted being more aware than before the pandemic of the human impact on the environment and its relationship to climate change, with 76% believing environmental issues are at least as important as human health issues. 70% noted that environmental issues should be either prioritized equally or more than economic recovery (Kachaner et al. 2020).

A 2021 report by the Economist Intelligence Unit in conjunction with the World Wide Fund for Nature (WWF) noted in an assessment of Google searches, where the searches were specific to sustainable goods, that sustainability-based searches increased by 71% globally since 2016. Further, demand for sustainable goods had appreciable gains, especially in high-income countries. Noted were the United Kingdom, the United States, Germany, Australia, and Canada. But this trend can also be seen in emerging countries. In emerging economies, 41% of consumers in China say they want eco-friendly products, and in India, a 13% increase in sales of organic products has been estimated for the period between 2021 and 2018 (Economist Intelligence Unit 2021). In an assessment conducted by Forrester, a market research company, the outcomes pertaining to the U.S. market revealed similar findings. Among consumers actively engaged in their purchasing behavior, 68% planned to step up their efforts to identify brands that reduce environmental impact, 61% intended to seek out energy-efficient labels when making purchases, and 47% responded that they regularly buy organic products (Forrester 2021). The Ernst and Young (EY) Future Consumer Index builds on these findings and notes that 43% of global consumers want to buy more from organizations that benefit society, even if their products or services cost more; 64% are prepared to behave differently if it benefits society. The report continues by noting that consumer actions are not necessarily in alignment with their values and intentions, opening the opportunity for consumer facing companies to literally capitalize on sustainability branding, making sustainability more accessible to the eco- and social justice-minded consumer (Rogers and Cosgrove 2021).

The consumer demand for sustainability is within the operational capacity of corporations. Tulder (2018) identified the rationale for corporate action, highlighting the following:

- Corporations show great ability to scale activities across sectors, borders and products;

- They are able to innovate through their ability and willingness to take risk;
- Companies—next to governments—are the largest investor in technology;
- They can develop new organisational practices alone, or in concert with others; they serve the fundamental desire of people to face individual challenges in an entrepreneurial manner, and take responsibility for costs and rewards;
- Corporations create jobs, products and services;
- Due to their often very powerful positions in networks, technologies, and sectors, they
- can be a formidable barrier to change if they are not involved in the change process;
- They can mobilise sizable and timely financial resources (either on the open stock
- market, or as part of other financial arrangements);
- They create efficiency, stimulated by competition, thus enabling cheaper solutions for
- existing products and services;
- They are aimed at investments rather than at subsidies (Tulder 2018 p. 73)

Given the consumer-driven brand value of sustainability, corporations have sought to integrate environmental and equity parameters in operations. This has been most easily observed in corporate social responsibility reporting (CSR) and environmental, social, and governance (ESG) classification seeking. As a result, SDG integration in corporations is typically reported in a company's CSR or ESG reporting. Interestingly the CSR approach is also supported by the UN Global Compact (Msiska and Kimmel 2021; Orzes et al. 2020; Voegtlin and Pless 2014) and is the attributed vehicle for addressing the SDGs (ElAlfy et al. 2020).

4 Corporate Sustainability Reporting

The Governance & Accountability Institute (G&A), a consulting firm specializing on corporate sustainability and ESG reporting, shared in their 2021 report that "92% of the S&P 500 companies published a sustainability report in 2020, up from 90% in 2019"; and "70% of the Russell 1000 companies published a sustainability report in 2020, up from 65% in 2019" (G&A 2021). From a financial product perspective ESG reporting, and classification is gaining investor interest. Financial product consumers have increased inflows into ESG labeled investments. At their current growth rate, ESG oriented assets are expected "to represent half of all professionally managed assets globally by 2024" according to a 2022 report from Deloitte (Taylor and Collins 2022).

The rationale for consumer interest in sustainability is complex. Though as survey evidence highlights the demand specific to consumer products may be focused on reducing the human footprint, this may not be the same for investment products. A consumer can be motivated by the sustainability platform and CSR profile of a company, and evidence addresses a correlation between consumer environmental and social justice values and purchasing behavior (Rodrigues and Borges 2015; Ali and

Sohail 2018; Vătămănescu et al. 2021). The focus on ESG is not simply sustainability but the perception that sustainability is aligned with the potential to generate higher financial returns. The academic and investment literature has promoted ESG as a growth driver but often with limited explanation as to how and why, the former has prompted increased investment in these assets, which may trigger their trading at valuations in excess of their worth, while also providing a market attribution of excess return (Hamlin 2022). Further, though correlations between ESG and risk management have been made as a basis for sustainability outperformance, there is an ambiguity with respect to financial performance of ESG with respect to non-ESG or sustainability-focused investments (Bhagat 2022; Polman and Winston 2022). The ambiguity obscures the role of risk management and sustainability with respect to fiduciary responsibility as presently defined, executed, and legitimized by Milton Friedman (Friedman 1970).

This chapter explores the relationship between firm fiduciary duty, consumer expectations and motivations, and the inclusion of sustainability in market activity. The data used in the discussion is taken from public sources, including company specific publications, United Nations Global Compact, Morningstar, and Forbes. The discussion is limited to the top 10 firms listed in the Fortune 500.

In the next section, the discussion addresses the relationship between financial returns and sustainability using the financial services sector as an example, arguably a natural experiment given the role of fiduciary responsibility in defining the sector. The discussion continues by addressing the superficiality of present sustainability reporting with a review of the inclusion of the SDGs in the sustainability platforms of the top 10 companies listed in the Fortune 500. The findings are provided, and the chapter concludes with a suggested catalyst for corporate SDG facilitation, sustainability marketing.

5 Fiduciary Duty and Sustainability

Looking at the alignment of the SDGs with business, sustainability is not a new concept. The origin of the concept of "sustainability" can be traced back to the eighteenth century. Its recent traction is attributed to the 1987 Brundtland Commission report (Lu et al. 2020). Sustainability embodies the sentiment of doing business in a self-regulating community where resources are acknowledged as being shared, utilization is symbiotic with the environment, and risk management is an integral and required holistic responsibility. Simply stated, at the core of sustainability is fiduciary responsibility and accountability that aligns financial return with a moral parameter.

5.1 *Natural Experiment: Financial Services*

The financial services sector would appear to have a comparative advantage with respect sustainability as it relates to the fulfillment of fiduciary duty, given that the definition of a fiduciary requires an agent to act in the best interest of the client and to disclose all conflicts of interest and costs. Additionally, by attribute, the sector has inherent to its activities, financial responsibility for client funds and therefore, would appear to have integrated the motivation for customer persistence by administering services under the highest ethical standards of doing business, as a fiduciary (Lydenberg 2014; Weber et al. 2016).

Unfortunately, the examples of the lack of fiduciary responsibility within the sector are abundant and range from individual corporate practices (Campbell 2019) to individual funds and money managers (Schoen 2017). Further, the realization of the need for a standard duty of care has had a slow enlightenment, being borne from one sector-based catastrophe after another and culminating in further regulatory involvement, scrutiny, and additional rules (Waitzer and Sarro 2014).

The question of why a sector, which provides financial services, products, and management would be able to persist and grow without a self-policed foundation in fiduciary responsibility should be at the core of any conversation on sustainability. Financial services are an integral component of the economy; the sector provides the liquidity and risk management services to enable the activities and growth of other sectors. The due diligence exercised by the sector and the requirements mandated with respect to fund access and availability, in many ways set the standard of practice of companies and individuals needing liquidity or risk sharing. As a result, given the publicly relayed risk assessment shortfalls, the sector may be inappropriately credited as a conduit of acceptable behavior and business ethics.

To some extent the assumption of the sector's fiduciary responsibility from the perspective of investors has provided it with trusting, ill-informed, return-seeking clients, whose very ignorance of finance and economics has propelled the growth of investment irresponsibility (Fairfax 2018). Furthermore, poorly informed bankers, insurers, and other financial services agents have compounded the issue by a lack of understanding of investment policy time horizon and over-reliance on the risk-return trade-off for justification in the solicitation of an individual investment's appeal. The limited exercise of own-education responsibility of both clients and agents has deterred the emergence of sustainability as a standard component of business as usual. However, at the very center of the problem may be the secular incentive system, the desire for immediate return, which has led to the exclusion of the evaluation of an investment's holistic impact on the environment, resources, and waste creation (Puaschunder 2016; Miller 2018).

Short-term investment horizons may by their very duration neglect long-term costs, resulting in the higher risk and reward outcomes at the expense of sustainable long-term income streams. This in turn may be responsible for higher valuations or lower risk assessments for specific operations or activities as experienced in the financial contagion preceding the Great Recession; and in cyclical flow, contributing to higher than warranted market returns, promoting, and prompting greater risk taking. Ultimately, this process leads to higher social costs, as resources and waste issues are eventually remediated. Unfortunately, the social costs, which may take the form of higher taxation or economic slowing are disproportionately borne by individuals who did not partake in the short-term returns; thereby, exhibiting a similarity to a regressive tax (Cushman 2015; Samuelson 2011). This latter aspect highlights the disproportionate impact of unsustainable practices and is similar in nature to the concept of common but differentiated responsibilities incorporated in the Paris Agreement specific to greenhouse gas emissions and resulting speed of climate change.

In looking at the overall mechanics described, there are many points where operationalized sustainability could have been implemented:

- Fiduciary duty to clients as part of business credibility
- Longitudinal assessment of an investment strategy or risk assumption
- Evaluation of long-term return to society, inclusive of social equity

However, as apparent and obvious as these may seem, these elements have yet to be included in an operationalized sustainability framework for many companies within the financial services sector. Companies still appear to be looking at the superficial elements of sustainability, limiting sustainability reporting to the marketing and communications function, culminating in a public relations, brand enhancing CSR report. This being noted, there are examples of financial services firms that have integrated sustainability in operations, determining what types of entities to finance and where financial resources will be deployed, this is observable in lending and investment practices. In the latter case, some sustainability initiatives have equated with risk management and resulted in positive correlation with returns (Wu and Shen 2013). For most sustainability-oriented actions, the time horizon for payoff may be more uncertain than the mitigation benefit. For example, UN Secretary-General António Guterres addressed insurers in 2021 noting their impact on global emissions reduction targets, "Your investments should not be contributing to climate pollution but should be directed towards climate solutions. Invest in renewables, low- and zero-carbon transport and climate-resilient infrastructure" (International Cooperative and Mutual Insurance Federation 2021). The action of insurers through investment of their own assets, as well as parameters related to underwriting are attributed significant impact in near-term mitigation given the size of the investment portfolio and the significance of insurance in supporting financial flexibility by risk sharing.

Unfortunately, to the extent that the consumer or investor is unable to distinguish between reporting and operations, CSR and ESG as a form of marketing can obscure the reality of sustainability. Company actions are aligned to financial return to stakeholders; fiduciary duty relates to maximizing shareholder value. To the extent that operationalized sustainability is not defined as a fiduciary responsibility, corporate actions may be in opposition with sustainability targets (Lashitew 2021). Simply stated, what is profit maximizing may not be sustainable. The discussion that follows reviews the top 10 companies listed in the Fortune 500, a listing of the largest public and private corporations filing financial statements in the United States, and provides insight on sustainability reporting, sustainability of corporate action, and ESG ranking.

5.2 Corporate Commitment

ESG has become a marketing tool to determine the environmental sensitivity of a company's operations. However, the lack of standardization of ESG evaluation, and definition, increases the ambiguity of the ESG relationship with operationalized sustainability. In this section Morningstar ESG Risk Ratings are used to assess the ESG risk associated with the top 10 Fortune 500 firms. "The Morningstar ESG Risk Ratings measure the degree to which a company's economic value is at risk driven by ESG factors or, more technically speaking, the magnitude of a company's unmanaged ESG risks. A company's ESG Risk Rating is comprised of a quantitative score and a risk category. The quantitative score represents units of unmanaged ESG risk with lower scores representing less unmanaged risk" (Sustainalytics 2021).

Table 2 provides a listing of the top 10 Fortune 500 companies and the tenure of their ranking on the listing. Additionally, tied to the UN focus on the UN Global Compact, the signatory status of the company and year of attainment is listed along with the type of sustainability reporting issued. Though there is no rationale for assuming that an ESG, CSR, or Sustainability report may be different is any significant manner, there is no basis for assuming their similarity as well. This highlights the lack of standardization in reporting and the challenge to attainment of the SDGs and sustainability, as the goal is not defined except by specific parameters. This creates ambiguity with respect to sustainability objectives: Are they focused on the short-term or long-term? Are they related to an aspect of operations or operations as a whole? Does sustainability assess the product(s) focus of the company or only the inputs? These issues are captured in the Morningstar ESG ratings, which arguably are limited by company ESG focus and self-reporting.

The ESG ratings highlight the relationship between risk and return, as well as the self-reported nature of ESG and its focus on metrics, which provides credit for carbon emission targets but is unable to assess qualitative adjustments toward business as usual. An example of the latter would be reduction in profit margins due to corporate initiatives to ensure fair wages and equitable supply chain standards. For this reason, though companies may provide disclosure, their communications and initiatives remain tied to fiduciary duty defined by financial returns, where perception of sustainability has greater financial reward through brand premium than reality, which would adversely impact investor enthusiasm through narrowing profit margins.

In its ESG Report, Walmart affirms its commitment to nine of the SDGs. However, the company has been cited for limited transparency and brown washing, where the latter is a new term referencing omission of information to promote a green platform (Bansal 2021). This aspect is important as the company operated in a price sensitive sector as a discount retailer. Profits are dependent on reducing costs, which is most likely to occur in the form of wages and purchasing of products for sale. These

Table 2 Fortune 500 Top 10

Company name	2021 revenue (billions dollars)	Employees	Number of years listed	Reporting (year signed, sustainability report type)	ESG risk rating
Walmart	559.15	2,300,000	27	UN global compact: 2019[a] ESG report	24.6 (medium risk)
State grid	386.62	896,360	21	UN Global Compact: 2006[b] CSR Report	37 (high ESG risk)
Amazon	386.06	1,298,000	13	UN global compact: 2021[c] CSR report	30.3 (high ESG risk)
China national petroleum	283.96	1,242,245	21	UN global compact: n.a.[d] CSR report	Not rated
Sinopec group	283.73	553,833	23	UN global compact: 2006[e] CSR report	38.5 (high ESG risk)
Apple	274.52	371,000	19	UN global compact: n.a.[f] ESG report	16.4 (low ESG risk)
CVS	268.71	76,000	26	UN global compact: 2018 [g] ESG report	22.9 (medium ESG risk)
UnitedHealth group	257.14	42,000	25	UN global compact: n.a.[h] Sustainability Report	17.5 (low ESG risk)
Toyota motor	256.72	156,500	27	UN global compact: 2013[i] Sustainability report	28.9 (medium ESG risk)
Volkswagen	253.97	72,000	27	UN global compact: n.a.[j] Sustainability report	29.7 (medium risk)

Source Fortune (n.d.); UN Global Compact (n.d.(c)). Morningstar (n.d.)
[a] Walmart Mexico
[b] State Grid Corporation of China; State Street Brazil Holding
[c] Amazon Fulfillment Poland
[d] No adoption
[e] China Petroleum and Chemical Corporation—Sinopec
[f] No adoption
[g] CVS Health
[h] No adoption
[i] Toyota Gibraltar Stockholdings Ltd
[j] No adoption

attributes are directly tied to exploitive behavior that can be obscured by supply chain procurement policies.

Similarly State Grid can take ESG credit for its energy efficiency and has even been lauded by the UN Department of Economic and Social Affairs (2020) for facilitating green recovery. However, though this seemingly aligns to SDG 7 there is no assessment of the sustainability impact of the corporation's infrastructure and the potential impact of its infrastructure roll-out on increasing energy use, meaning a misalignment with SDG 12. The case of Amazon with respect to SDG 7 mimics this, as the focus on emissions in one aspect of the company's business may be neglecting emerging and realized adverse sustainability impacts in another. For example, transportation-based emissions may obscure the impact of waste creation in packaging and natural carbon sequestration elimination and habitat impacts due to the use of cardboard.

China National Petroleum highlights its focus on green energy, decent work, and even poverty alleviation in its ESG report that highlights governance. However, though details are provided limited information is given on implementation of policies and no auditable data is available to track progress. For Sinopec Group, the focus is only on ESG with governance being the driver. Given the self-reported nature of ESG and its check the box validation, reliance on governance is a limited perspective of sustainability operationalization.

Looking at Apple, the reality of e-waste, design for obsolescence business model, and supply chain violations are largely outside of the discussion of the company's ESG score though they are noted in the company's environment page, "The products you love also save the planet" (Apple 2022). The company addresses SDG 12 but has no metrics to justify its sustainable production. Further, the energy dependence related to product use is rarely if ever integrated into the sustainability discussion. The latter can also be attributed to auto manufacturer Toyota Motor and Volkswagen, where the life cycle impacts of their products and their direct an indirect relationship (plastics) with fossil fuels are not incorporated in the discussion of their sustainability footprint. Volkswagen highlights initiatives related to all SDGs with the exception of SDG 1, 2, and 14. However, the SDGs are addressed solely with their connection to ongoing corporate initiatives with no detail on implementation or progress provided. This is also true for Toyota. The company provides an interactive webpage devoted to the SDGs but no details on implementation or related metrics are included (Toyota n.d.).

Highlighting the relationship between healthcare and sustainability specific to UnitedHealth and CVS, the methods related to the provision of care and the use of single-use plastics is not addressed. Given the focus on carbon emissions related to energy production from the fossil fuel sector, plastics are a growth engine for fossil fuels, meaning that as emissions are reduced in one area, they may be increased in another. UnitedHealth addresses the SDGs from the perspective of SDG 3 but provides no details other than its progress toward universal health coverage, conferring legitimacy for sustainability primarily from its UN Global Compact signatory status (UHC 2022). Lack of data is also the case for CVS, which highlighted its focus on SDGs in its CSR, but with no metrics, the announcement aligns with a desire to be perceived as sustainable than a reality of the same.

In viewing all ten of these companies it is clear that an operational focus on carbon emissions distorts the interconnectedness of the overall operations of the company with respect to indirect channels that promote carbon emissions. Further, the variation in targets, implementation, and limited detail and transparency on sustainability initiatives limits information flow, arguably a barrier for investor and consumer action to promote sustainability. This once again highlights a topic discussed significantly in the sustainability literature, what is the goal of sustainability? If implementation of the SDGs are not regulated, standardized, or mandated, how is there assurance that corporate action holistically is aligned to their attainment? However, what is even more neglected is that pricing provides the opportunity to transform products from wants to need. As costs are externalized leading to lower pricing, prices become a barrier for change, as consumers have assumed financial viability of their present consumption standards. So how can stakeholders be engaged in promoting the corporate attainment of SDG goals when they benefit from their non-attainment? Can we move beyond the hypocrisy of sustainability attributable to a focus on carbon?

6 Conclusion: Promoting Sustainability Traction

Consumption choices in general are based on the dynamics of demand and supply of a good and are identified with satisfying a need or a want. The impact of consumption decisions can be significant when there is asymmetry of information and alternatively transparency; fundamentally, there is a relationship between economic and environmental outcomes and consumption choices. Purchases affect labor and environmental resource use. However, most purchase decisions are made through a market mechanism, where the consumer is not aware of the entire production process and waste is not a factor in the consumption decision. This limitation in information transparency often creates a disconnect between the social and environmental justice sensitivities of a consumer, and the realities of their consumption choice in enabling and maintaining the values that they espouse. It is for this reason that marketing has a significant role in an economy. Subject to any prevailing regulatory constraints, in a consumption-led, GDP focused economy, marketing, through establishing demand via marketed wants, promotes purchasing behavior.

Consumers rely on product marketing as a source of information on a product and often default to certifications and brand image as a decision-making tool (Kahn 2012; Overbeek 2019). For social good and sustainability to be achieved or enhanced through marketing, marketing itself needs to move beyond just the selling phase to the responsibility stage. Present CSR, ESG, and Sustainability reporting is limited to marketing for brand positioning rather than operational alignment with the rationale for the UN Global Compact and the SDGs.

These reports can provide a channel for consumer and investor education. In lieu of superficial sustainability brand marketing, marketing can be educational, consider science, including environmental uncertainties, and take a holistic stance in its relationship with products especially when the product image is sustainability (Whiteman

1999). Sustainable marketing requires that marketing practitioners "spend an equal, if not greater, amount of time and energy on what happens after the individual or business 'consumes' the product...[M]arketing's responsibilities and opportunities do not end with the sale" (Murphy 2005). Noting these parameters for the discipline, there is a need to enlarge the discission of sustainability to include a product life cycle and essentially assess the sustainability of a product holistically across the multiple supply chains that are incorporated within it.

In the present global economy, there is an obvious and urgent need to incorporate sustainability into all activities but there is an even more important goal to align all sustainability objectives to the common goal of reducing the human planetary footprint through behavioral change. This change will require a new relationship with the planet that will need awareness, knowledge, and action. Marketing, if implemented responsibly, can be instrumental in facilitating all three, while also aligning to financial fiduciary responsibilities.

For operationalized sustainability, change in operations will likely require stakeholder engagement to recondition senior management to the active facilitation and operationalization of a sustainability framework. Specifically, until a threshold is reached whereby enough stakeholders actively promote sustainability as described as being a part of senior management fiduciary responsibility, assessment of quantifiable carbon emissions, waste generation and electricity use will continue to be the simplified interpretation of sustainability. Arguably this aligns to carbon mitigation but is only a necessary not sufficient step toward sustainability operationalization as it only addresses one of the SDGs, SDG 7, independent of the other 16. Further, rules to facilitate fiduciary responsibility will not be sufficient without stakeholder impetus; rules alone will not foster the sentiment of the regulation but instead will lead to perpetuation of minimal fulfillment strategies. The most significant factor for sustainability integration is the stakeholder; the necessity to enable stakeholder vigilance is education and understanding of the risks related to limited or non-existent sustainability integration. For this reason, sustainability marketing is significant but what and who is involved in dissemination will determine the outcome.

References

Ali F, Sohail M (2018) Effects of corporate social responsibility on consumer purchase intention. Pakistan J Human Soc Sci 6(4):477–491. https://doi.org/doi.org/10.52131/pjhss.2018.0604.0059

Apple (2022) The products you love also save the planet. https://www.apple.com/environment/

Bansal T (2021) Walmart appears to be greener than Costco. But is it really? forbes. https://www.forbes.com/sites/timabansal/2021/12/20/walmart-appears-to-be-greener-than-costco-but-is-it-really/?sh=4238f83a63b2

Bhagat S (2022) An inconvenient truth about ESG investing. Harvard Business Review. https://hbr.org/2022/03/an-inconvenient-truth-about-esg-investing

Campbell JL (2019) Self-responsibility gone bad: institutions and the 2008 financial crisis. Am Behav Sci 63(1):10–26. https://doi.org/10.1177/0002764218816801

Cushman T (2015) The moral economy of the great recession. Society 52:9–18. doi.org/https://doi.org/10.1007/s12115-014-9852-4
Economist Intelligence Unit (2021) An Eco-Awakening. https://explore.panda.org/eco-wakening#full-report
ElAlfy A, Palaschuk N, El-Bassiouny D, Wilson J, Weber O (2020) Scoping the evolution of corporate social responsibility (CSR) research in the sustainable development goals (SDGs) era. Sustainability 12(14):5544
Fairfax LM (2018) The securities law implications of financial illiteracy. Virg Law Rev 104(6):1065–1122. https://www.jstor.org/stable/26790703
Forrester (2021) Empowered consumers call for sustainability transformation. Forbes. https://www.forbes.com/sites/forrester/2021/01/21/empowered-consumers-call-for-sustainability-transformation/?sh=4af6d0a62042
Friedman M (1970) The social responsibility of business is to increase its profits. New York Times Magazine 13 September 1970:122–126
Global Justice Now (2018) 69 of the richest 100 entities on the planet are corporations, not governments, figures show. https://www.globaljustice.org.uk/news/69-richest-100-entities-planet-are-corporations-not-governments-figures-show/
G & A (2021) 92% of S&P 500® Companies and 70% of Russell 1000® Companies Published Sustainability Reports in 2020, G&A Institute Research Shows. https://www.globenewswire.com/news-release/2021/11/16/2335435/0/en/92-of-S-P-500-Companies-and-70-of-Russell-1000-Companies-Published-Sustainability-Reports-in-2020-G-A-Institute-Research-Shows.html
Hamlin J (2022) With Too much love from institutional investors, companies with high ESG ratings may be overvalued. Activists Are Taking Note. Institutional Investor. https://www.institutionalinvestor.com/article/b1xqgjgywp9yfm/With-Too-Much-Love-From-Institutional-Investors-Companies-With-High-ESG-Ratings-May-be-Overvalued-Activists-Are-Taking-Note
Jain S, Hagenbeek O, Shogren B (2021) Global Sustainability study: what role do consumers play in a sustainable future? Simon-Kucher & Partners. https://www.simon-kucher.com/en/blog/global-sustainability-study-what-role-do-consumers-play-sustainable-future
Kachaner N, Nielsen J, Portafaix A, Rodzko F (2020) The pandemic is heightening environmental awareness. Boston Consulting Group. https://www.bcg.com/publications/2020/pandemic-is-heightening-environmental-awareness
Kahn R (2012) Fair trade activists in the United States. In Linton A (ed), Fair Trade from the Ground Up: New Markets for Social Justice (pp. 101–119)
Kim IS, Milner HV, Bernauer T, Osgood I, Spilker G, Tingley D (2019) Firms and global value chains: identifying firms' multidimensional trade preferences. Int Stud Quart 63(1):153–167. https://doi.org/doi.org/10.1093/isq/sqy055
Lashitew AA (2021) Corporate uptake of the sustainable development goals: mere greenwashing or an advent of institutional change? J Intern Bus Policy 4:184–200. https://doi.org/doi.org/10.1057/s42214-020-00092-4
Lu J, Liang M, Zhang C, Rong D, Guan H, Mazeikaite K, Stremikis J (2020) Assessment of corporate social responsibility by addressing sustainable development goals. Corp Soc Responsib Environ Manag 28(2):686–703
Lydenberg S (2014) Reason, rationality, and fiduciary duty. J Bus Ethics 119(3):365–380. http://www.jstor.org/stable/42922011
Miller S (2018) The global banking sector: corruption, institutional purpose and economic justice. Bus Profess Ethics J 37(1):13–44. http://www.jstor.org/stable/45149309
Msiska M, Ng A, Kimmel RK (2021) Doing well by doing good with the performance of United Nations global compact climate change champions. Human Soc Sci Comm 8:321. https://doi.org/doi.org/10.1057/s41599-021-00989-2
Morningstar (n.d.) Company ESG Risk Ratings. Sustainalytics. https://www.sustainalytics.com/esg-rating
Murphy P (2005) Sustainable marketing. Bus Prof Ethics J 24(1/2):171–198

Orzes G, Moretto AM, Moro M, Rossi M, Sartor M, Caniato F, Nassimbeni G (2020) The impact of the United Nations global compact on firm performance: a longitudinal analysis. Intern J Prod Econ 227:107664. doi.org/https://doi.org/10.1016/j.ijpe.2020.107664

Overbeek A (2019) Examining the efficacy of fair trade and alternative consumption on environmental sustainability and human rights in developing countries. Consilience 21:158–171

Polman P, Winston A (2022) Yes, investing in ESG pays off. Harvard Business Review. https://hbr.org/2022/04/yes-investing-in-esg-pays-off

Samuelson RJ (2011) Rethinking the great recession. The Wilson Quarterly (1976-) 35(1):16–24. http://www.jstor.org/stable/41001064

Rodrigues P, Borges AP (2015) Corporate social responsibility and its impact in consumer decision-making. Soc Respons J 11(4):690–701. https://doi.org/doi.org/10.1108/SRJ-02-2014-0026

Puaschunder JM (2016) On the emergence, current state, and future perspectives of Socially Responsible Investment (SRI). Consilience 16:38–63. http://www.jstor.org/stable/26188773

Rogers K, Cosgrove A (2021) The CEO imperative: make sustainability accessible to the consumer. EY Future Consumer Index. https://www.ey.com/en_us/consumer-products-retail/make-sustainability-accessible-to-the-consumer

Schoen EJ (2017) The 2007–2009 Financial crisis: an erosion of ethics: a case study. J Bus Ethics 146:805–830. https://doi.org/doi.org/10.1007/s10551-016-3052-7

Sustainalytics (2021) ESG risk ratings—methodology abstracts. https://connect.sustainalytics.com/hubfs/INV/Methodology/Sustainalytics_ESG%20Ratings_Methodology%20Abstract.pdf

Taylor TL, Collins S (2022) Ingraining sustainability in the next era of ESG investing. Deloitte. https://www2.deloitte.com/us/en/insights/industry/financial-services/esg-investing-and-sustainability.html

Toyota (n.d.) Toyota sustainable development goals. https://global.toyota/en/sustainability/sdgs

Tulder R van (2018) Business & the sustainable development goals: a framework for effective corporate involvement. Rotterdam School of Management. Erasmus University, Rotterdam. https://www.rsm.nl/fileadmin/Positive_Change/SDGs/Business_and_Sustainable_Development_Goals_-_Positive_Change_0_Rob_van_Tulder.pdf

UHC (2022) UHC 2030, Our mission. https://www.uhc2030.org/our-mission/

UN (n.d.) Poverty eradication. https://sdgs.un.org/topics/poverty-eradication

UN Chronicle (n.d.) The UN Global compact: finding solutions to global challenges. https://www.un.org/en/un-chronicle/un-global-compact-finding-solutions-global-challenges#:~:text=With%20more%20than%209%2C500%20companies,role%20in%20improving%20our%20world

UN Global Compact (n.d.(a)) The SDGs Explained for Business. https://www.unglobalcompact.org/sdgs/about

UN Global Compact (n.d.(b)) Our Mission. https://www.unglobalcompact.org/what-is-gc/mission

UN Global Compact (n.d.(c)) See who's involved. https://www.unglobalcompact.org/what-is-gc/participants

UN Department of Economic and Social Affairs (2020) State Grid Integrated Energy facilitates the green recovery of MSEs by five comprehensive services. https://sdgs.un.org/partnerships/state-grid-integrated-energy-facilitates-green-recovery-mses-five-comprehensive

Vătămănescu E-M, Dabija D, Gazzola P, Cegarro-Navarro JG, Buzzi T (2021) Before and after the outbreak of Covid-19: Linking fashion companies' corporate social responsibility approach to consumers' demand for sustainable products. J Clean Prod 321:128945. https://doi.org/doi.org/10.1016/j.jclepro.2021.128945

Venkatesan M (2017) Sustainability and civic engagement: a communications engagement and education plan. In Walter Leal Filho, John Callewaert and Robert Marans (Eds.) Handbook of Sustainability and Social Science Research. London, England: Springer-Verlag

Voegtlin C, Pless N (2014) Global governance: CSR and the role of the UN global compact. J Bus Ethics 122:179–191. https://doi.org/doi.org/10.1007/s10551-014-2214-8

Waitzer EJ, Sarro D (2014) Fiduciary society unleashed: the road ahead for the financial sector. Bus Lawy 69(4):1081–1116. http://www.jstor.org/stable/43666019

Weber O, Feltmate B (2016) History of sustainable banking. In Sustainable Banking: Managing the Social and Environmental Impact of Financial Institutions (pp. 25–44). University of Toronto Press. http://www.jstor.org/stable/https://doi.org/10.3138/j.ctv2fjwx89.5
Whiteman G (1999) Sustainability for the planet: a marketing perspective. Conser Ecol 3(1)
World Economic Forum (2016) How do the world's biggest companies compare to the biggest economies? https://www.weforum.org/agenda/2016/10/corporations-not-countries-dominate-the-list-of-the-world-s-biggest-economic-entities/
Wu M-W, Shen C-H (2013) Corporate social responsibility in the banking industry: Motives and financial performance. J Bank Fin 37(9):3529-3547. doi.org/https://doi.org/10.1016/j.jbankfin.2013.04.023

The Quest for Impact Research: Position, Strategies and Future Directions

Kent A. Williams, Loretta Baidoo, Tony R. Walker, Binod Sundararajan, and Alexander Davis

Abstract There is an urgent global call for sustainable development (SD) action to meet the United Nations 2030 Agenda. Business schools play a significant role in facilitating engagement and enabling action on the global commitment for impact. Such institutions strive for responsible and meaningful impact by fostering the sustainable development goal (SDG) commitments, where research is a core element of this impact lexicon. However, research commitments in business schools are challenged by the pressures of academic life and its progressions that scholars face, often hindering pathways towards impact research. As an interdisciplinary group of researchers embarking on a major research project focused on SD, this positional paper considers why impact research is important, what impact research is, and how we as researchers can ensure we best do impact research that is embedded in purposeful shared partnerships and collaborations. To continue the impact research dialogue and enable other researchers, we share strategic pathways and community considerations for conducting impact research that can be applied, measured, and put into action towards the SDGs and the 2030 Agenda.

Keywords Research impact · Impact research · Sustainable development · 2030 agenda

1 Introduction

> How selfish soever man may be supposed, there are evidently some principles in his nature, which interest him in the fortune of others …
>
> —Adam Smith (1790), Theory of Moral Sentiments

Business schools were first developed with the intention to enable professionalism, or a standard level of statesmanship (Khurana 2007). The idea was business schools would develop capacity and capability for leaders to elevate their ideals and contribute

K. A. Williams (✉) · L. Baidoo · T. R. Walker · B. Sundararajan · A. Davis
Faculty of Management, Dalhousie University, Halifax, NS, Canada
e-mail: Kent.Williams@Dal.ca

T. Wall et al. (eds.), *Sustainability in Business Education, Research and Practices*, World Sustainability Series, https://doi.org/10.1007/978-3-031-55996-9_8

to the success and flourishing of their community. In the early days of the Tuck Business school the aim for their program was stated, "to so broaden the minds, and raise the ideals of its graduates that it will do something to elevate the business community above the plane of merely money-getting (Khurara 2007, p. 110). In 1919, Harvard Business School's Dean Donham argued the core responsibility of the school was to develop and strengthen the mindset of its learners to be socially minded, being able to embrace emerging social challenges of the day (Khurana 2007). What happened to these ideals? Led by Milton Friedman in the late 20th-century, and held high by capitalist doctrine, shareholder primacy was taken as the fiduciary responsibility of organizations and businesspeople alike. This is still the dominant theme today. Coming from the business school's early beginnings, it can be suggested the moral compass and developing the social mindsets of leaders through business school education and research has deteriorated over the years an has left and unfulfilled promise (Khurana). Capitalism today drives the depletion of the earth resources, which has made the earth-system dangerously unstable (Haraway 2016; Moore 2015). There are some that suggest capitalism is broken, and it is time for something different before it is too late (Varoufakis et al. 2020). Our entire economic paradigm is a threat to the earth-human system that supports human and non-human flourishing (Moore 2015). Part of the responsibility for this lies with the business school institution in how we have positioned our research and teaching practice (Hoffman 2021). It can be argued the research agenda of the business school has held an unprecedented 'looking away' from the onrushing challenges facing society, that our research is not creating impact for ongoing challenges.

The focus of this text is impact research (IR), to answer the call to accelerate learning about practices of change and transformation through a participatory, reflective, and action-based forms of research (Sharpe et al. 2013). More to the point, as a global network of business schools, how do we take response-ability through IR and practice to co-inspire and influence a collective way of knowing, being, and doing? In this positional paper we thoughtfully think about steps for taking 'response-ability' for business research and practice. In the following narrative, we will explore: (1) what impact research means; (2) why impact research is crucial; and (3) how to thoughtfully do impact research to positively impact society and the biosphere; and, (4) how might we measure impact by the utilizing tools like the United Nation's (UN) sustainable development goals (SDG) or Indigenous knowledge frameworks (IKF). Tied to the latter exploration, through our recent research grant work, our working group responds to the invitation to explore novel and emergent approaches to collaborating, co-inspiring, and co-creating impact through an inclusive network of relationships and a collective transdisciplinary approach in and around the university community. Using imagination and seeds of emerging ideas we consider what IR could look like that has the potential to open portals to thriving futures through embracing societal challenges.

2 Cracks in the Foundation

The global turmoil of recent years often leads us to perceive the world through a dystopian lens, filled with much uncertainty through the recent pandemic, the war in Ukraine, wide-spread famine in Africa, and the grand challenge of the climate crisis. The visions of Cormac McCarthy's (2006) fictional dystopian narrative, *The Road*, seems not so far away from today's living reality. Forest fires, ongoing drought, and onrushing and increasing rates of natural disasters are becoming iconic to our living times—the suggested epoch of the Anthropocene. This period is defined by humans as a geophysical force that is wreaking havoc on the earth-human system (Rockström et al. 2009). Coupled with this, we are witnessing increased levels of fear, anxiety, division, and distrust on a global geo-political scale (Haraway 2016). Humankind's patterned way of living is highlighted by bounded individualism and a knack of unprecedented looking away from the catastrophes staring us in the face (Haraway 2016). Our present way of life, the *business-as-usual* mindset, is disrupting planetary boundaries that support human development (Steffen et al. 2015). Scientific truth shares the possibility of meeting global goals of keeping the Earth temperature below the 1.5 °C threshold is no longer possible (Rockström et al. 2024). Even as steps for a sustainable future are taken, our global carbon emissions continue to creep up. *What does this all mean?* It means stable living conditions for humankind on the planet are no longer guaranteed, and positions society for an uncertain and turbulent future. Examples of this are witnessed every day by the increasingly frequent catastrophic weather events, war and conflict over resources, loss of biodiversity, growing inequality, and human migration patterns. The progression in Earth temperatures could spell the endgame for humanity and other life on the planet, pointing towards extinction scenarios (Bendell 2018; Kemp et al. 2022). With increasing knowledge of climate scenarios, we are sensing an urgency by global leaders and institutions to take immediate action towards societal transformation that can embrace a changing planet through mitigation and adaptation strategies (Sachs et al. 2019).

It is evident that humankind and more specifically our approach to business scholarship is at a crossroads. Humanity needs to move towards immediate transformational living on the planet through SD as a species. The alternative is to continue the present course trajectory that continues to push us outside the planetary living conditions that support life on the planet (Rockström et al. 2009). This makes a compelling argument for researchers to uphold our part through IR that imbues response-ability, meaning we must think and act thoughtfully. Evidence suggests that it is behavior change and awareness that enables transformative change to occur (Kegan 1982). It is through education by using reflexivity and know-how that humans can build capacity and capability to embrace conflict and challenges to get into the rhythm of sustainable futures (O'Brien et al. 2013). Through the calls for transformative action, it is recognized there is limited understanding on how to go from talking or writing about it, to moving to action and results (Bentz et al. 2022). Researchers have always taken on humankind's challenges, leading explorations into the unknown and opening new pathways for knowing, being, and doing in the world (Gardner et al.

2021; Wittmayer and Schäpke 2014). However, we are beginning to understand, and academia is being called out for falling into the trap of producing research that does not really matter (Harris 2021). 'Not mattering' in the fact the research is not contributing positively to tackling societal challenges. Often the pressure to produce research is not primarily centred on contributing towards positive impact for society and beyond, but seemingly more focused on meeting tenure promotional requirements (Alperin et al. 2019). Consideringly, how do we become more aware and purposeful in our research explorations and ensure that our focus is on impact? Critically, what does impact mean and what does it look and feel like?

3 The Intentionality to Impact

Discourses on impact initiate thoughts on legitimacy and contribution, with many institutions including business schools criticized for not staying relevant to their primary goal to lead in the development of leaders and for "producing research considered as an expensive waste" (Lejeune et al. 2019, p. 88). Business schools purport to produce community leaders and better decision makers, however they are critiqued to being disconnected from society, dehumanizing their contribution to only a set of skills, and not personalized virtues (Petriglieri and Petriglieri 2015). This critique is not only subject to business schools but other components of the university system, where institutes of higher education have been critiqued as disconnecting with community priorities (Benneworth 2013; Harris 2021). The route to career growth as required in business schools and other departments or colleges of higher education, where academics are expected to publish as a prerequisite for growth, underscores the question of what impact means and who and what interests does produced research respond to. Are these interests of community members, peer-reviewed journal publishers or academic standard boards, and how do all these different actors synergize to ensure actual impact? These concerns on how impact ensues, also drives in the argument of whether the impact of business schools, transcends scholarly impact (Lejeune et al. 2019; Pettigrew and Starkey 2016). An attempt to respond to these questions and concerns brings up the question of what impact is? One important factor to note is the meaning of IR is not finite, however the constant is that amidst the array of definitions, is the consensus that IR only fulfills its intended and not so intended purpose when it transcends the academic sites of business schools.

A dialogue on IR by business schools brings up pointers on reflexivity, critique, and collaboration, and respond to what knowledge is being produced, for whom and with whom. Although we agree that IR has an infinite connotation and definition, we attempt to define by considering the effect or contribution research makes to society (ERSC 2022; York n.d.). Lejeune et al. (2019) classify IR as how research contributes to the economy, how it is produced and transferred, and what responsible actions are taken to ensure ethical, sustainable, and community driven outcomes. Therefore, in business schools we need to be intentional on achieving impact research by being intentional in our strategies for impact, especially where business schools represent multiple actors with diverse perspectives (Pettigrew and Starkey 2016).

Reed et al. (2022) conceptualizes IR as a final outcome or series of intermediary outcomes leading to a final one. IR from a simplistic perspective is research that informs change or is intended, to ensure a shift, from a neutral or low, to a positive state. Against this conceptual background, research that is consciously intended to make an impact fits the description of IR. Impact is also measured by reach, quantity, extent, and representation and IR can be assessed through these measuring lenses, mostly interconnected, paying attention to the significance the intended research has on beneficiaries, and/or how far it covers beneficiaries in terms of numbers as well as diversity (Reed et al. 2018, 2021). Another consideration in defining IR comes from Steffen et al. (2015), and what is referred to as the *wedding cake model* (see Fig. 1—*Wedding Cake Model*). This model enables a nested view that presents both a holistic systems and *relational turn* (West et al. 2020) perspective of biosphere, society, and business interacting through interdependence and reciprocity, which are the foundations of Indigenous knowledge frameworks (Cajete 2000). This model places the priority for impact at the biosphere level, understanding without a healthy biosphere there cannot be a healthy society, and in a cascading effect—no healthy business operations. Reflectively, IR efforts should intentionally consider its positive effect from a relational perspective and the interdepend effects on the biosphere, society, and business. Researchers can identify the interconnected SDGs framework to guide action driven IR.

In our deliberations to understand IR for business schools, we explore the formulation of objectives, methods used and our own contributions in terms of how our positionality as researchers informs the research's approach and outcome. At the center of being intentional lies the interest of the beneficiary actors, not as recipients of research produced, but rather as co-creators, answering the earlier questions for, and with whom is knowledge produced? Participation of actors in co-creating research ultimately responds to the question of what knowledge is produced and whose interests are addressed, as the rationale, process and effect of the research is critically taken into consideration by all actors to ensure the conduct of meaningful research. Although not highlighted in the earlier tenets of assessing impact—numbers, reach among others- it is pertinent to note, and we would like to explicitly indicate that the kind of partnerships formed when conducting research also informs its impact. This point facilitates the intersection of divergent interests of actors.

Intentionality to research by business schools therefore does not only stop at conceptualizing objectives but also how we can reflexively critique acknowledging our biases and privileges as researchers and academics and how these positions influence outcomes. How do we inclusively involve multiple actors, what methods can we incorporate to ensure that we realize a collaborative goal for impact? How do we move away from researching on participants or running business schools as producers of skills set, alienating the human-centeredness, and thus losing our focus on impact. How are we certain the impact intended is representative of community needs, and construed by our lenses informed by scientific epistemic and technical gaps? We talk about humanizing business school activities to impact considering ethics, sustainability, and community needs, but how do we move away from individual self-love or selfish interests to focusing on group impact (Reed et al. 2022, p. 7)? Questions such as these may not be conclusive but are initial points to really start thinking about impact, with phronesis in perspective. By being critically practical we rethink impact with focus on context, collective voice, and power (Reed and Rudman 2021). The context the research is being developed, engaged, and actualized; whose voice is heightened and how; and how power is yielded and utilized are important signifiers that will inform the impact we desire. Considering being intentional, and reflexively critiquing our activities, we document processes by acknowledging our own influences in its minutest detail to ensure transparency. We acknowledge the array of definitions that may come up when attempting to define impact, but leaning on a pluralistic worldview is our guide to maintain practical knowledge, recognizing co-producers as effectively responsible and present in all areas of the research process. Business schools as a reflexive critical action can explore their role not as producers or developers but rather as facilitators, encouraging collaboration and accounting for inclusiveness and representativeness to achieve impact. By having these thoughts and acting on them we can therefore begin to be intentional about creating impact.

4 The Why for Impact

The obvious and primary reasoning for IR is to transform it into action towards societal challenges. However, this has not always been the case. Focusing on research exploration of SD, it is evident that humankind is at a fragile precipice living in the Anthropocene as shared above (Steffen et al. 2009). As the name, Anthropocene, suggests, one single species, humankind, constitutes the largest driver of environmental change on the planet (Kemp et al. 2022). Our species is now a geological physical force that is negatively affecting life supporting planet systems (Rockström et al. 2009). Scientific truth is suggesting this progression in Earth temperatures could spell the endgame for humanity and other life on the planet, pointing towards extinction scenarios (Bendell 2018; Kemp et al. 2022). With this increasing knowledge of possible climate scenarios, we are sensing an urgency by global leaders and institutions to take immediate action towards societal transformation that can embrace a changing planet through mitigation and adaptation strategies (Sachs et al. 2019). It is evident in this 'moment of truth' that humanity is at a crossroads. Humankind needs to move towards immediate transformational living on the planet through SD as a species. The alternative is to continue the present course trajectory that continues to push us outside the planetary living conditions that support life on the planet (Rockström et al. 2009). This makes a compelling argument for researchers to uphold our part through IR that imbues action. From a global perspective we are hearing urgent calls by many voices for a radical shift and re-thinking of how we live in the planet (UN Climate Change Conference UK 2021).

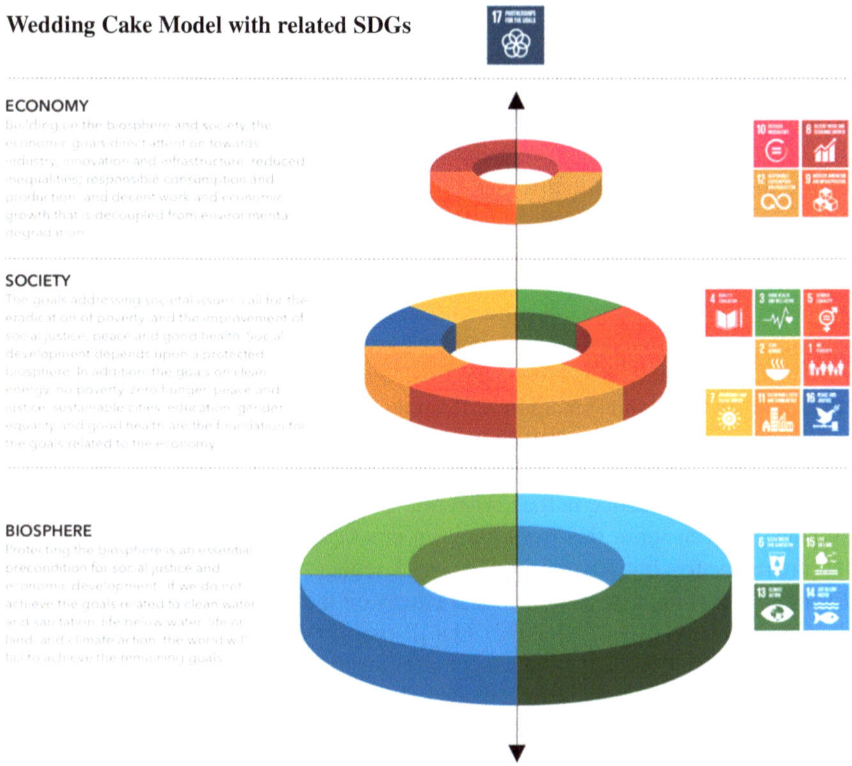

Fig. 1 Shows the interdependency and reciprocal relationality of the holistic social-ecological system and interconnecting relationship of the SDGs. *Credit* Adapted from Azote Images for Stockholm Resilience Centre

IR is critical to human flourishing on the planet. Humankind's history gives witness that science research has been the foundation for societal innovation and well-being (Price-Smith 2002; Wilson 1999). More than ever, there is a need for research to inquire and explore critical questions and challenges that humanity faces (Bentz and Shapiro 1998). Examples of these challenges are many, from climate change, increase in infectious disease prevalence (Price-Smith 2002), deforestation, world conflict, pollution, food insecurity (Hawkins 2007), poverty, resource exhaustion (Moore 2015), loss of species (Ceballos et al. 2015), and cultural diversity (Davis 2009).

4.1 *Mobilizing Our Research in the Classroom*

In our business school institutions, over the past decades we have positioned our graduates well to prosper in organizational life, to be able to enter the workforce and be productive leaders, contributing towards the bottom-line. However, we have left them vulnerable to the changing complexities of our world. With existential threats like the grand challenge of climate change, many students today are experiencing eco-anxiety and are not feeling hopeful for the future (Ojala 2017; Pihkala 2022). The Covid-19 pandemic has only heightened these feelings. This is leaving students feeling they cannot make a difference and is affecting their mental health (Ágoston et al. 2022; Lovero 2022; Scheirich 2020). As learning institutions, we need to question whether we are really preparing our graduates for organizational and community life. The business school's aim should be to enable graduates with the capacity and capability to be a positive influence on society and embrace complexity. The good news is our institutions are waking up to this need, and we are seeing guiding bodies for business education, such as the Association to Advance Collegiate Schools of Business (AACSB) and Principles of Responsible Management Education (PRME), making societal impact the foundation of their goals and principles. As researchers practicing IR, we become exemplars for our students. Our IR informs and filters into our teaching practice in the classroom, as we share emerging narratives of our findings. We can sense this from our own practice in the classroom, whether it is sharing innovative initiatives on plastic waste reduction, to novel practices in equity, diversity, inclusion, assess-ability or practices on the use of SDGs in the organizational value chain (Tiller et al. 2022). Our IR enables students to understand there are pathways and strategies to embrace complex societal challenges. This also engages student voices, where they relate it to their everyday practices. These shared insights can open new angles of perspectives that can inform the researcher and learners to open deeper meaning-making. Through this dialogic process theories and practice start to inform action; we gain understanding of societal challenges from the bottom up, rather than the top-down, enabling new moving conclusions (Bolm and Biedermann 1999; Chandler 2014). Through mobilization of our research in the classroom these actions open feelings of hopefulness and helpfulness for students and further meaning making for the researcher (Hicks 2014).

4.2 Theory as a Bridge to Agency

Reflecting further on the importance of 'why' IR, we sense the theory or insights that are produced from IR can act as seeds and scenarios for others to act out new pathways. On mass, there is no argument, humanity needs a transformation in mindset and behavior in how we live in our communities to address emergent challenges and calls for action. Knowingly, we acknowledge through societal growth, transformation is slow and painful, as it is human nature to resist and hang onto resilient habits and mindsets (Kegan and Lacey 2011; Senge et al. 2008). The mounting awareness and knowledge of potential existential threats to humankind tends to weigh heavily on individuals and society and creates many emotions (Frankl 1972). One emotion, fear, can dominate and enable thoughts of a dystopian world, leading to feelings of meaninglessness, hopelessness, and helplessness (Frankl 1972; Ojala 2017). This hopelessness and helplessness can act as chains that imprison us in our old mindsets and actions. On an individual level, many of us often feel we cannot make a difference on the grand scale or planetary level (Hicks 2014).

Theory produced through IR could potentially be a liberating force to the fear and emotions that bind us. Bell Hooks (2014) reminds us that theory can be a healing place to the patterned ills and pains of the world. Hooks shares, "How I envied Dorothy, her journey in the *Wizard of Oz*, that she could travel to her worst fears and nightmares only to find at the end that *there is no place like home" (p. 61).* We have a history of examples of theories and frameworks that liberate new ways of thinking, e.g., the UN's SDGs. This global framework has enabled us to reimagine and co-create nascent strategies for living more sustainably on the planet (United Nations 2015). As researchers exploring existential threats and SD, when we apply critical thinking, reflexivity, and analysis to the lived experience of the research, it can open creative portals of understanding. These pathways can lead to new rhythms of embracing emergent challenges in our world. In sharing our work to audiences through the knowledge mobilization process, it can educate through a transmission or practice of freedom, enabling others agency to use their imagination to create desired futures (Freire 2000; Escobar 2020). Considering this, producing IR opens possibilities and potential for liberating our hopefulness, helpfulness, and meaning for the future.

4.3 Collaborating for Impact and Common Good

In our pursuit of IR, it is clear, it needs to be embedded in a collaborative effort that has purpose. According to Antonacopoulou, "collaboration when founded on *Isotima* (equal worth) places individual and collective dignity as the linchpin that fuels mutual care, respect and trust in each other to appreciate each partner in the collaboration" (p. 3). *Philotima* is the commitment, through respect and dignity, that we make to each other as researchers to co-create something of worth that serves the common good (Antonacopoulou). Most times research is collectively done through an intentional effort that aims to co-create something worthwhile. However, often research intentions are misaligned and co-creating meaningful impact is missed. Rodenburg et al. (2022) share common reasons for misaligned IR, "perverse effects on faculty and decanal priorities have been identified, including incenting behaviors that are at odds with the achievement of the United Nations Sustainable Development Goals (SDGs) and the *2030 Agenda*" (p. 1).

Fostering networks and collaborations for productive research is being recognized as a key starting point for research activity that promotes impact outcomes (Reed et al. 2022). Bringing people together in a collective thinking process enables dialogue and diverse perspectives to be shared in exploring ongoing challenges. Thoughtful research processes that explicitly weigh out impact outcomes through a pluriverse or multitude of engaged citizens has the potential to gain deeper meaning-making around ongoing challenges and to uncover emergent seeds, scenarios, and pathways to embrace the unfolding complexity of the situation (Pereira et al. 2018; Reed et al. 2018). The collaborative approach can be seen as counter-intuitive to the often-siloed structure and culture of the university research setting. However, studies are strongly suggesting that the interdisciplinary and transdisciplinary approach to conducting research holds the most promise for embracing the ongoing challenges of our rapidly changing world (MacGregor and Phipps 2020; Sellberg et al. 2021).

Albert Einstein once stated, "Not everything that can be counted counts, and not everything that counts can be counted." When considering IR another challenge that needs to be addressed is the institutionalized phenomenon of rating the quality of research according to what journal it is published in. For example, the publication list of the top 50 journals endorsed by Financial Times (FT50) has been demarcated as the golden standard of quality and prestige by business school and faculty (Rodenburg et al. 2022). However, a recent study of FT50 articles strongly suggests that 90% of the publications sampled, using the SDGs as an impact measure, had no explicit relationship or perceived impact (Rodenburg et al. 2022). More commonly, researchers are rewarded for tenure and promotion for their tally of publications, not on whether it has impact, but considering the rating of journals that are often misaligned to social-ecological imperatives (Adler and Harzing 2009; Mu and Hatch 2021). What business schools have rated as the most prestigious journals are suggested to be no better than what is seen as lower tier journals in representing valued research (Starbuck 2005).

The lack of intentional focus on impact questions academia's cultural approach towards publications. Explicitly, we need to critically ask, are we publishing for morally 'good reasons' or more from self-serving pursuits? Starbuck (2007) highlights that we know academia has exceptional research capability and capacity to co-create the world anew in the ideas we present through our work, but surprisingly only five to ten percent of researchers are focusing on someone or something other than themselves. This questions academia's methodological approach to publishing research. In our ethical reviews and methodological approaches, we need to explicitly consider our impact on others and the relationships connected to the research. As researchers we need to hold ourselves accountable to produce IR that matters and demonstrates care towards otherness. We can do this by explicitly defining, identifying, and rewarding research in our institutions that impacts social-ecological system challenges that are measurable, e.g., aligns to SDGs or Planetary Boundary frameworks. Additionally, not only is this the 'good' approach but accrediting bodies, such as AACSB and PRME in business schools, and institutional values and mission are becoming aligned to SDGs and are setting these goals as criteria for membership and accreditation.

5 Exploring Strategies for Impact Research

As a group we want to be intentional and explicit that our goal is to conduct research that matters, not for goals of the business school or university per se, but for the common good of society. Researchers set out to apply their research in some matter of form or action. However, the facts do not lie. Most research published is rarely applied to the outside world or provides positive influence for social outcomes. Through our literature research and praxis, it is evident IR living in today's seriously messy and tangled world needs to imbue collective thinking and be more aligned with transdisciplinary approaches. This enables researchers to consider the many lines of perspective to be able to embrace the ongoing challenges. IR is complex and subjective in nature (Read and Rudman 2022) with many overt and hidden dynamics that can amplify or dampen the impact result. In our case study, the goal is to do applied research to solve real life challenges that connect to the SDGs and *2030 Agenda.* As an effective team of researchers, to best ensure we deliver IR in our efforts, the starting point is to explicitly express our intention for IR by providing an open strategy for conducting and delivering IR. In this section we share our intention and strategy for IR, which are guided by measuring the following elements: (1) Awareness as researcher; (2) Purpose of research; (3) SDG Framework; (4) Inclusive Collaboration; (5) Embracing Conflict/Challenges; and, (6) Mobilization of Results. We share details of our strategy below.

5.1 Researcher Awareness

Awareness is a key element in the human experience for developing and flourishing. This principle also correlates to the researcher, where we must continue to develop a reflexive muscle to shine a spotlight on ourselves and the world around us. First, as a research team we must reflexively practice bracketing our research bias. In this process, we continually inquire and address our subjective biases and examine the power dynamics that we are embedded in. Questions for the researcher to consider are, who is setting the goal of the research, and whose lens is driving the direction of the research? For example, is it a western lens, and does it include the marginalized voices that intersect with the research? Also, relationships between academic and non-academic partners can be complex and clash, with hidden power dynamics and moving conclusions that can derail and disrupt pathways for impact (Fritz and Meinherz 2020; Reed and Rudman 2022). In developing inclusive partnerships for IR, we need to be aware of the biases and positionality of how we are showing up in the research through a bracketing process. How do we ensure a collaborative role with participants, so they too are researchers, and we are participants intermingling in the search for meaning making and impact? Through this process it is also important to understand why efforts to generate impact fail or lead to negative impacts (Reed and Rudman 2022).

As researchers it is valuable to explicitly espouse and commit to best practices for research assessment that focuses on impact outcomes. The San Francisco Declaration of Research Assessment (DORA) is an example of commitment that researchers can sign and enable their awareness on research assessment. DORA recognizes the need to improve the ways in which researchers and the outputs of scholarly research are evaluated (DORA 2023). Through signing the declaration, you become part of a collective transdisciplinary network that subscribes to a set of principles that seeks to improve the ways that the output (impact) of scientific research is evaluated by funding agencies, academic institutions, and other parties. These principles move away from traditional and dominate research practices that are focused on promotion, hiring, and publication (DORA 2023).

5.2 Purpose of Research

Axiology is the critical examination of the values that govern thinking, behavior, and actions of humans. "Axiology marks the worthiness that each partner honors in themselves and the others they collaborate with, so that their partnership itself is not only a practice and values work but marks the worthiness of their shared purpose" (Antonacopoulou 2022, p. 7). In group work, including research collaboration, we do not often explicitly consider the values and purpose that are driving the group forward. However, if we intend to co-create impact with our research it is imperative, we understand the underlying values and purpose that drives the work—to understand the worthiness of the shared purpose. As our transdisciplinary research team explores sustainable pathways in our community, we need to ask what the worthiness of our shared purpose is. More specifically, what is the axiology that guides us in research exploration? What purpose and whose purpose does the research serve? Explicitly, does the purpose intersect with the SDGs, specifically which ones, and how?

5.3 Framework of Social Theory, SDGs, and UNDRIP

Since antiquity, theory has guided researchers in their quest for emergent epistemology (Gross 1958). With the urgent call by global leaders to tackle the 'messy challenges,' more than ever, researchers are compelled to be guided by moral frameworks to embrace global challenges like the climate crisis. Since 1948, the *Declaration of Human Rights* has given us moral guidance in how to guide human policy and action on how to live in the planet. We have seen these rights evolve through the progression of the Millennial Goals and to the present day with the SDGs. The growth towards the SDGs has given humankind a moral framework that considers and is strengthened and imbues human rights. The SDG framework shows researchers a foundation and measuring stick to test the impact of their work. The good news is researchers are acquiring tools that consider the impact of their work as correlated to the SDGs. These tools will give researchers guidance on where they can publish to gain the most impact. Cabells International is an example of a publication evaluation system that shares a complete source for journal information, evaluation metrics, and submission details for academic institutions (Cabells 2022). This evaluation source enables researchers to define their audience and find the right fit for a journal. Cabells has recently been awarded membership with the *United Nations SDG Publishers Compact*, and through this membership is committed to supporting research and publications focused on sustainable solutions. They recently announced they will soon to be releasing a new metric shortly—*SDG Impact Intensity*™. This will be a valuable tool for researchers by being the first academic journal rating system for evaluating how journals contribute to positively impacting the SDGs (Cabells 2021).

The United Nations Declaration of Rights for Indigenous Peoples (UNDRIP) is another framework consideration for researchers. UNDRIP is the most comprehensive international instrument on the rights of Indigenous Peoples and establishes a universal framework of minimum standards for the survival, dignity and well-being of the Indigenous Peoples of the world (United Nations (General Assembly), 2007, art. 5). UNDRIP elaborates on existing human rights standards and fundamental freedoms that are covered by the SDGs, as they apply to the specific situation of Indigenous Peoples globally. By thoughtfully considering this framework upfront in the researcher's lexicon, this avoids valuable Indigenous perspectives being an afterthought in the research process.

5.4 *Inclusive Collaboration*

In truly inclusive IR, we need to decolonize and ensure marginalized and oppressed voices are heard and participate as researchers in the effort. Poorly designed and executed engagement in research fails to account for elite capture, various forms of cultural imperialism, the exacerbation of social inequities and opportunity costs for those who engage for little or no benefit (Reed 2022). Escobar (2020) advocates the *pluriverse* approach is imperative in delivering impact, as this approach enables a multi-world view of knowing, being, and doing in the world. Through this approach, the researcher participates in a web of co-existing and reciprocal relationality and communality. According to Escobar (2020):

> Relationality signals that life is interrelationship and interdependence, always and at every level. Everything exists because everything else exists, as the southern African principle of *Ubuntu* tells us. There are no intrinsically existing objects, subjects, or actions, as modernity has taught us to presume since at least the time of Descartes. In other words, the real is not made of isolated objects that interact with one another; the observer does not preexist what she observes; there is no external world for us to cling to; everything living is always an integral part of the ever-changing pluriverse. (P. 92)

Therefore, recognizing the relationality and communality that exists is paramount for IR; to accurately identify, welcome, and validate a multiple world lens that touches on the research question being asked. Fritz and Binder (2020) identify in transdisciplinary research the ideals of co-leading and balancing of power in the research process between those involved is important for symmetry producing robust knowledge. However, holding research design to these ideals is often challenging with power imbalances in the research process, especially with polarizing topics, such as sustainability and climate change that has many different meanings for those involved. It becomes important for researchers to recognize the entanglements of power and participation in the research process (Bieluch et al. 2017). In our research project, it is critical that we understand, welcome, and validate all community members who are interested, concerned, and want to share their voice on pathways for a sustainable future. In the research process we need to continually ask, who holds the power in relationships? Whose voices might be oppressed or marginalized and how might we welcome them into the process? Also, how do we design research that explicitly examines power, and enables collaborative approaches to unleash latent agency of participants? It is important that these reflexive questions are continually asked through the research design and process.

Antonacopoulou (2022) highlights the call for research that builds inclusive partnerships and collaborative networks that are not focused on narrow or individual goals but on holistic network goals. In the case of our exploration of SD pathways, we seek to engage in 'networks of networks' to sense the impact on the social-ecological system and relationality that intersects our university with interconnected communities and the biosphere. However, in this inclusive collaboration, we will continue to respect and celebrate the diversity of perspectives that participates in the mutual co-mingling of the research process. For example, in our research how do we acknowledge the stolen and sacred land of the Indigenous Peoples and celebrate them as exemplars for sustainability in how they stand as stewards of the land, embodying reciprocity and interdependence. "This positioning of inclusiveness transcends the place and perspective different partners hold in their respective fields and invites honoring equality amongst partners whilst celebrating the diversity and uniqueness each has" (Antonacopoulou, p. 8). Being intentional with these elements in mind will enable us to design and best ensure IR through inclusive partnering.

5.5 Embracing Conflict and Challenges

We acknowledge we live in a world defined by volatility, uncertainty, complexity, and ambiguity; the VUCA world acronym used in today's literature (Bennett and Lemoine 2014). With the rapid flux and conflict that present today, we often hear how we need to 'fight' the changing complexity of our world, whether it is climate change or pandemics. This mindset separates us from the world around us, not enabling us to consider our relationship to the phenomenon. A characteristic of being human is that we are habitual in nature, and we are mostly resistant to change (Kegan and Lahey 2011). This is evident in how we have been born and patterned in our "groves of thought" that no longer work in today's world of complexity (De Quincy 2002, p. 154). Mechanistic and positivist methodological approaches, from the legacy of Descartes, have dominated and separated us from the living world and no longer serve us best in adapting to emerging challenges. As researchers we must embrace the conflicts and challenges that face us, understanding that we are not outside observers or impartial bystanders.

When conducting research, we also become participants and bring in our biases to the work. It is important that we own and bracket these biases, but also use them to drive deeper meaning making. Kant argued that humans must be subject to critique, that none of us can reject scrutiny, therefore we need to be open to exploring change (O'Shea 2012). Additionally, we should be compelled to move away from the patterned hubris mindset that the primary goal of research is to 'crack the code' or find the solution to problems. We suggest the goal of our work, and that of others, is to get into the rhythm of understanding the inter-relationality that gives complex systems their vitality. Through this approach we can sense the interconnectivity and reciprocal relations at play, and through this approach we can contribute more fully to the epistemology and ontology of our evolving world. In the research process we can lean into the polarities and conflict to open creative portals of understanding, prototyping new methodologies, e.g., Indigenous and/or expressive art-based methodologies, that can provide new angles and lines of perspectives. The intention should not be to discount our past research approaches, but to add to our pursuit of new knowledge and ways of *being* through emerging processes.

5.6 Knowledge Co-production and Mobilization

There is an increasing awareness that researchers no longer hold the traditional position as the sole provider of knowledge, and a concern for the outcomes, how it is applied, and the impacts of the research (Sellberg 2021). In facing these challenges, we need to set the intention to create IR that is actionable. To address social-ecological system challenges, we need to sense we are performing social-ecological systems research, this is research that is focused on informing SD practice and governance, and opening resilient pathways for communities (Folke et al. 2016). Williams and Whitman (2019a) argue, the "primary research aim must not be to see how we can build theory out of a crisis, but rather how our organizational and management research and theories can contribute concretely to helping humanity prepare for and respond to these shocks and build long-term societal resilience" (p. 2). This positions the urgent need to conduct IR, research that can be applied to open pathways for SD—to embrace social-ecological system challenges.

In understanding the complex nature of our social-ecological system challenges, researchers are engaging in transdisciplinary processes. According to Roux et al. (2017), "A key aim of transdisciplinary research is for actors from science, policy, and practice to co-evolve their understanding of a social–ecological issue, reconcile their diverse perspectives and co-produce appropriate knowledge to serve a common purpose" (p. 711). This transdisciplinary approach enables dialogue and interplay between academics, non-academics, and other concerned citizens to embrace social-ecological systems changes (Roux et al. 2017). In our transdisciplinary research process the goal is for the co-production of knowledge. Considering this, participation is a key element and consideration in our research discourse, and this encompasses mobilizing what is learned through the collaborative process (Collin and Ison 2009). As researchers we will carry out a knowledge mobilization plan with the traditional route of publishing and presenting at conferences. However, in being conscious of our work's transdisciplinary nature, we will share outcomes with the various citizens and community members who participated in the research. We will deeply listen to their needs to understand how the outcomes can be mobilized to enable the co-produced knowledge to be applied to open pathways for sustainable futures for those the project touches. Also, we will continue to network with colleagues, researchers, and practitioners, to assess the impact of our work, e.g., DORA and Impact Research Canada. We believe this will best enable us to find seeds that can create pathways for SD that can co-inspire our university and local communities and provide glimpses for other interconnected communities to explore sustainable futures.

6 Conclusion

This is no doubt the business school approach to research and education needs to change. An approach that for decades has been dominantly focused on an individualistic and profit-first scenario. This capitalistic approach is placing tremendous pressure on the planetary living conditions and leaving humans and all other living species in a vulnerable place, and overall threatens flourishing life on the planet. Through the cacophony of our rapidly changing world, we have heard repeatedly the need to shift away from business as usual, and this is no different for researchers (Hoffman and Jennings 2018). We need to break-out of the patterns and habits of doing research for the self-interest of profile and promotion. As Alder and Harzing (2009) remind us, "Tenure, in fact, was institutionalized as a way to support scholars in asking important (and often controversial) questions and reporting even their most unpopular findings without jeopardizing their livelihood" (p. 92). For decades we have heard the empty echo for urgency and the need for change in how we are living in the planet—to take on the grand challenges before us. Considering this, it is action that shows our intention and seriousness to embrace and tackle the challenges before humanity. As scholars we need to hold an unwritten oath towards worthiness of shared purpose to support the common good, to embrace and conduct research that matters—to do IR.

As we have shared in this text, it is imperative that we conduct IR correlated and measured through the SDGs and other frameworks that consider impact on society and the biosphere. This is work that needs to be done explicitly and with intention towards a shared purpose. Through our meaningful knowledge co-production of nascent theory and case experience as researchers we provide pathways for others to travel. As researchers we become geographers of the human condition, where we open new pathways, where before others did not have direction or sense of where to travel; through our research we provide guideposts. These beacons enable hopefulness and helpfulness for the future, to sense, as Hooks (2014) metaphorically shares, that 'Dorothy can make her way home safely through the tornados and volatility of the world'. Through our collective thinking as researchers, we can provide seeds and scenarios that lead to viable and desired pathways for others to travel and adventure. Through IR, we pass the baton to others as we embrace challenges collectively, through this ongoing process we enable conditions for multispecies flourishing in our one and only planet.

Acknowledgements This study was funded by the Social Sciences and Humanities Research Council (SSHRC) Insight Grant Number **1174082** to Kent Williams.

References

Acker S, Webber M (2016) Discipline and publish: The tenure review process in Ontario universities. In: Shultz L, Vickzo M (eds) Assembling and governing the higher education institution: democracy, social justice and leadership in global higher education. London: Palgrave Macmillan, pp 233-255

Adler NJ, Harzing AW (2009) When knowledge wins: Transcending the sense and nonsense of academic rankings. Acad Manag Learn Educ 8(1):72–95

Aguinis H, Glavas A (2012) What we know and don't know about corporate social responsibility: a review and research agenda. J Manag 38(4):932–968

Ágoston C, Csaba B, Nagy B, Kőváry Z, Dúll A, Rácz J, Demetrovics Z (2022) Identifying types of eco-anxiety, eco-guilt, eco-grief, and eco-coping in a climate-sensitive population: a qualitative study. Int J Environ Res Public Health 19(4):2461

Alperin JP, Nieves CM, Schimanski LA, et al. (2019) How significant are the public dimensions of faculty work in review, promotion and tenure documents? eLife 8:e42254. https://doi.org/10.7554/eLife.4225

Antonacopoulou EP (2010) Making the business school more 'critical': reflexive critique based on phronesis as a foundation for impact. Br J Manag 21:s6–s25

Antonacopoulou EP (2022) Partnering for impact: a grand challenge and design for co creating a just, resilient and flourishing society. J Appl Behav Sci 58(4):571–594

Bendell J (2018) Deep adaptation: a map for navigating climate tragedy. In: Institute for Leadership and Sustainability (IFLAS) Occasional Papers, University of Cumbria, Ambleside, UK. 2:1–36. http://insight.cumbria.ac.uk/id/eprint/4166/ (accessed 15 November 2022)

Bennett N, Lemoine GJ (2014) What a difference a word makes: understanding threats to performance in a VUCA world. Bus Horiz 57(3):311–317

Benneworth P (2013) University engagement with socially excluded communities. In: Benneworth P (ed) university engagement with socially excluded communities. Springer, Dordrecht. https://doi.org/10.1007/978-94-007-4875-0_1

Bentz J, O'Brien K, Scoville-Simonds M (2022) Beyond "blah blah blah": exploring the "how" of transformation. Sustain Sci 17(2):497–506

Bentz VM, Shapiro JJ (1998) Mindful inquiry in social research. Sage Publications

Bieluch KH, Bell KP, Teisl MF et al. (2017) Transdisciplinary research partnerships in sustainability science: an examination of stakeholder participation. preferences. Sustain Sci 12(1):87–104. https://doi.org/10.1007/s11625-016-0360-x

Bohm D, Biederman C (1999) Bohm–Bierderman Correspondence: Volume one-Creative and science. London: Routledge
Cabells International (2022) Scholarly Analytics. https://www2.cabells.com (accessed 17 November 2022)
Cajete G (2000) Native science: natural laws of interdependence
Chambers JM, Wyborn C, Klenk NL et al. (2022) Co-productive agility and four collaborative pathways to sustainability transformations. Glob Environ Change 72:102422
Economic and Social Research Council (2022) Defining impact. https://www.ukri.org/councils/esrc/impact-toolkit-for-economic-and-social sciences/defining-impact/ (accessed 10 November 2022)
Chandler D (2014) Resilience: the governance of complexity: Routledge
Collins K, Ison R (2009) Jumping off Arnstein's ladder: social learning as a new policy paradigm for climate change adaptation. Environ Policy Gov 19(6):358–373
De Quincey C (2002) Radical nature: the soul of the matter. Park Street Press, Rochester, USA
Escobar A (2020) Pluriversal politics: the real and the possible. Duke University Press
Evensky J (2005) Adam Smith's theory of moral sentiments: on morals and why they matter to a liberal society of free people and free markets. J Eco Persp 19(3):109–130
Fernández-Ríos L, Rodriguez-Diaz J (2014) The "impact factor style of thinking": a new theoretical framework. Intern J Clin Health Psychol 14(2):154-160
Fritz L, Meinherz F (2020) Tracing power in transdisciplinary sustainability research: an exploration. GAIA-Ecol Persp Sci Soc 29(1):41–51
Frankl VE (1972) The feeling of meaninglessness: a challenge to psychotherapy. Am J Psyc 32(1):85–89
Freire P (2000) Pedagogy of freedom: Ethics, democracy, and civic courage. Rowman and Littlefield Publishers
Haraway DJ (2016) Staying with the trouble: making kin in the Chthulucene: Duke University Press
Harris MS (2021) The soft underbelly of universities as anchor institutions: the disconnect between university and community research priorities. High Educ Policy 34(3):603–621
Hicks D (2014) Educating for hope in troubled times: climate change and the transition to a post-carbon future. London: Institute of Education Press
Hoffman AJ (2021) Management as a calling: leading business, serving society. Stanford University Press, Redwood City, UNITED STATES
Hoffman AJ, Jennings PD (2018) Re-engaging with sustainability in the Anthropocene era: an institutional approach. Cambridge University Press
Hooks B (2014) Teaching to transgress: education as a practice of freedom. Routledge
Gardner CJ, Thierry A, Rowlandson W et al. (2021) From publications to public actions: the role of universities in facilitating academic advocacy and activism in the climate and ecological emergency. Frontiers in Sustainability 42
Kegan R (1982) The evolving self: problem and process in human development. Harvard University Press, Boston
Kemp L, Xu C, Depledge J et al. (2022) Climate Endgame: Exploring catastrophic climate change scenarios. Proceedings of the National Academy of Sciences 119(34):e2108146119. https://doi.org/10.1073/pnas.2108146119
Korten DC (2010) Agenda for a new economy: for phantom wealth to real wealth. Berrett-Koehler Publishers Inc., San Francisco, CA
Krivine H (2015) The earth: from myth to knowledge. Verso, London, UK
Kolbert E, Wilson EO and Lovejoy TE (2017) Living in the Anthropocene: Earth in the Age of Humans. Smithsonian Institution
Lejeune C, Starkey K, Kalika M, Tempest S (2019) The impact of business schools: increasing the range of strategic choices. Manag Int 23(2):88–98
MacGregor S, Phipps D (2020) How a networked approach to building capacity in knowledge mobilization supports research impact. Intern J Educ Policy Leade 16(6)

Moore JW (2015) Capitalism in the web of life: ecology and the accumulation of capital: Verso Books
Mu F, Hatch J (2021) Becoming a teacher scholar: the perils and promise of meeting the promotion and tenure requirements in a business school. J Manag Educ 45(2):293–318
Lovero (2022) Climate change: its impacts on microbial communities and mental health. eScholarship, University of California
Ojala M (2017) Hope and anticipation in education for a sustainable future. Futures 94:76–84
O'Shea J (2012) Kant's critique of pure reason: an introduction (1st ed.). Routledge. https://doi.org/10.4324/9781315730097
Petriglieri G, Petriglieri JL (2015) Can business schools humanize leadership? Acad Manag Learn Educ 14(4):625–647
Pettigrew A, Starkey K (2016) From the guest editors: the legitimacy and impact of business schools—Key issues and a research agenda. Acad Manag Learn Educ 15(4):649–664
Pihkala P (2022) The process of eco-anxiety and ecological grief: a narrative review and a new proposal. Sustainability 14(24):16628
Pereira L, Bennett E, Biggs R, Peterson G, McPhearson T, Norström A … Vervoort J (2018) Seeds of the future in the present: Exploring pathways for navigating towards "good" Anthropocenes
Price-Smith AT (2001) The health of nations: infectious disease, environmental change, and their effects on national security and development. Mit Press
Potts K, Brown L (2015) Becoming an anti-oppressive researcher. In: Strega S, Brown L (eds) Research as resistance: revisiting critical, indigenous, and anti-oppressive approaches. Toronto: Canadian Scholars Press, pp 17-41
Rich A (2002) Arts of the possible: essays and conversations. WW Norton and Company
Reed MS, Bryce R, Machen R (2018) Pathways to policy impact: a new approach for planning and evidencing research impact. Evid Policy 14(03):431–458
Reed MS, Ferre M, Martin-Ortega J et al (2021) Evaluating impact from research: amethodological framework. Res Policy 50(4):104–147
Reed MS, Rudman H (2022) Re-thinking research impact: voice, context and power at the interface of science, policy and practice. Sustainability Science, 1–15
Rockström J, Steffen W, Noone K et al. (2009) A safe operating space for humanity. Nature 461(7263):472
Rockström J, Kotzé L, Milutinović S, Biermann F, Brovkin V, Donges J ... Kim R (2024) The planetary commons: A new paradigm for safeguarding earth-regulating systems in the Anthropocene. Proceedings of the national academy of sciences, 121(5):e2301531121.
Rodenburg K, Rowan M, Nixon A et al. (2022) The Misalignment of the FT50 with the achievement of the UN's SDGs: a call for responsible research assessment by business schools. Sustainability 14(15)
Roux DJ, Nel JL, Cundill G, O'farrell P, Fabricius C (2017) Transdisciplinary research for systemic change: who to learn with, what to learn about and how to learn. Sustain Sci 12:711-726
Sachs JD, Schmidt-Traub G, Mazzucato M et al. (2019) Six transformations to achieve the sustainable development goals. Nat Sustain 2(9):805–814
Sellberg MM, Cockburn J, Holden PB, Lam DPM (2021) Towards a caring transdisciplinary research practice: navigating science, society and self. Ecosyst People 17(1):292–305. https://doi.org/10.1080/26395916.2021.1931452
Senge P, Smith B, Kruschwitz N et al. (2008) The necessary revolution: how individuals and organizations are working together to create a sustainable world New York: Doubleday
Scheirich C (2020) Coping with the Climate crisis: exploring art therapy for sustainable mental health
Starbuck WH (2005) How much better are the most-prestigious journals? The statistics of academic publication. Organiz Sci 16(2):180–200
Starbuck WH (2007) Living in mythical spaces. Organiz Stud 28(1):21–25
Steffen W (2022) The earth system, the great acceleration and the anthropocene. Sustainability and the New Economics. Springer, Cham, pp 15–32

Steffen W, Richardson K, Rockström J, Cornell SE, Fetzer I, Bennett EM, De Wit CA (2015) Planetary boundaries: guiding human development on a changing planet. Science 347(6223):1259855
Tiller SJ, Rhindress AP, Oguntola IO, Ülkü MA, Williams KA, Sundararajan B (2022) Exploring the impact of climate change on arctic shipping through the lenses of quadruple bottom line and sustainable development goals. Sustainability 14(4):2193. https://www.mdpi.com/2071-1050/14/4/2193
Tong S, Bambrick H, Beggs PJ, Chen L, Hu Y, Ma W … Tan J (2022) Current and future threats to human health in the anthropocene. Environ Intern 158:106892
UN Climate Change Conference UK 2021 (2021) COP26 World Leaders Summit – Statement on The Breakthrough Agenda. https://ukcop26.org/cop26-world-leaders-summit-statement-on-the-breakthrough-agenda/
United Nations (General Assembly) (2007). Declaration on the Rights of Indigenous People
United Nations (2015) Transforming our world: The 2030 agenda for sustainable development. *New York: United Nations, Department of Economic and Social Affairs.* https://www.york.ac.uk/staff/research/research-impact/impact-definition/
University of York (n.d.) Research Impact: What is research impact
Varoufakis Y, Brooks A, vanden Heuvel K, Brooks D (2020) Is Capitalism Broken?: Simon and Schuster
West S, Haider LJ, Stålhammar S, Woroniecki S (2020) A relational turn for sustainability science? relational thinking, leverage points and transformations. Ecosyst People 16(1):304–325
Williams A, Whiteman G (2021) A call for deep engagement for impact: addressing the planetary emergency. Strateg Organ 19(3):526–537
Wittmayer JM, Schäpke N (2014) Action, research and participation: roles of researchers in sustainability transitions. Sustain Sci 9(4):483–496

Environmental Impact Reduction Micro-Placements: An Opportunity to Overcome Business Management Students' Reluctance to Undertake a Long-Term Work Experience Placement Whilst Stimulating Environmental Improvements in Businesses?

Kay Emblen-Perry

Abstract Work placements are generally accepted as a valuable route into work experience, which provides personal, professional, and academic benefits to the placement student and resource benefits to the placement employer. These placements have traditionally been with larger organizations for 12 months; often termed 'a year in industry'. However, this year in industry is now becoming less attractive for students and a challenge for employers. In response, a student-led consultancy programme of micro-placements designed around environmental impact improvements within SMEs was established for business management students. This paper presents the pilot micro-placement project and explores the benefits and challenges it gave the placement student and placement employer. This paper will be of interest to educators and placement coordinators seeking to develop innovative work experience opportunities in Higher Education.

Keywords Student-led consultancy · Micro-placement · Work experience

1 Introduction

The traditional route into work experience for both undergraduates and postgraduate students is through a long-term, 12-month work placement; frequently called a 'year in industry'. It is generally accepted that placement students gain academic, personal, and professional benefits such as the opportunity to acquire skills which cannot be

---

K. Emblen-Perry (✉)
University of Worcester Business School, City Campus, Castle Street, Worcester, UK
e-mail: k.emblenperry@worc.ac.uk

T. Wall et al. (eds.), *Sustainability in Business Education, Research and Practices*, World Sustainability Series, https://doi.org/10.1007/978-3-031-55996-9_9

learned in the classroom and to test potential future careers (Elarde and Chong 2012; Ismail 2018,). In turn these enable placement students to become more likely to obtain a job with a higher salary (Saniter and Saniter 2014). This focus on work experience and work readiness emanated from the mass marketisation of Higher Education, which resulted in competition between Higher Education Institutions (HEIs) and the use of employment destination league tables (Green et al. 2009). There has also been a transfer of responsibility for graduates' employment skills from in-company graduate training to the recruitment of work ready graduates, such that HEIs are now expected to prepare students to operate effectively in their future workplaces (Docherty 2014). Consequently, there has been an increasing emphasis on the importance of placements as part of the broader educational experience (Saniter and Saniter 2014).

Despite students considering a degree as a passport to employment, most employers no longer consider a degree sufficient for entry into their workforce (Jackson et al. 2017; Temmerman 2019). Employers now require graduates to be empowered reflective lifelong learners, with appropriate attitudes and interpersonal skills, gained through practical, relevant work experience (Harvey 2000; Ismail 2018; HR News 2022). Business degrees, unlike health and engineering courses, frequently lack a core element dedicated to gaining work experience in a real-world business. Consequently, many students graduate unable to demonstrate their preparedness for employment (Smith et al. 2014).

To overcome this, educators encourage students to undertake a work experience placement that enables them to not only develop the required attitudes and interpersonal skills, but to explore career options early enough to change their intended career direction before graduation, become familiar with workplace behaviours and expectations, recognize their skills and skills gaps, and make informed choices about their next steps (Education and Training Foundation 2021; Institute of Student Employers 2021).

The value of a placement to the student is well recognized as being a fundamental opportunity to enhance their learning and softer employment skills (Bullock et al. 2009; Institute of Student Employers 2021). However, the benefits for the employing organization from recruiting a placement student are less well recognized. Previous research has generally framed potential benefits as generic outcomes such as placement students providing lower cost resource (Maertz et al. 2014; Jackson et al. 2017) and has provided little specific detail of the practical benefits for individual organizations or determined the potential value of placements for the placement employer.

This study contributes to closing this gap. It highlights the potential value of placements to the placement employer by presenting the specific benefits received through participation in the Student-led Environmental Consultancy Placement Programme. The pilot placement project is used to explore the opportunities provided to both students and businesses through this novel micro-placement opportunity; a short-term alternative to the traditional 12-month 'year in industry'. The study suggests this student-led environmental consultancy placement may enable the student to bring diversity, creativity, and innovation to the business within a cost-effective skilled resource provision. It also appears to have engaged the student in a more manageable work-based learning experience that may help to overcome the challenges of traditional a placement. This can help develop students' resources in the context of the real-world job market (Finch et al. 2016), whilst maintaining the familiarity of friends and family as they undertake a work placement whilst studying. This approach may overcome the declining interest in and reluctance to undertake a 'year in industry' placement.

2 Engagement with Placements

2.1 Higher Education Institutions' Engagement with Placements

In UK the effectiveness of placements is largely embedded in the measures of employability published at the university level (Clarke 2018). Employability success is established by the number of students in employment six months after graduation, which measures the success of the individual HEI rather than the students' personal success. (Cheng et al. 2022). Despite this it is generally accepted that work experience prior to graduation can help students obtain a graduate level role (Mason et al. 2009). Consequently, offering placements is seen as a way to address the policy makers' and graduate employers' demands for work experience activities (Docherty 2014; Brooks and Youngson 2016), and develop a greater awareness of the importance of core employability skills amongst students prior to graduation (Patrick and Crebert 2004). If successful, this may contribute to a positive placing in employment destination league tables (Green et al. 2009) and respond to the growing expectations that HEIs produce work ready graduates (Woodfield and McIntosh 2022).

In addition, as working with under-represented groups is high on universities' agendas, work is frequently undertaken to encourage placements for under-represented groups (Woodfield and McIntosh 2022). This can encourage company diversity as well as helping students build confidence and resilience (Institute of Student Employers 2021).

2.2 Student Engagement with Placements

The oversupply of graduates requires individuals to differentiate themselves if they are to be successful in the hyper-competitive UK graduate job market (Jackson et al. 2017). In 2022, 800,000 graduates competed for just 300,000 graduate roles; more than 50% of applications were rejected during the first review (Milkround 2022). One of the most effective ways to achieve differentiation is to gain real-world work experience (Smith 2021). Traditionally, this has been a through 'year in industry', in which an undergraduate takes a year out of studying, to work full-time in an organization. It is well established that placement students not only gain personal, professional, and academic benefits; for example, they are better prepared for the workplace as they understand the importance of meeting deadlines, time management, and working in teams, skills which cannot be taught in the classroom (Elarde and Chong 2012; Ismail 2018). Placement students are also considered to develop self-efficacy, which can promote work readiness as it enables an individual to make career decisions, face challenges and accept risks (Nisa et al. 2022). Consequently, students who have undertaken a placement generally have an easier transition from study to work (McLennan and Keating 2008).

From the student's perspective the placement may also result in significant benefits, for example, the offer of a job to return to after graduation and/or the ability to command a higher salary (Saniter and Saniter 2014; Williams 2020). In a separate study into graduate work readiness undertaken by the author, a student identified that the confirmation of a post-university job removed the stress of the final year of study; it removes the need to complete job applications whilst studying and graduate employers generally accept a lower degree classification than students previously expected.

2.3 *Employer Engagement with Placements*

It is generally accepted that an employer engaging in a long-term student placement will receive demonstrable benefits including access to lower cost resource (e.g., Blackwell et al. 2001; Bullock et al. 2009; Paisley and Paisley. 2010; Allen et al. 2013; Edwards 2014; Finch et al. 2016; Brooks and Youngson 2016; Crawford and Wang 2016; Jackson et al. 2017; Romgens et al. 2020). However, this perspective neglects the benefits that can help make a business more sustainable, for example, the provision of innovative and diverse resources, and the opportunity to engage the organization in fresh ideas and the latest thinking (Placements Scotland 2019). In addition, placement students can also form part of a longer-term recruitment or Corporate Social Responsibility (CSR) strategy.

2.3.1 Placement Students as a Temporary or Long-Term Resource

Placement students provide an opportunity for organisations to gain access to fresh talent with fresh ideas at a lower, indirect rather than fixed cost (Jackson et al. 2017; Placements Scotland 2019; Bespoke HR, 202; Indeed 2022). They can bring innovation, creativity, and new perspectives, as well as providing an additional, temporary short-term resource that closes a skills gap (Bespoke HR 2021). This can be particularly useful for a specific project where appropriate resource or a specific skillset is not available internally; having one person with appropriate skills to focus on one piece of work for a fixed period can be very productive (Institute of Student

Employers 2021).This can also help reduce recruitment costs in the short term as placements generally receive applications from a pool of talented individuals who are likely to possess the skills the employer hopes for (Education and Training Foundation 2021); frequently for SMEs such roles are advertised through a university free of charge (Institute of Student Employers 2021).

Longer-term, a recruitment pipeline may be created, as placement students frequently re-join the organization after graduation and/or encourage their friends to apply to the organization (Placement Scotland 2019; Bespoke HR 2021). Further, this can facilitate staff retention in two ways. Firstly, hosting a placement student can promote the development of supervisory skills such as coaching and mentoring for existing staff as they are required to manage the placement. In turn, this can boost morale and job satisfaction and encourage individuals to stay within the organization (Bespoke HR 2021). Secondly, graduating placement students are less likely to leave their first role than non-placement students (Institute of Student Employers 2021).

2.3.2 Placement Students' Contribution to a CSR Strategy

Recruiting a placement student can offer an organization an opportunity to contribute to their CSR practices, which in turn can offer a public relations benefit by increasing the attractiveness of employment and providing communication content (Placement Scotland 2019). Employing a placement student can enable the business to give something back to the student community whilst narrowing the skills gaps in their industrial or service sector (Education and Training Foundation 2021). It is also widely accepted as an effective way to enable future graduates to both obtain and retain a job (Van der Heije and Van der Heije 2006) and help the economy by preparing students for the world of work (Education and Training Foundation 2021). Thus, employing a placement student can enable a business to contribute to the Sustainable Development Goals, specifically Goals 8 (Decent Work and Economic Growth) and 9 (Industry Innovation and Infrastructure), the world-wide programme that drives government, business, and individuals to make the world a better place (United Nations, nd).

3 Challenges of Placements and an Innovative Response

3.1 Challenges of Placements

Traditionally placements for students have operated as a 'year in industry' within an undergraduate or postgraduate programme, with Higher Education being perceived as the preparation for work readiness and employability (Hillage and Pollard 1998; Hodgkinson 2005; Eraut 2009). However, despite the well-recognized personal, professional, and academic benefits for students that can promote success in the hyper-competitive graduate job market, there has been a significant downward trend in number of students opting to go on such a long-term placement (Finch and Peacock 2015; Education and Training Foundation 2021; Institute of Student Employers 2021). This may be due to a lack of confidence and/or self-awareness, a lack of understanding of the benefits of taking a placement, personal responsibilities, unwillingness to leave friends and family, unwillingness to leave an existing part-time job, or desire to complete formal education as soon as possible (Education and Training Foundation 2021; Institute of Student Employers 2021).

Employing organisations also face challenges in offering a placement. In smaller organisations where staff perform multiple functions there may be little time to train and support a placement student, and long-term placements may be too costly (Jackson et al. 2017). It is frequently difficult to determine the extent of administration, coordination and support required so the placement can be difficult to cost (Clark et al. 2014). In addition, many employers' have little understanding of how to get involved, identify a suitable student, identify a suitable role or project for the placement student, generate resources to coordinate the placement and/or how to supervise and mentor the student (Jackson et al. 2017; Dalrymple et al. 2014). Ultimately, some placements may not add value or become unsuccessful as there is a misalignment between the employing organisation's expectations and arrangements, and the purpose of the placement expected by the university (Patrick et al. 2009). To overcome this, HEI placement coordinators, workplace supervisors, and students must cooperate effectively to maximise the placement's value (Dalrymple et al. 2014).

3.2 Micro-Placements as an Innovative Response to Placement Challenges

To address these challenges and encourage more students to undertake a work experience placement, the author has developed an innovative approach to placements: a student-led environmental impact consultancy micro-placement. Within this micro-placement the student works a maximum of 20 hours per week over a period of 12 weeks whilst maintaining their studies. It may therefore be more attractive to students who are reluctant to undertake long term placements. The placement is undertaken during semester or a holiday period and is undertaken with a local SME. This overcomes students' challenges proposed above by enabling them to maintain their university relationships, social circles, family commitments, etc. whilst gaining work experience in the real world. The programme does not require students to relocate, give up part-time jobs, or fall behind their friends in obtaining a degree. It does, however, allow them to and develop their competitive advantage and personal resources in the highly competitive graduate job market and explore a potential future career which is vital for future career success (Van der Heije and van de Heije 2006; Finch and Peacock 2015; Institute of Student Employers 2021).

In this micro-placement the student acts as a consultant and focuses on a bespoke project that will develop an understanding of environmental impacts for the placement organization. The project is designed as a collaboration between the author and the placement organization to focus on a key area of their recognized need, starting with an environmental audit.

As the Business School is in an area dominated by SMEs and extensively recruits graduates from the local areas, most students progress into employment with smaller organizations after graduation. To develop familiarity with their most likely future work environment, as well as helping local businesses to access innovative skillsets, SMEs are targeted for these placements. The environmental focus was chosen as SMEs are increasingly facing demands for environmental impact improvements and reporting from customers and legislation, or a desire to become more sustainable; many do not have the knowledge and skills to deliver these changing demands. The author, who designed this programme, works in a sustainable and ethical Business School which has a clear focus on responsible management education and possess the expertise and resource to support the students and companies participating in short-term placement micro-placement projects.

To maximize the positive outcomes of these student-led micro-placements the author trains the students in data collection, analysis and presentation techniques based on tried and tested environmental auditing and reporting processes such as those utilized in ISO14001, ISO50001 and the Global Reporting Initiatives.

Alongside the potential environmental improvements offered, this approach may benefit the placement employer as it provides a cost-effective skilled resource for around 12–15% of the cost of a year-long placement and 30% of the cost of an established business consultant.

4 The Study's Context and Approach

4.1 The Context of This Study

The study presented in this paper evaluates the pilot student-led consultancy micro-placement which focused on a carbon emissions audit. It was established with a specialist plastic manufacturing and distribution company which is located near the Business School and supply the National Health Service (NHS) with a range of specialist products.

The placement employer in the pilot project had a clear rationale for recruiting a placement student for a micro-placement; they face a growing demand for carbon emissions data from customers and recognized the need to establish Scope 1 and 2 emissions but lacked the resource to do so. Many customers now demand carbon emissions data within tender submission documents; UK public tendering bodies such as the NHS, may exclude potential suppliers from the tendering opportunities if they do not provide emissions data and a formal carbon management strategy.

The inclusion of carbon emissions data within tender submissions is soon to become a mandatory requirement for UK public bodies' tenders. In addition to obtaining carbon emissions data that supported their future tendering, the placement employer wanted to voluntarily adopt good carbon management practices but struggled to find sufficient and appropriate internal resource to undertake the required data collection and analysis.

After initial discussions with the placement employer's project lead, the author who is a Principal Lecturer and the Sustainability Lead for the Business School, proposed a placement role designed to establish Scope 1 and 2 emissions for the placement employer. The planned output was a carbon emissions audit, which could be used for tender documentation and provide the baseline from which to develop a carbon management strategy, and the identification of areas of opportunity to reduce carbon emissions. It was agreed that this would be presented as a formal written report and presentation to the senior management team of the company.

This agreed role was formally written into a job description by the Placements Coordinator for the Business School to prime the recruitment process. The student-led consultancy role was then edited and approved by the placement employer and the author before being publicized through the Business School's online placement portal. Ten applications were received. After a first review and filtering of the submitted CVs, five potential candidates were forwarded to the placement employer's project lead. Interviews were conducted by the project lead and an offer made to a candidate, an international postgraduate student.

The placement employer's project lead and student agreed to run the project online as it largely involved data collection and analysis. Weekly meetings were held between the project lead and student to monitor progress and facilitate support required. Guidance from the author was provided on an ad hoc basis as required by the student and project lead. In practice minimal guidance was needed as the task was well defined and the placement student managed the project independently. The project lead had successfully recruited placement students in previous year long placements so was confident of the opportunity a student micro-placement could provide the company.

4.2 *The Means of Evaluating the Pilot Micro-Placement*

To establish the benefits and challenges from this micro-placement approach for both the placement employer and placement student, the placement employer's project lead and the placement student were asked to reflect on their experiences of working on the placement once it had been completed. These reflections were sought around three areas, project process, project outcomes and project impact so that the findings of this study could be captured and shared.

These project reflections, which are presented below, were facilitated by separate project debrief sessions between the author and the placement employer's project lead, and between the author and placement student and then written up by the project lead and student in their project reports via an online Microsoft Forms document. In line with the project debrief, the written reflections were structured around the three areas of reflection. This two-phased approach was implemented to enable the project lead and student to develop their thoughts beyond the initially recognized benefits and challenges. Additionally, the project lead and student were asked in the written reflection to identify those parts of the project that worked well and to suggest potential improvements that could enhance future student-led consultancy projects.

5 Impact of the Student-Led Consultancy Project

5.1 Benefits for the Company Provided by the Placement Student

The placement organization had a clear motivation for establishing the micro-placement, which enabled a clear structure and goal to be established. For example,

> *The requirements for reporting CO_2 for medical device companies is changing. This required us to make changes or risk losing out to competitors for tenders. This need was the driving force for organizational buy in for the project from all levels of management within the company.*

Whilst the placement organization obtained focused lower-cost resource, as recognized by Jackson et al (2017) and Indeed (2022), and the innovative resource that engaged the company in the latest sustainability thinking and fresh ideas proposed by Placements Scotland (2019) and Bespoke HR (2022), this study identified that the placement organization also obtained more specific, valuable outcomes above and beyond the generic benefits previously proposed (e.g. Blackwell et al. 2001; Bullock et al. 2009; Paisley and Paisley 2010; Allen et al. 2013; Edwards 2014; Finch et al. 2016; Brooks and Youngson 2016; Crawford and Wang 2016; Romgens, Scoupe and Beasaert 2020). These are explored below.

The success of the project has been accepted by those involved: the placement employer, the student, and the study's author. The organization's placement lead, recognized:

> *The most valuable learning experience was what a great resource students can be in this situation. It solved a problem for us, and it offers the students some real-world work experience. It really is win-win and somewhat of an under-utilized resource. Other than that, obviously the output of the student's work has been incredibly useful and has put us on a road to being a more sustainable company.*

The outcomes of the micro-placement project were presented in the end of placement written report which detailed the environmental impacts from the company's carbon emissions and the source of the emissions. This was also presented by the student to the senior leadership team. The report provided a baseline understanding for the organization and a tool for corporate communication. The Chief Executive reported the findings of the report to all staff members and has adopted its findings to develop a corporate sustainability vision and an environmental impact reduction strategy; the senior leadership team committed to move forward by setting energy consumption reduction and emissions reduction targets following the end of placement report presentation. The senior leadership team were impressed by the student's work and have published project report to their staff, customers, and LinkedIn community, and have begun to act upon the evidence presented and recommendations suggested.

Key benefits of the placement student's work for the placement employer were the provision of evidence that demonstrated that carbon emissions are a problem for the organization, and that the extent of the problem was greater than previously thought. This was established by the student's collection, analysis, and presentation of emissions figures from each of the company's three sites for both emissions Scopes 1 and 2.

The student's work highlighted the company's corporate culture as a contributing factor to the emissions problems established through the project. The student demonstrated that the company had simply considered that emissions were everyone's problem. However, through this project the organization realized that because of this, they had allowed emissions to be no one's problem. At the project review meeting the senior leadership team committed to enhance sustainability practices by setting targets to reduce energy consumption and reduce emissions. In addition, the project prompted the company to undertake a review of their sustainability organization, staff responsibilities and communication. This involved communicating the student's report with all staff and rethinking the roles and responsibilities of sustainability to establish a formal sustainability structure within the business. Whilst Placement Scotland (2019) recognized that a student's contributions to the placement organization's CSR and sustainability practices as a potential public relations benefit, this study suggests the contribution can also change internal practices and enable the management team to rethink corporate structures and accountability for sustainability.

The data analysis conducted by the micro-placement student presented information with which to make evidence-based decisions on sourcing, logistics and stock holding practice within the organisation. For example, understanding the emissions impacts of the current airfreighted products from the USA has enabled the company to evaluate the opportunity to utilise sea freight and local stock holding to reduce carbon emissions.

Following the placement student's calculation of Scope 1 and 2 emissions and proposal of emissions reduction interventions, work is now underway to understand carbon offsetting opportunities and sources of lower carbon electricity. The placement student not only provided information with which to make evidence-based decisions, it helped to highlight and close skills gaps as previously recognized by the Education and Training Foundation (2021) and Placement Scotland (2019). For example, the student's work identified that the three company sites had local purchase agreements for electricity, with the three sources providing a range of carbon intensive electricity. It also raised additional issues that the organization have now appointed owners to address; why the three different sites consume fuel differently, e.g., differences in fuel efficiency of heating systems, different fuel tariffs, etc., how can fuel efficient be improvement and what cannot be done?

The data analysis of electricity undertaken by the student highlighted an opportunity to reduce emissions by sourcing electricity from one supplier who provides lower carbon power, e.g., EDF from nuclear generation or Octopus from wind generation. This objective is now built into the contract renewal process. In addition, at the project review meeting a member of the senior leadership team was appointed to explore carbon offsetting as a way to address emissions that cannot be reduced through energy consumption reduction and the use of lower carbon sources of electricity.

The project lead confirmed the evidence provided by the placement student through the emissions audit will now be used to support the organization's EcoVardis supply chain accreditation (https://ecovadis.com/) which is required for NHS tenders. This was facilitated by the student's skillset and completion of a focused task, previously recognized as a placement benefit by Bespoke HR (2021). It will protect future corporate growth and development. The value of potential student-led projects for the placement organization was highlighted:

The student certainly delivered on all the tasks we asked her to carry out. Beyond that she also highlighted "easy" areas we could improve in regards to our CO2 emissions. Another extra benefit was simply the fact it has opened out eye to how useful student led projects like this can be to the company.

5.2 *Benefits for the Student Provided by the Micro-Placement Project*

The placement student had a clear goal in undertaking this micro-placement project,

The main reason to undertake this project was to gain practical knowledge of Sustainability along with the academic knowledge.

Whilst the study shows clearly that the student-led consultancy project was of considerable benefit to the placement organization, it appears from the student's reflections that it was also of meaningful personal and professional benefit to her as well. For example, the student reflected on professional skills development.

I have gained professional experience working on this project. I could correlate my academic knowledge with the tasks, and it gave me new insights and an understanding of sustainability.... this opportunity opened up new avenues for me on the professional front and added weight to my CV.

This mirrors the expectations of the Institute of Student Employers (2021) and Education and Training Foundation (2021) who advocate placements as an opportunity for the student to recognize their skills and close skills gaps. In addition to recognizing their development of specific skills, the placement student also acknowledged the impact of the consultancy project on her personal behaviours. For example,

It's not only business or corporations' responsibility, but each individual is also responsible to act. I learned how small changes can also make an impact and can contribute to emissions reduction. For instance, I started to unplug or turn off the devices when not in use to reduce electricity consumption, use water efficiently, etc.

The student recognized that the project helped to develop sustainability knowledge, for example UK environmental regulations. In addition, she reflected upon her development of skills for the collection, collation, analysis and presentation of data, the use of IT, and presenting the data in visual formats to enhance its' impact. All of these are the real-world skills recognized as vital to ensure graduates are better prepared for the workplace by Elarde and Cheong (2012), Ismail (2018) and Smith (2021). However, perhaps the most important learning point that the student understood was,

I learned to communicate my ideas and thoughts effectively.

The author recognized the student's development of self-efficacy and confidence during the project as suggested by Nisa et al. (2022). Following the placement, the student successfully applied for a data analyst role within the University's Library. The real-world experience gained through the consultancy placement enabled her to demonstrate the real-world learning and preparedness for employment and overcome the common graduates' skills gap identified by Smith, Ferns and Russell (2014).

From the reflections on the project provided by the placement student, the recognition of professional skills development, subject learning and communication skills have been beneficial from a personal and professional level. This suggests student-led consultancy projects can provide the opportunities for students to develop both the learning and softer employment skills that are not available in the classroom (Elarde and Cheong 2012; Ismail 2018; Smith 2021). These are recognized by Bullock, Gould and Hejmadi (2009) and Institute of Student Employers (2021) as key preparation for the workplace. The project lead and senior management team were so impressed by the student's engagement, effort, and professionalism they offered her a full-time job after-graduation, a positive outcome of placements previously recognized by Saniter and Saniter (2014) and Williams (2020).

6 Good Practice and Challenges in the Student-Led Consultancy Project

The organization's project lead was also asked to reflect on what contributed to the success of the project. Good practice came from many design and practice elements within the project which will be of use to other placement practitioners.

Establishing a single project lead who operates as the placement student's mentor was highlighted as good practice. In this project the project lead had previously run student placements and worked in academia so knew how to support the student. However, whilst this was beneficial, this previous experience was not considered a pre-requisite for a consultancy placement; staff from the HEI can coach and support the placement lead(s) as well as the student when required. The project lead highlighted the project's limited scope and defined objectives as good practice,

> *The best approach is to have a very set task for the student to carry out. The short time frame of these projects don't allow much time for the student to get up to speed with anything out of their comfort zone. Therefore, have a set project in mind, with clear milestones and deliverables that allow the student the best chance of getting the work done in the time allotted.*

Project success was considered to have been promoted by basing it on a single problem as advocated by the Institute of Student Employers (2021). The clear problem statement set by the placement employer for the placement student enabled all participants to have a clear focus, which gave the student a clear vision from the start. It also defined the timeframe and what the employer expected from the student. This expectation was fully met by the placement, for example the project lead reflected,

> *We found that basing the work on a "set" problem worked really well. There was a problem statement and a timeframe to which the student could work from. This clear vision from the start made it straight forward for the student and made it clear what we were expecting from them.*

The opportunity for a student to provide additional, lower cost resource (Jackson et al. 2017; Placement Scotland 2019; Indeed 2022) was identified by this study. For example, the project lead recognized good practice from,

> *The ability to get a task/project completed in a concise timeframe to a high level, while not needing to add workload to current staff or having to go through a full recruitment process to find someone who can do the work.*

Although the micro-placement was considered to be a success by the placement employer, the student and the author, a challenge has been established within the project. The major concern, although it proved to be easily managed, was the virtual operation of the project. Working online was required as the company is not served by public transport and the student did not drive. However, despite this initial challenge, the project lead recognized,

> *Communication is always the issue when working remotely. But this was managed well and didn't cause any issues. The benefit was that the student could concentrate on the task she was working on, and not worrying about a difficult commute.*

To address this challenge, the student and project lead established a formal communication process with a daily communication via Teams messaging, a weekly project update session and monthly project reviews all of which were held face-to-face online through the Teams portal. In addition, the author provided ad hoc guidance. Both the student and the project lead recognised the value of the support available and highlighted this as good practice in driving both the successful outcomes of the project and facilitating the effective running of the project.

7 Limitations of the Study and Next Steps

The study into the student-led consultancy project presented here reflects the pilot project conducted. Consequently, research into the outcomes of further projects is required to enable generalizations to be made about the value of micro-placements for both the placement employer and the student. If this proves that micro-placements focused on environmental impact improvements are successful in driving change within SMEs, the approach could be used be extended to additional topics. In addition, the student-led consultancy approach could be adapted to offer shorter projects (both part-time or full-time) that may further engage students in real-world work experience and SMEs in more effective sustainability practices.

8 Conclusion

From the reflections on the short-term student-led consultancy project presented above it is clear a valuable impact was made for the both the placement employer and the student; for the organization this did not end on conclusion of the placement project. Whilst the placement student delivered the planned emissions audit, gaining valuable personal, academic, and business skills, additional benefits were also delivered for the placement organization beyond those expected. For example, the student demonstrated that the placement organization needed to manage the company culture and staffing differently to achieve the carbon management focus they desired. The Scope 1 and 2 emissions data and opportunities to mitigate emissions impacts also enabled the company to make evidence-based decisions, develop sustainability. understanding and skills for environmental impact reduction. It will also facilitate EcoVardis accreditation which protects future business opportunities. Overall, the student-led consultancy project has provided the placement employer with an emissions baseline and opportunities for environmental improvement from which to financially and strategically benefit whilst contributing to the Sustainable Development Goals and mitigation of climate change.

References

Allen K, Quinn J, Hollingworth S, Rose A (2013) Becoming employable students and 'ideal' creative workers: exclusion and inequality in higher education work placements. Br J Sociol Educ 34(3):31–452

Bespoke HR (2021) Small business guide to hiring placement students. https://www.bespokehr.com/small-business-guide-to-hiring-placement-students/ (Last Accessed 18 November 2022)

Blackwell A, Bowes L, Harvey L, Hesketh A, Knight P (2001) Transforming work experience in higher education. Br Educ Res J 27(3):269–285

Brooks R, Youngson P (2016) Undergraduate work placements: an analysis of the effects on career progression. Stud High Educ 41(9):1563–1578

Bullock K, Gould V, Hejmadi M, Lock G (2009) Work placement experience: should I stay, or should I go? High Educ Res Dev 28(5):481–494

Cheng M, Adekola O, Albia J, Cai S (2022) Employability in higher education: a review of key stakeholders' perspectives. High Educ Eval Develop 16(1):16–31

Clark L, Rowe A, Cantori A, Bilgin A, Mukuria V (2014) The power dynamics and politics of survey design: measuring workload associated with teaching, administering and supporting work-integrated learning courses. Stud High Educ 41(6):1055–1073

Clarke M (2018) Rethinking graduate employability: the role of capital, individual attributes and context. Stud High Educ 43(11):1923–1937

Crawford I, Wang Z (2016) The impact of placements on the academic performance of UK and international students in higher education. Stud High Educ 41(4):712–733

Dalrymple R, Kemp C, Smith P (2014) Characterising work-based learning as a triadic learning endeavour. J Furth High Educ 38(1):75–89

Docherty D (2014) Universities must produce graduates who are ready for any workplace. https://www.theguardian.com/higher-education-network/2014/may/22/universities-must-produce-graduates-who-are-ready-for-workplace

Education and Training Foundation (2021) The value of work experience to students and employers. https://www.et-foundation.co.uk/t-level-professional-development-tlpd/the-value-of-work-experience-to-students-and-employers/. (Last Accessed 21 February 2023)

Elarde J, Chong F (2012) The pedagogical value of "eduployment": information technoogy internships in rural areas. In Proceedings of the 13th Annual conference on Information Technology Education. pp 189–194. https://dl.acm.org/doi/pdf/https://doi.org/10.1145/2380552.2380607 (Last Accessed 20 November 2022)

Eraut M (2009) Transfer of knowledge between education and workplace settings. In Daniels H, Lauder H, Porter J (eds) Knowledge, values and educational policy. A critical perspective, pp 65–84. Routledge: London

Finch D, Peacock M, Levallet N, Foster W (2016) A dynamic capabilities view of employability. J Educ Train 58(1):61–81

Green W, Hammer S, Star C (2009) Facing up to the challenge: why is it so hard to develop graduate attributes? High Educ Res Dev 28(1):17–29

Harvey L (2000) New realities: the relationship between higher education and employment. Tert Educ Manag 6(1):3–17

Hillage J, Pollard E (1998) Employability: developing a framework for policy analysis. DfEE, London

Hodkinson P (2005) Reconceptualising the relations between college-based and workplace learning. J Work Learn 17(8):521–532

Indeed (2022) What is labor cost? Definition, direct vs. indirect costs and examples. https://www.indeed.com/career-advice/career-development/labor-cost (Last Accessed 20 November 2022)

Institute of Student Employers (2021) 5 reasons why employers are running micro-placements. https://insights.ise.org.uk/work-experience-internships/blog-5-reasons-why-more-employers-are-running-micro-placements/ (Last Accessed 18 November 2022)

Ismail Z (2018) Benefits of internships for interns and host organisations. https://opendocs.ids.ac.uk/opendocs/bitstream/handle/20.500.12413/13848/Internships.pdf?sequence=1&isAllowed=y (Last Accessed 21 February 2023).

Jackson D, Rowbottom D, Ferns S, McLaren D (2017) Employer understanding of work integrated learning and the challenges if engaging in work placement opportunities. Stud Cont Educ 39(1):35–51

Maertz C, Stoeberl P, Marks J (2014) Building successful internships: lessons from the research for interns, schools and employers. Career Dev Int 19(1):123–142

Mason G, Williams G, Cranmer S (2009) Employability skills initiatives in higher education: what effects do they have on graduate labour market outcomes? Educ Econ 17(1):1–30

McLennan B, Keating S (2008) Work-integrated learning (WIL) in Australian Universities: The Challenges of Mainstreaming WIL. ALTC NAGCAS National Symposium

Milkround (2022) How competitive are graduate schemes? https://www.milkround.com/advice/how-competitive-are-graduate-schemes (Accessed 22 February 2023)
HR News (2022) Employers reassess value of degrees in face of worker shortages. https://hrnews.co.uk/employees-reassess-value-of-degrees-in-face-of-worker-shortages/ . (Last Accessed 21 February 2023)
Nisa N, Sugiharto D, Awalya A (2021) The relationship between creative thinking, problem solving skills, and self efficacy with work readiness. J Bimbingan KonseLing 10(1):8–13
Paisey C, Paisey N (2010) Developing skills via work placements in accounting: student and employer views. Account Forum 34(2):89–108
Patrick C, Crebert G (2004) The contribution of work placement to generic skills development. Creating flexible learning environments: proceedings of the 15th Australasian conference for the Australasian association for engineering education and the 10th Australasian women in engineering forum, Australasian association for engineering education
Patrick D, Peach C, Pocknee C, Webb F, Fletcher M, Pretto G (2009) The WIL (Work Integrated Learning) report: a national scoping study, final report. Brisbane: Queensland University of Technology
Römgens I, Scoupe R, Beausaert S (2020) Unraveling the concept of employability, bringing together research on employability in higher education and the workplace Stud High Educ 45(12):2588–2603
Saniter N, Saniter T (2014) Door opener or waste of time? The effects of student internships on labor market outcomes. IZA Discsusion p 8141, Institute for the Study of Labor (IZA)
Placements Scotland (2019) What are the benefits of undertaking a placement through e-Placement Scotland? https://www.e-placementscotland.com/our-blog/view/26/what-are-the-benefits-of-undertaking-a-placement-through-e-placement-scotland.aspx (Last Accessed 1 March 2023)
Smith C, Ferns, S, Russell L (2014) The impact of Work integrated learning on student work-readiness, The impact of Work integrated learning on student work-readiness: Final Report, Curtin University of Technology, LSN Teaching Development Unit
Smith J (2021) Work placements. https://www.prospects.ac.uk/jobs-and-work-experience/work-experience-and-internships/work-placements (Last Accessed 18 February 2023)
Temmerman N (2019) When a degree just isn't enough, also offer upskilling. University World News. https://www.universityworldnews.com/post.php?story=20190304094729650 (Last Accessed: 18 February 2023)
United Nations (no date) The 17 Goals. https://sdgs.un.org/goals. (Last Accessed 20 November 2022)
Van der Heijde C, Van der Heijden B (2006) A Competence-based and multidimensional operationalization and measurement of employability. Hum Resour Manage 45(3):449–476
Williams V (2020) Career Focus: Top 6 reasons you should do a work placement. https://www.kaplanpathways.com/blog/top-6-reasons-to-do-a-work-placement/ (Last Accessed: 18 February 2023)
Woodfield L, McIntosh R (2022) Employability Blog series: three policy turning points that changed the higher education employability agenda. https://www.hepi.ac.uk/2022/03/11/employability-blog-series-three-policy-turning-points-that-changed-the-higher-education-employability-agenda/ (Last Accesses 27 February 2023)

Responsible Global Business Graduates: Closing the Gap Between Strategy and Practice

Jeanine Gregersen-Hermans, Marcus de Vries, and Connie Heiligers-Duckers

Abstract Inspired by the United Nations Sustainable Development Goals (UNSDGs), the mission and vision of the International Business School Maastricht (IBSM) is to educate resilient business leaders with a global mind. Sustainability and internationalisation are core to this mission. At IBSM we found that our students and lecturers had difficulty in connecting the dots between these two educational initiatives and the various business disciplines. This paper describes the introduction of an innovative and radical pedagogy 'Student in the Lead' and how ethics, responsibility, and sustainability and international, intercultural, and global dimensions of learning are being positioned as a foundational lens for the core business disciplines of Marketing, Finance, Supply Chain Management, and Organisation and People. In addition, the paper reports on the first experiences of students and lecturers with Student in the Lead, which point towards their appreciation of the new pedagogy. Both students and coaches indicate they are still finding their way in what is expected of them and what the focus on the UNSDGs implies for them as International Business professionals. The paper concludes that the radical change trajectory is not only about developing new skills but also about sense making and transforming professional identities. The IBSM lecturers have to develop into reflective practioners who understand the fine line between coaching and their own positionality on the values underpinning ethical, responsible, and sustainable dimensions of business challenges and their international, global, and intercultural contexts and conditions. The IBSM experience also inspired the initiation of a new Eramus + Cooperation Partnership in Higher Education (2022–2025), titled "Innovating Business Education for Responsible Global Minds".

J. Gregersen-Hermans (✉) · M. de Vries · C. Heiligers-Duckers
The International Business School Maastricht, Zuyd University of Applied Sciences, Heerlen, The Netherlands
e-mail: jeanine.hermans@zuyd.nl

M. de Vries
e-mail: marcus.devries@zuyd.nl

C. Heiligers-Duckers
e-mail: connie.duckers@zuyd.nl

T. Wall et al. (eds.), *Sustainability in Business Education, Research and Practices*, World Sustainability Series, https://doi.org/10.1007/978-3-031-55996-9_10

Keywords UNSDGs · Education for sustainable development · Internationalisation of the curriculum · Business education · Student centred learning · Experiential learning · Curriculum development · Transformational change

1 Introduction

The International Business School Maastricht (IBSM) of Zuyd University of Applied Sciences in the Netherlands has been involved in a curriculum development initiative, which implied radical change in its bachelor programme International Business (IB). This paper highlights some of the lessons learned from the IBSM's change trajectory in which it introduced a student-centred learning approach, referred to a Student in the Lead (SitL), whilst at the same time embedding the United Nations Sustainable Development Goals (UNSDGs) at the core of its curriculum.

IBSM's change in focus and pedagogy of its curriculum illustrates a worldwide trend of Higher Education Institutions (HEIs) which state they are improving the quality of education by enhancing the alignment of their curricula with their wider societal and environmental contexts. With an understanding that graduates need to operate in an increasingly volatile and uncertain world, they report they are working to integrate the United Nations Sustainable Development (UNSDGs, n.d.) in their teaching and learning (e.g., Michel et al. 2020). Often Business Schools are assumed to take the lead for developing a new type of graduates who can facilitate the transformation of companies and other organisations to more sustainable and socially just business models. This might be particularly true for Business Schools, like the example of IBSM, who have signed up the Principles of Responsible Management Education (PRME). However, whilst highlighting the role of management education 'as agents in moving the UNSDGs forward' (Weybrecht 2017, p. 91) in the training of current and future business leaders this author concludes that the increasing awareness of the UNSDGs 'in most cases has not yet translated into engagement, and that engagement has been slow, inconsistent and haphazard' (Weybrecht 2022, p. 8).

Although carefully planned, the implementation of IBSM's curriculum development initiative also has not been straightforward. In this paper we explore the process of change and organisational learning within IBSM to deliver on the educational innovation. First, we will describe the IBSM context and the need for change. Second, the SitL approach and the distinctive IB lens are described what this implies for students and lecturers. Third, we will share the first experiences with this curriculum development project and highlight blockers and enablers that have come to the fore in this process so far. In the fourth section we will reflect on these experiences from the perspective of change processes and organisational learning. The concluding section addresses the next steps for IBSM's curriculum innovation and its involvement in the Eramus + collaborative partnership project 'Innovating Business Education for Responsible Global Minds' (IBE-ReGloMi), a collaboration of five continental European Business Schools: Kedge Business School (Fr.), Kozminski University (PL),

Technische Hochschule Koln (Ge), Babes Bolyai U (Ro), Zuyd IBSM (NL.)) and the European Foundation for Management Development (EFMD).

2 The IBSM Context and the Need for Change

IBSM includes an EFMD programme accredited Bachelor of Business Administration (BBA) in International Business, with specialisation opportunities in International Marketing, Finance, Management and Organisation, and Supply Chain Management. Further students can opt for a graduation project in Entrepreneurship. The school accommodates over six hundred students and further includes a research centre for Sustainable International Business (SIB). The mission and vision of IBSM expresses its commitment to the SDGs and PRME in the following way:

> In this fast-changing world we, at International Business School Maastricht, guide our young professionals to become resilient business leaders with a global mind. (IBSM 2022)

To implement this mission and vision in its curricula, a curriculum development trajectory was initiated in 2018 (IBSM 2022). From a predefined curriculum with a focus on the themes of various business fields, the IB programme changed to a real life challenged based education, with a focus on student learning. Although the theme-based curriculum included intended learning outcomes (ILOs) for ethics, sustainability, and corporate social responsibility (jointly ERS), the associated teaching and learning activities were scheduled as separate modules from the core business disciplines such as Marketing, Finance, Supply Chain Management, or Organisation and People. Moreover, sustainability was only considered an as one of the possible options for approaches to business. Economic growth still underpinned these business areas as a core value. Further, programme modules for international, global, and intercultural (jointly IGI) learning were scheduled in another pathway referred to as Intercultural Business. Inspired by the model of Global Competences developed by the Organisation for Economic Development (OECD 2018), the associated ILOs in this pathway addressed the SDGs as well and drew on similar underlying competences such as ethical awareness, social responsibility, and critical thinking. This led to difficulty for students and lectures alike in connecting the dots between ERS and IGI and with the specific business disciplines (Gregersen-Hermans and Boonen 2021).

To ensure a more comprehensive approach the IBSM community of students and staff realised that a more radical approach to the pedagogy was needed, which also allowed for an integrated view on ERS and IGI and the business disciplines (Gregersen-Hermans 2021). To this end, a SitL approach was introduced, which was piloted in the fast track of the IB programme from 2019 onward and fully implemented in the programme as a whole in the academic year 2020–2021. The next section further discusses this educational innovation of the IB programme.

3 Innovation of the International Business Programme

The need for innovation of the IB curriculum resulted from the awareness at IBSM that traditional learning models which focused on the delivery and assessment of knowledge were less fit for purpose to deliver graduates who matched the profile of IBSM's mission and vision and who could fulfil the current and future needs of the international business world and society at large. Increasing volatility, uncertainty, and complexity globally, and the need to addresses the challenges of current and future times lead businesses to search for a new type of professional. Success in business in this new context requires autonomous graduates, who are resilient, agile, and capable of managing themselves in a rapidly changing world (Van der Sluijs et al. 2009). These new types of graduates are often referred to as changemakers. They are aware of the growing scarcities in natural resources that come with human overpopulation and can operate in new ways of economic thinking which focus on planetary health and well-being in support of the UNSDGs. The that time Dutch Minister of Education, Culture and Science, Robert Dijkgraaf referred to the required curricular change in his Kohnstam Lecture as the responsibility of education for 'developing students' uncertainty capabilities and personal interaction skills on a personalized learning journey." (Dijkgraaf 2022).

The next two sub-sections address how IBSM's innovated its curriculum to educate these changemakers and embedded the PLOs for ERS and IGI as the defining lens for student learning and the SitL pedagogical in its IB programme.

3.1 IBSM's Integrated Approach to the Programme Learning Outcomes

An important element of the SitL approach was the introduction of the idea of the International Business worlds (IB worlds). The IB worlds build on the Dutch national framework for International Business programmes (Sijben et al. 2017), which lists four general domains (ways of thinking, ways of working, living in the world, and tools for working and management) and 24 programme learning outcomes (PLOs). In an attempt to move beyond the siloed approach of the 24 PLOs in the national framework, the IB worlds group these PLOs and connect the separate PLOs for ERS and IGI dimensions to the core business areas and to real life business challenges and global trends. They describe in a student friendly way how the general domains and PLOS are derived from actual and future business challenges and needs. The nine IB worlds are listed in Table 1. In addition to the nine worlds there is a free study space, in which students can study any subject of their choosing.

These IB worlds were complemented in second year of implementation (2021–2022) with a conceptual model in the form of the IB lens (de Vries et al. 2022). To ensure that the ERS and IGI PLOs are holistically embedded in the core business worlds, this lens arranges the IB Worlds and the PLOs in a particular way to support

Table 1 The nine international business worlds of IBSM. *Source* IB Worlds (2020)

Nine IB Worlds
Research based consultancy
Culture and the international business context
Communication
Personal & professional development
Collaboration
Marketing
Finance
Operations & supply chain management
Organisation & People

the students and lecturers to connect the PLOs to real life business challenges in an integrated way and to the UNSDGs. The lens in Fig. 1 visualises this conceptual model.

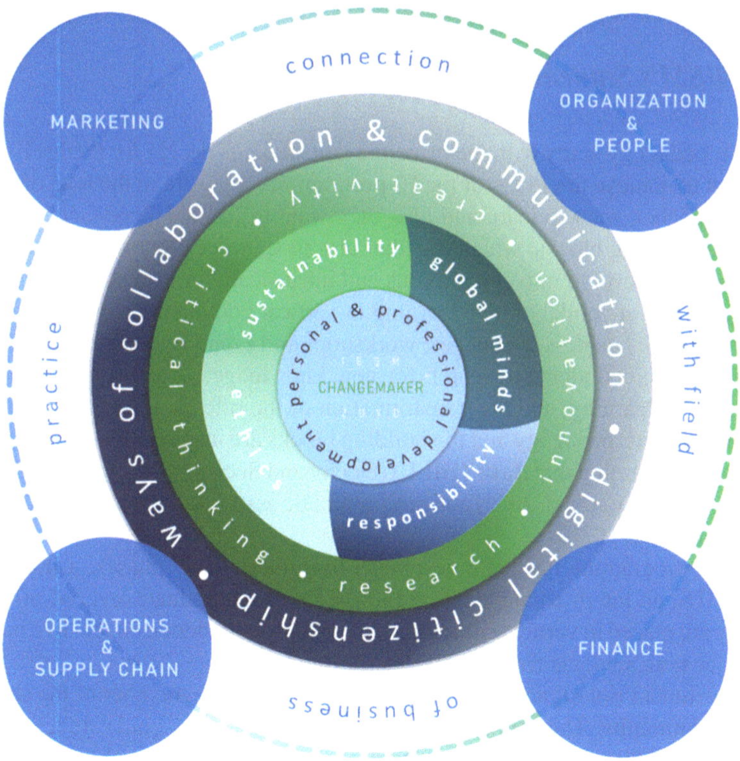

Fig. 1 The IB lens in support of developing change makers in international business. *Source* de Vries et al. (2022), used with permission

At the heart of the lens sits the student and his or her professional and personal development as a changemaker. The first ring includes the PLOs for ethics, responsibility, and sustainability (ERS), and international, intercultural, and global dimensions (Global Minds). This ring is aligned to the mission and vision of IBSM. It refers to a value-based approach to business challenges grounded in the UNSDGs.

The second ring points to the student's development into an autonomous critical thinker, and includes PLOs related to a research-based attitude to business challenges, innovation, and creativity. The third ring focuses on ways of working and refers to the PLOs for collaboration, communication, and digital citizenship. Together these three rings form the lens which characterise IBSM's distinctive and value-based approach to business challenges. The fourth and outer rings consist of the international business practice and the disciplinary international business fields (marketing, finance, management and organisation, and supply chain management). Applying this IB lens to business challenges supports students to connect the dots between ERS, IGI dimensions and the disciplinary international business fields, enhancing their capability to address the current and future needs of companies and their employability. (Gregersen-Hermans and de Vries 2022 (as cited in Kalogeras 2022)).

3.2 IBSM's Student in the Lead Approach

The SitL pedagogy starts with the principle that students decide for themselves how they want to achieve the PLOs of the IB programme. They are in the lead of what to learn, when they want to learn and how they want to learn. The pedagogy consists of several building blocks, which together aim to spur deep learning and challenge the students to develop into autonomous and intrinsically motivated learners. These building blocks are collaborative learning in vertical communities of mixed student cohorts; real life business challenges; workshops and events by lectures and business experts to inspire students to search for new knowledge; a focus on feedback and assessment as learning, and the transformation of the role of lecturers into business coaches. The SitL pedagogy and the building blocks are inspired by the principles of High Impact Learning that Lasts (HILL) (Dochy and Seegers 2018), challenge-based learning (e.g., Leijon et al. 2022) and programmatic assessment (van der Vleuten et al. 2015; Sluijsmans 2020).

The IB programme has been designed along three study phases, a junior level of 60 ECTS, an intermediate level of 120 ECTS and a graduate level of 60 ECTS. Students are placed in vertical communities, and they collaborate on real live business challenges with students from all levels.

Every student is assigned a personal coach who is also a member of the student's specific community. The role of the coaches is twofold; they guide students during their learning journey, and they provide academic and professional input related to their own area of expertise. The coaches in each community function as a multidisciplinary team. These teams are equipped to address all business areas at the junior and intermediate level. Moreover, to ensure crossovers between the communities, students are encouraged to gather business expertise and knowledge from coaches connected to the other communities. For graduate level expertise students must consult with the lecturer-coaches who are specialised in specific business areas. To enhance deep learning, lecturing is on demand and related to the business challenges students are working on.

In line with the aspiration to develop intrinsically motivated graduates, the SitL pedagogy focusses on assessment as learning. Students develop e-portfolios for the business challenges they are working on and connect these to the relevant IB worlds and the PLOs. They must collect regular feedback from a variety of sources at the beginning, during and at the end of a business challenge and document in reflective writing how they processed the feedback. For summative assessment and allocation of credits, a six-point rubric per study phase has been developed that includes requirements pertaining to the distinctive IBSM lens and the PLOs, to increasing learner autonomy, and to complexity and the international context of the business challenges. This rubric is grounded in the Dublin descriptors of the three-cycle system of the European Higher Education Area (EHEA, n.d.) and the European Qualifications framework (Europass, n.d.). The rubric is included as supplementary material.

In support of SitL, an electronic student portfolio system with interactive feedback capability has been implemented for documenting student progress and providing feedback, feedforward and feed up. Further, a Community Office has been created to provide administrative support to the lecturers and students. The Community Office has operational oversight and bridges the engagement of students with the business world.

In summary, embedding ERS and IGI dimension as the defining lens of the IB programme and implementing the SitL pedagogy implied a radical shift from the traditional way of teaching and learning at IBSM. To get all lecturers on board and to deliver on the profile of the future graduates to business and society, the innovation of the curriculum therefor needed to be approached as a change management trajectory. The next section describes IBSM's change trajectory.

4 The Change Trajectory

The development and implementation of SitL can be divided in three major phases: design and testing, implementation and training, continuous improvement and development. In the design and test phase of this pedagogical innovation the principles of HILL, challenge-based education (e.g., Leijon et al. 2022) and programmatic assessment were blended into the SitL pedagogy, with learner autonomy and realistic business projects at the core. For the pilot and testing phase a team of innovation minded lecturers was recruited from the existing pool of IBSM lecturers, who represented a well-balanced blend of experienced educators with a business background. They were given carte blanche to pilot the SitL pedagogy in an initial learning lab with a cohort of the IB short track students, facilitated by their own physical community space. The pilot further developed the SitL pedagogy into a proof of concept for learning in vertical communities, coaching, and assessment as learning. During the pilot and test phase several team-wide and individual continuing professional development (CPD) opportunities were offered to prepare the full team of IBSM lecturers for the role out to the full IB programme.

Since the SitL approach implied major changes in the role of the lecturer, the commitment and preparedness of the wider team of IBSM lecturers was essential to go ahead with the implementation. With agreement of the majority of the IBSM team members, the SitL pedagogy was implemented from the academic year 2020–2021 onwards across the whole IB programme for about six hundred students.

To strengthen IBSM's ties with the business world and to allow sufficient time for coaching of individual students, six seasoned professionals from the business world were added to the team. Furthermore, time had been allocated to lecturers' workload schedule for training into their new roles as IB coaches and for learning on the job. This primarily consisted of regular peer to peer consultation during calibration sessions and involvement in the further refinement of the SitL pedagogy. Community teams of IB coaches held weekly calibration sessions that dealt with the day-to-day issues the coaches experienced in their community. This could range from student attendance and progress, their new role as coach and how to provide formative feedback and summative assessment, or current business trends. Team-wide training and calibration sessions were organised monthly to allow learning across the community teams, in which individual coaches and coach teams would present student cases for mutual inspiration, reflection and the development of a joint approach. Systemic issues with the SitL pedagogy that came forward from these calibration sessions were included in the quality assurance processes of IBSM. Finally, each business coach had the opportunity to contribute to the further refinement of the pedagogy in project teams on each of the SitL building blocks.

The quality assurance of this change trajectory was transferred from the initial pilot team to a new management body, the Educational Support and Advisory Committee (ESAC). The function of ESAC is to advice the IBSM leadership, to guide the process of educational innovation, assure the quality of education through its plan-do check-act management cycle (PDCA), and arrange further CPD for the team. As part of its role, ESAC coordinates the further fine-tuning of the SitL pedagogy and addresses teething problems and concerns of business coaches and students. The IB lens, and the six-point assessment rubric as part of the IBSM Assessment Plan were among the deliverables of the ESAC during the first year of implementation. Further, in collaboration with the Community Office and current students an onboarding programme has been created to facilitate the transition of new students into IBSM's SitL pedagogy.

As demonstrated by the change trajectory of IBSM, the educational innovation implied continuous engagement between the leadership, ESAC, business coaches, students, and other university stakeholders. To this end, the experiences of students and business coaches have been and are continuously researched through surveys and focus groups. In the next section we will report on the first findings of this research.

5 First Experiences of Students and Business Coaches

To include all voices in the further development of the IB programme, further research was conducted in January 2023. The research included five focus groups with students and twenty-five interviews with IB coaches. The focus groups were held with students in each community and were facilitated by the IB coaches. The protocol for the focus groups was developed by the Quality officer of IBSM (Heiligers-Duckers 2023). Further, the first experiences of the IB coaches themselves were collected through individual interviews that addressed their new role as IB coach and their individual learning needs. The interviews were conducted by the two IBSM programme directors.

5.1 Student Experiences

Three topics were selected for the focus groups: planning and structuring your own programme, academic content and learning resources, learning in communities. For each topic small groups of four to five students discussed a single statement that was based on earlier feedback provided by students. The role of the IB coaches was to clarify the procedure and the statements if necessary. All five communities participated in the sessions, which lasted 45 min. The sessions were recorded, transcribed, and analysed using thematic content analysis. The key points of the report (Heiligers Duckers 2023) are summarised below.

- Students generally appreciated the freedom to design their own learning pathway. They valued the room for discovery and creativity and to study at their own pace, without the stress of exams.
- Students reported they struggled with organising their studies, especially in the first of months of their programme. They felt the IB coaches needed to be more proactive in helping them to bridge the gap between their pre-bachelor education and the SitL pedagogy. Moreover, they sometimes were confused by the variety in the feedback that they felt was not always consistent.
- Students preferred guidance on how and where to find business knowledge and expertise to succeed in their business challenges over lecturing. Many of them start to understand the importance of critical thinking and that the insights they develop through their business challenges make the fabric of expertise.
- Students would appreciate however to gain a better understanding and knowledge of the core business principles, so they can study towards that. In their view, larger scale events and workshops should be offered that engage the whole community. They would like to have more opportunities to engage in activities that facilitate the interaction with the business and finding suitable challenges.
- Students indicated a need to learn how to discover their talents, personal qualities, and interests; how to present themselves towards businesses and challenge providers and how to network to acquire challenge projects.

Some students further mentioned the focus on the UNSDGs. They understood these are important but stated that the way it is pushed to the students tends to make it unattractive. As a starter, IBSM itself should collaborate more with sustainable companies and make its own operations more sustainable.

5.2 *Staff Experiences*

To ensure a personal approach and a confidential setting, individual interviews were held with the IB coaches by one of the two programme directors. Each interview lasted approximately 45 min. Five topics were addressed during these interviews: the attitude and behavior towards students, the bonding within the communities, the SitL pedagogy and the focus of the IB programme on the SDGs, collaboration between colleagues, and the need for professionalisation. The key points are summarised below (ref internal report).

- Coaches appreciated their new role and the genuine nature of the coaching relationship, not only with their own coachees but also with other students. They mentioned the value this professional relationship has for the students in terms of their personal development and for building their confidence. The student is not a number in this new set-up.
- Coaches mentioned the importance of pro-active coaching, especially during the early phases of the IB programme, thereby referring to the misconception they sometimes saw with their colleagues that student-in-the-lead means leaving the student to his or her own device.
- Coaches struggled with the group dynamics within their communities and how to enhance the bonding between the students. Students quickly formed their own subgroups of friends to work on their business challenges and often stayed within these subgroups.
- Coaches mentioned the challenge of appropriately balancing the art of asking coaching questions to students with the habit of explaining and lecturing. This was especially the cases when coaches were concerned the students needed business knowledge to solve a challenge.
- Coaches stressed the importance of understanding the big WHY of the new pedagogy and the focus on the SDGs for themselves and for the students. In their view, this also needed more attention in the induction of new students and new colleagues.
- Coaches felt the need for sharing and closer collaboration with each other on how to implement the SitL pedagogy, how to ensure fair and transparent summative assessment, and how to interpret and work with the IBSM lens. In this context they indicated a need for further professionalisation regarding the working in communities, assessment, and coaching.

Although coaches felt important improvements were made since the role out of the educational innovation, more structure within SitL was needed.

In conclusion, the first experiences of students and coaches point towards their appreciation of the SitL approach and how this supports students in their development of autonomous and critical professionals, with a values-based approach to international business contributing towards the UNSDGs. Both students and coaches indicated they are still finding their way in what is expected of them. For students, this implies developing the attitudes and skills to successfully work within the IBSM pedagogical approach. The IB coaches are in the process of redefining their role and acquiring new skills. Next to developing new skills, coaches need time and certain degrees of freedom to make sense of this educational innovation. The focus on the UNSDGs is understood by both the students and the business coaches. However, how this currently plays out in the practice of business challenges indicates more works needs to be done to embed an integrated perspective on ERS and IGI as visualised by the IB lens in their hearts and minds. Critical for the further development of the SitL approach is the continued dialogue between and among students and business coaches on how to collaborate with each other and with business and challenge providers towards delivering graduates with responsible global minds.

5.3 *Limitations*

To build a balanced case story on IBSM's journey of educational innovation, the perspectives of all stakeholders, and in particular those of students and IB coaches have been carefully considered. We are aware however of our own positionality in this change trajectory. Based on the research underpinning the SitL pedagogy and the first experiences, we are confident the initiative has come at the right time and responds to the need of transforming business education towards the UNSDGs. We are also aware that some students and colleagues may state we paint a too optimistic or a too critical picture.

6 Discussion and Next Steps

The process of change IBSM has embarked on is approached as a team driven change. To turn IBSM's mission and vision of educating resilient business professional with a global mind in practice for the IB programme as a whole, the pilot team realised that the larger IBSM team needed to make sense of the radical change in its own way. Radical change requires a deeper level of change than merely adopting a new teaching practice. It deeply impacts personal and professional beliefs and values, for instance related to how to assess, or what constitutes educational quality. Further, offering ready made solutions from the top 'helps to kill creativity and causes people to wait and see' (Boonstra 2022, p. 18). These insights have two major implications for IBSM's change trajectory. First, the leadership needs to allow time and resources for the IBSM lecturers to develop from experts assessing what they teach in a series of well-designed lectures towards business coaches engaging in reflective feedback dialogues with students along a student's learning journey. In other words, investment needs to be made for continuing professional development and acquisition of new skills. Second, it would be a misconception to belief the radical change trajectory of IBSM is a linear process that can be managed through consecutive steps as suggested by some scholars (e.g., Kotter 1999). The change trajectory is also deeply personal and IBSM's coaches need certain degrees of freedom to make sense of the educational innovation, their changing professional identity, and their relationship with students. What is needed is a safe environment to experiment, discover and learn together in the direction of the desired changes, which is the further development of the SitL approach and embedding ERS and IGI competences as the defining lens for the problem solving of business challenges. For team learning to happen a sense of belonging to the team is essential, with social activities and room for mistakes.

Further, IBSM's educational innovation also implies a normative change (Wals 2014) by its positioning of ERS and IGI dimensions as foundational in its programme learning outcomes. As suggested elsewhere, the knowledge and skills of the IB coaches needs to be extended to include sustainability, global and intercultural competence development (Gregersen-Hermans 2021). Furthermore, coaches have to understand their own biases, and assess how their own values and assumptions influence their guidance of students, as this otherwise may be perceived as indoctrination (Boeve – de Pauw et al. 2015). In other words, the IBSM coaches need to further develop their understanding of the UNSDGs and what these imply for the coaching of international business students. Most importantly they have to develop into reflective practioners who understand the fine line between coaching and their own positionality on the values underpinning ethical, responsible and sustainable dimensions of business challenges and their international, global and intercultural contexts and conditions.

One of the next steps for IBSM on this journey is its participation in the Eramus + co-funded collaborative partnership project Innovating Business Education for Responsible Global Minds (Gregersen-Hermans et al. 2022) This partnership consists of six European partners and a wide range of global and local associated partners. They have started an initial international community of practice to co-create business graduate profiles with students and business with a focus on the intersections between ERS, IGI and Social Innovation and Entrepreneurship (SIE) as the defining lens for the various business disciplines. Moreover, the IBE-ReGloMi project will deliver principles for innovative pedagogical approached for embedding ERS, IGI and SIE in the curriculum and a CPD programme in support of lecturers and coaches. The ambition is to move beyond the statements about enhancing the quality of student learning and in Weybrecht' s (2022) terms to act as agents who deliver graduates who can function as changemakers for a sustainable and just global society and world.

References

Boeve - de Pauw JB, Gericke N, Olsson D, Berglund T (2015) The effectiveness of education for sustainable development. Sustainability 7(11):1-25

Dijkgraaf R (2022) Hoe talent op zijn plek valt. Kohnstamm lecture. https://www.rijksoverheid.nl/documenten/toespraken/2022/11/21/toespraak-van-minister-dijkgraaf-kohnstammlezing-2022

Dochy F, Segers M (2018) Creating impact through future learning: The high impact learning that lasts (HILL) model. Routledge

EHEA (n.d.). http://ehea.info/page-three-cycle-system

Europass (n.d.). https://europa.eu/europass/en/description-eight-eqf-levels

Gregersen-Hermans J (2021) Towards a curriculum for the future: synthesizing education for sustainable development and internationalisation of the curriculum. J Stud Int Educ 25(4):1–21. https://doi.org/10.1177/10283153211031033

Gregersen-Hermans J, Boonen J (2021) The international business school Maastricht: from an educational strategy to an internationalized educational practice. In: Leask B, de Wit H, Torres Hernández M, Bustos-Aguirre M (eds) Reimagining the internationalization of the curriculum. Best practices and promising possibilities. University of Guadalajara

Gregersen-Hermans J, Cosma S, Cordos GJ, Gijsberts G, Gudkova S, Faget M, Kalogeras N, Khalil M, Kraak J, Kowalski S, Maillard J, Terrasse C, Thuerbach K, Wojtowicz D (2022) Innovating business education for responsible global minds (2022–2025). Grant application for the Erasmus+ KA220-HED Cooperation Partnerships in Higher Education

Heiligers-Duckers C (2023) Student Feedback Sessions. IBSM Internal Report

IB Worlds (2020) The 10 International Business Worlds of the BBA. Internal document

IBSM (2022) EFMD Programme Accreditation: Self-Assessment Report. Internal Report

Leijon M, Gudmundsson P, Staaf P, Christersson C (2022) Challenge based learning in higher education–a systematic literature review. Innov Educ Teach Int 59(5):609–618

Michel JO, Holland LM, Brunnquell C, Sterling S (2020) The ideal outcome of education for sustainability: transformative sustainability learning. New Direct Teach Learn 2020(161):177–188

OECD Organisation for Economic Development (2018) Global Competence. https://www.oecd.org/pisa/innovation/global-competence/

Sluijsmans AD (2020) Toetsing als motor voor leren: naar succeservaringen voor alle leerlingen. https://wij-leren.nl/toetsing-motor-voor-leren-succeservaringen.php

van der Sluis L, Van de Bunt-Kokhuis S, van der Bunt S (2009) Competing for talent

United Nations Sustainable Development Goals (n.d.) https://www.un.org/sustainabledevelopment/

Van Der Vleuten CP, Schuwirth LWT, Driessen EW, Govaerts MJB, Heeneman S (2015) Twelve tips for programmatic assessment. Med Teach 37(7):641–646

de Vries M, Gregersen-Hermans J, Keulers F, Seriere J, Benedik S (2022) The IB Lens. Internal policy document International Business School Maastricht

Wals AE (2014) Sustainability in higher education in the context of the UN DESD: a review of learning and institutionalization processes. J Clean Prod 62:8–15

Weybrecht G (2017) From challenge to opportunity–Management education's crucial role in sustainability and the sustainable development goals–an overview and framework. Intern J Manag Educ 15(2):84–92

Weybrecht G (2022) Business schools are embracing the SDGs–But is it enough? How business schools are reporting on their engagement in the SDGs. Intern J Manag Educ 20(1):100589

Integrating Environmental Sustainability in Construction and Real Estate Management Education

Erwin Mlecnik, Queena Qian, Ad Straub, Aksel Ersoy, Hilde Remoy, Vincent Gruis, Fred Hobma, Remon Rooij, Herman Vande Putte, Gerard van Bortel, and Monika Roeling

Abstract It is recommended to integrate specific management competencies in academic education to support the transition towards environmentally sustainable practices, particularly in the construction and real estate sector. This paper explores how architectural management education can integrate environmental sustainability within its current university programmes. In recent years TU Delft explored and experimented with various education initiatives to bring forward environmental sustainability knowledge and to connect with policy, societal and industry practices. This paper describes what we learned from both bottom-up and top-down initiatives implementing environmental sustainability in construction and real estate management education. Bottom-up educational initiatives show that knowledge about transition policies, stakeholder experiences, business models and management practices from a European perspective can help students across the globe to apply knowledge into their local context, reflecting on the overarching management principals across stakeholders, institutions, academic disciplines and cultures. Top-down initiatives show that the university has a vision on integrating sustainability in its curriculum, but that integrating environmental sustainability in construction and real estate management education is still challenging. Adapting the academic curriculum to integrate building and portfolio responses to environmental challenges might be a way forward and the experiences from numerous elective courses and educational initiatives can be a basis to identify the development of a future standard curriculum.

Keyword Education · Environmental sustainability · Real estate management · Architecture · Buildings · Construction · Urban development

---

E. Mlecnik (✉) · Q. Qian · A. Straub · A. Ersoy · H. Remoy · V. Gruis · F. Hobma · H. V. Putte · G. van Bortel
Department Management in the Built Environment, Faculty of Architecture and the Built Environment, TU Delft, P.O. Box 5043, 2600 GA Delft, The Netherlands
e-mail: e.mlecnik@tudelft.nl

R. Rooij · M. Roeling
Faculty of Architecture and the Built Environment, TU Delft, P.O. Box 5043, 2600 GA Delft, The Netherlands

T. Wall et al. (eds.), *Sustainability in Business Education, Research and Practices*, World Sustainability Series, https://doi.org/10.1007/978-3-031-55996-9_11

1 Introduction

Multiple scholars and experts (Grosseck et al. 2019) consider that students and teachers need to develop key competencies in sustainability, which can also guide academic program development and course design (Wiek et al. 2015; Wiek and Redman 2022). There is a broad agreement that students need to be enabled and empowered to effectively contribute to sustainability problem-solving (Brundiers et al. 2021). Particularly in engineering disciplines students need to know how to act on deep knowledge in the outside world on the fringes of technological sophistication, societal acceptance and politics (Kamp 2020). Particularly knowledge is needed regarding "environmental sustainability", interpreted by Morelli (2011) as an expansion of our common perception of the nature of *human activity* so as to more clearly connect it with the ecological concept of interdependence and to serve *as a goal for managers*.

Particularly, formulating and managing sustainable development in the form of 17 internationally accepted Sustainable Development Goals and 169 associated targets (UN 2015) present a formidable 'to do' list for the integration of learning objectives in the development of education (Vare et al. 2022).

In 2018, the buildings and construction sector accounted for 36% of final energy use and 39% of energy and process-related carbon dioxide (CO_2) emissions (IEA 2019). It is therefore particularly essential to deal with such goals in construction and real estate management education. At the same time, European policies (EPBD 2010; EPBD 2018; EED 2018; RED 2018; Green Deal 2019; Renovation Wave 2020; eu2020.de 2020) and international environmental sustainability goals (UN 2015; Paris Agreement 2015; UN 2017; UNFCCC 1997) drive countries towards the development of nearly zero-energy buildings including energy efficiency and renewable energy systems, as well as sustainable urban transformation. Furthermore, the sector also recognizes the need for circular, socially just and economically sustainable initiatives, as well as related business innovations. However, most challenges ahead require more of a management perspective instead of a design or policy perspective.

In architecture education, most attention is paid to designing energy-efficient or nearly zero-energy buildings and a low-carbon built environment (Gething and Puckett 2019; Oliveira et al. 2018; Altomonte 2009; Sassi 2006). But reflecting on the original definition from the report *Our Common Future* (Brundtland and Khalid 1987) one could consider in a more holistic manner the need for "managing" the needs of the present generation without compromising the ability of future generations to meet their own demands. From a sustainable housing perspective, this would for example suggest a need to identify housing management strategies in such a manner that this entails equal social and economic opportunity for present and future households, reflecting also on the ability of the already existing building stock to absorb the housing needs and the environmental limits to economic growth in housing portfolios.

To deal with the needed rapid innovation cycles, students need to be trained to provide a fast response to urgent dilemmas (Kamp 2020). They need to be enabled to navigate between conflicting stakeholder viewpoints, and manage change, and be willing and able to recognise and pursue opportunities for new value creation and problem solving in any organisational setting (Feyen et al. 2019). In the field of construction and real estate management therefore adapted education is needed to be able to manage the transition towards a sustainable built environment.

This paper aims to contribute to a current gap in knowledge and information regarding the degree of implementation of environmental sustainability in higher education, as well as an understanding of what challenges and opportunities universities have faced during attempts at implementation (Wright 2002), particularly for construction and real estate management education. After explaining the research approach, we explore experiences of integrating environmental sustainability in the education at the TU Delft, Faculty of Architecture and the Built Environment, using an integrated framework of competencies for advancing sustainability transformation (Wiek and Redman 2022). In a following section, experiences from some key bottom-up initiatives are identified that either were abandoned or led to a further development of integrating environmental sustainability in design, construction and real estate management education. In the section thereafter this paper scopes what happened simultaneously top-down on the higher level in the development of Bachelor and Masters programmes. After that, the paper discusses further on the specific implementation patterns and challenges that could lead to the further institutionalization of curricula including environmental sustainability (Weiss et al. 2021).

2 Research Approach

The research question put forward is: *How can construction and real estate management education curricula integrate environmental sustainability?*

The question is explored with a case study on the development of education at the Department of Management in the Built Environment at the Faculty of Architecture and the Built Environment of the TU Delft is chosen. The department of Management in the Built Environment (MBE) works towards a built environment where the interests of the end user and other stakeholders are key. "Management in the Built Environment" education at TU Delft focuses on integrating design, managerial, organisational, economic, financial, legal, technological, sustainable and policy aspects into development and operation processes aimed at adding value during the entire life cycle of buildings, areas and housing stock (BOSS 2023).

This research does not aim to give a complete picture on how education for construction and real estate management can implement and manage integrating environmental sustainability in university curricula. Nevertheless, the case is representative to research some key challenges and opportunities for integrating environmental sustainability in education as education at MBE focuses on providing the next generation of managers in the built environment, particularly with knowledge about solutions for the development and management of buildings, portfolios and urban areas (MBE 2023). The organisational sectoral classification includes three key domains: Design & Construction Management (DCM), Real Estate Management (REM), Urban Development Management (UDM). In this paper we focus on the development of education at REM and DCM and we illustrate also how educational initiatives emerging from UDM and other departments influence integration of environmental sustainability in REM/DCM education. Using DCM and REM as the overarching concepts allows for clearly positioning some of the courses discussed as either oriented towards mutation or towards operation, positioning environmental sustainability as part of engineering and procuring assets or operating and maintaining the assets portfolio.

Education for sustainable development can be implemented from the bottom-up, from the top-down, or both (Weiss et al. 2021). Therefore the following sub questions are addressed in the following sections:

- *What opportunities were detected during attempts of implementing environmental sustainability bottom-up?*
- *What challenges are faced when integrating environmental sustainability in the standard curriculum top-down?*
- *What challenges lie ahead for integrating bottom-up and top-down initiatives?*

3 Bottom-Up Education for Environmentally Sustainable Construction and Real Estate Management

Construction and real estate management education is strongly linked to social, economic, policy and environmental developments. This paper exemplifies experiences from some TU Delft courses that aimed to integrate environmental sustainability and that are standard for, or popular amongst, students construction and real estate management. Besides these developments there are also numerous initiatives to integrate social and economic sustainability, which are not covered in detail in this paper; in a later section this paper discusses the embedding. Earlier initiatives to integrate environmental sustainability in construction and real estate management education at TU Delft started from the notion that the development of sustainability in the built environment is inherently linked with changing technology management practices as used by demand, supply and policy actors. Therefore, education for sustainable housing developed a close relationship with–and knowledge from–these stakeholders and their practices to bring forward needed transitions. We exemplify some experiences on three levels: 1. Changes in the courses in the standard curriculum; 2. Development of elective courses, and 3. Development of other educational initiatives.

3.1 *Examples of Implementing Environmental Sustainability in the Standard Curriculum*

In most courses from DCM/REM there is attention for the sustainability agenda. Here we illustrate some courses that put environmental sustainability more central as part of the course goal.

3.1.1 Re-Design: From Area to Building Block

Re-design is a MSc course that aims to bring students' knowledge to analyse and re-design cases in a demanding, realistic context. Students are introduced to the structure and processes of management at multiple, interconnected levels involving individual buildings, larger complexes, the areas they belong to, the portfolios they belong to, relevant policies and governance. The course focuses on the development of essential cognitive and practical skills for applying management methods and techniques at these levels. The focus in this course is on understanding the effectiveness, efficiency, consistency, comprehensiveness and reliability of processes including reflection on various types of political, economic, social, technological, legal and environmental conditions.

In Re-design students are requested to examine a particular case also from various viewpoints, including environmental sustainability. Environmental sustainability is one of the public values that local and national governments focus on (in process and product) and that play a role in the design and development of real estate. Although environmental sustainability is not at the core of the course, it's importance is increasing and is seen as a value—next to societal and financial values.

3.1.2 Housing: Policy, Management and Sustainability

The MSc course "Housing" was the first in the Real Estate Management department that reserved specific attention for environmental sustainability as well as social sustainability and affordability. In this course, students learned about the way in which housing is provided for, and sometimes by, a variety of users and stakeholders, including (local) governments, professional organizations such as housing associations, individual households and self-organised groups. The core study topics were the housing market, user preferences, housing policies and institutions, the strategic portfolio management of housing providers; policies, measures and processes to increase the environmental sustainability of the housing stock; and policies and practices to improve the social sustainability of housing and neighbourhoods. Students defined sustainable housing challenges on various scales and in relation to social, economic, policy and environmental developments. As a learning goal, they were enabled to analyse and develop management strategies, particularly also from the perspective of various stakeholders and professions.

To support such learning objectives teaching activities explained and illustrated for example:

- Policy developments, particularly showcasing the effect of recent housing and energy policy measures on housing management practices;
- Existing housing management strategies from an environmental sustainability perspective, particularly analysing factors influencing the adoption of measures and strategies;
- Housing developments from a management perspective, particularly analysing the role of various stakeholders in implementing sustainability goals.

The course received excellent student ratings. A few years ago, when the course merged with the Real Estate Management course that had been running parallel with it for years, environmental sustainability was positioned therein as a goal and context of overall real estate management, next to safety, reliability, adequateness, sufficiency and efficiency. Several specific techniques to achieve environmental sustainability were added to the supporting management tools. Some parts of the Housing course were moved to a new elective course on social sustainability.

3.1.3 Urban and Infrastructure Redevelopment Game

The MSc course Urban and Infrastructure Redevelopment Game focuses on understanding the changing context of the global and local environment and economic, social and cultural elements that contribute to various urban problems. Students study the context, content, players and means of implementation during the cyclic phases of urban area development. To be able to do that, they play roles in project simulation and learn how to assess, analyse, research and improve operation practice in the field or urban development. The course aims to train students to grasp an integral approach when managing urban (re)development both at the urban area scale and at the portfolio and object scale. Through a role-playing simulation project, students are given design assignments that drive them to (re)develop a complex urban location with both residential and non-residential elements.

Environmental sustainability is an important part of the course as it runs as a red thread through the four dimensions of urban development, namely place, process, product and person. The term Product refers to the physical outcome of an area-based urban development project, i.e. the real estate, infrastructure, and/or public space that the project seeks to realize. Place refers to the geographic situation of a project, often sharply defined in spatial plans, briefs or contracts and signified by a specific history and distinct locational qualities. Person relates to the individuals involved in an intervention, acting either for themselves or in behalf of a larger group or organization (i.e. 'actors'), such as government bodies, real estate developers, investment agencies, businesses, pressure groups, or public private partners. The decisions and actions of these actors shape the content of an area-based project, and the negotiations between them are conceptualized as an on-going, dynamic, and interactive Process (van Bueren et al. 2016). Moreover, there are specific roles students play as part of their simulation game that focuses on environmental sustainability at the core. They are named as environmental law adviser and climate change and resilience officer.

3.1.4 Graduation Laboratory Management in the Building Environment

The Graduation Laboratory Management in the Built Environment was recently reconfigured to introduce specific thematic graduation labs, some of which also explicitly deal with environmental sustainability. The labs focus for example on management for a circular built environment or for achieving a faster energy transition of the existing building stock. So far, these specific labs attracted a few dozens of students. However, it is observed that for current student generations, sustainability is no longer a special theme. They include the topic in most of their reflections and course work. Even if environmental sustainability is not the main theme of a paper, thesis or discussion, it is often present as part of the goal or process.

3.2 Development of Elective Courses with a Focus on Environmental Sustainability

3.2.1 Solar Decathlon Europe

The Faculty of Architecture and the Built Environment of the TU Delft also regularly participates in the Solar Decathlon, which is offered as an elective management course. The course enables students to work together in an interdisciplinary manner with teachers and students from multiple departments and faculties. Initiated in 2002 by the United States Department of Energy (US DOE, n.d.), the Solar Decathlon is a university-level student competition for resource-responsible and energy-efficient architecture and engineering in the building sector. The competition challenges universities from all over the world to design, build and operate a solar-powered and energetically self-sufficient home equipped with advanced technologies, and designed to the highest standards of sustainability (SDE 2021). Award criteria such as reflections on affordability, viability, and sustainable innovation are typically developed and covered by students from Construction and Real Estate Management. Influenced by construction and real estate management strategies, student teams chose for radically different thinking compared to newly-built home construction, for example, the demonstration of a nearly zero-energy home renovation, the transformation of an office building unit and the development of topping-up units for responding to the increasing housing needs.

3.2.2 Climate Proof Sustainable Housing Renovation

In this course the renovation process of a housing complex is unravelled as a challenging social-technical system. Students gain knowledge of sustainable renovation from different perspectives and apply gained insights and knowledge in conducting a feasibility study. The interests of all involved stakeholders in housing renovation have to be merged to result in a successful renovation, the process as well as the end result. This course is organised as in-depth lectures covering all perspectives of integral climate proof adaptations of housing and a tender game for the renovation of a housing complex. The award criteria of the tender cover stakeholder involvement and process management, energy performance, environmental performance, health & comfort, financial case for housing association and tenants, and the integration of all the perspectives.

3.2.3 Social Sustainability in Human Habitats

In this newly established elective course, students learn about the role of housing to achieve socially sustainability human habitats in different geographical contexts. This elective course focuses on maintaining and improving social inclusion and wellbeing in housing and living environments and teaches four core dimensions: (1) the social preconditions for sustainable development (i.e., values, habits and rules); (2) the equitable distribution and consumption of housing; (3) the quality of social relations in housing and living environments; and (4) the physical conditions or liveability of housing and living environments. In parallel to the introduction of this course there was also a development track to establish an elective on Environmental Sustainability in Human Habitats.

3.2.4 Energy-Friendly Renovation Processes

This case-based elective blended course was based on developing process management competencies for high quality and energy ambitious renovation projects within the existing housing stock. The goal of the course was amongst other to guide students through interpreting and managing the complexity of the processes and taking a position from different stakeholder perspectives. The course pays attention to technical and financial factors, however the main focus lies on other success factors like the skills to work with various parties with different interests on ambitious goals. Students develop a real case solution, reflect in an on-line environment with involved professionals, and hold face-to-face meetings with stakeholders.

The course thus delivered student competencies to identify and eliminate barriers for sustainable housing renovations, particularly looking at:

- The impact of changing policies on decision-making, with a focus on the implementation of innovative measures and processes to increase the environmental sustainability of the housing stock;
- The business models of stakeholders with a focus on solving split-incentive (tenant-landlord) problems and identifying financially viable business cases that are applicable for market up-scaling;
- The influence of stakeholder interests on management practices and the opportunities for stakeholder dialogue;
- Real-life examples of social practices to improve the development, affordability and sustainability of housing renovation at various scales.

Although the course was highly relevant and rewarded by the sector as outstanding blended educational initiative, in the end it did not attract enough students any more to pursue the needed yearly educational effort.

The elective courses are situated in a highly competitive framework where students can also choose courses from other departments and faculties, some of which also focus on environmental sustainability. For example, other TU Delft elective courses are available regarding climate change, ocean energy, fuel cell systems, wind energy, waterpower, solar energy, renewable energy systems, geothermics, geo-design for a circular economy, aesthetics of sustainable architecture, green building, zero energy design, recycling engineering materials, environmental physics, urban sustainability, innovation and sustainability, sustainable urban engineering of territory, architectural sustainability, ecological engineering, water management, sustainable business models, strategic and sustainable design, sustainable innovation and social change, sustainable innovation and transitions, climate ethics, sustainable energy innovation and transitions, and so on. The sheer number of electives illustrates the complexity of course alignment. In reality—besides their standard curriculum–students only have a limited amount of time to grasp the complexity of environmental sustainability issues.

3.3 New Educational Approaches with a Focus on Environmental Sustainability

3.3.1 Summer School Sustainable Housing from a European Perspective

The department manages a summer course where, during two weeks, students gain insight into the European perspective on the development of sustainable housing stock, and speeding up energy transition goals, using European and national goals, policies, regulations, assessment tools, process innovations, behavioural, institutional and management approaches. The course also embeds reflection on emerging economies and on the role of institutions and urban contexts in less developed countries. The course mixes Bachelors, MScs, PhDs and professionals from all around the world using groups based on a mix of nationality, educational background and gender. Interaction with practice was also provided with a visit to a project example involving residents, professionals and staff and a building innovation centre.

Within the Summer School, the GO2Zero Game introduces a simulation environment to support understanding transition processes in a safe playful environment for students and experts to jointly experiment with process challenges (Bekebrede et al. 2018). The Game particularly represents the transition towards a renewable energy system in urban districts as strongly embedded within society, therefore requiring a thorough analysis of the stakeholder complexities (Bekebrede et al. 2018). Playing the game helps students to formulate stakeholder viewpoints and collaborative strategies for collaboratively achieving energy transition goals. The simulation game was introduced for further experimentation as part of the curriculum of a summer school.

Winning a Tender was developed to be a core objective of the whole Summer School. It was developed as an active teaching method for gradually improving environmental sustainability in management processes. In a competitive environment, groups of students are asked to develop and defend a plan for a sustainable housing renovation responding to a "tender" with political, economic, social, technological perspectives. Proposals prepared by the students are regularly reflected by peers, tutors (who also sometimes act as social media influencers) and a professional 'jury', that takes up stakeholder roles for evaluating predefined award criteria. Students finally also reflect lessons on their own national context, preparing take-home messages for local development.

As an experiment a one-week summer school was also tailor-made for Chongqing University, China at TU Delft and later on site at Xi'An University. In this course students were also challenged to connect with local residents to understand their renovation needs, and to interact with speakers during a high-end policy development workshop. Bringing awareness to local graduates and professionals and using a local project case in particular, made a significant impact not only on the way students perceived sustainability challenges in their own context, but also at a societal level on how they could change the construction culture locally.

3.3.2 Massive Open Online Courses (MOOC)

The Faculty of Architecture and the Built Environment of the TU Delft also provides successful massive online courses, some of which pay specific attention to environmental sustainability and might attract students from REM/DCM. The courses *Zero Energy Design* and *Advanced Zero-Energy Design* teach international students who possess basic technical knowledge how to make a building more sustainable and net zero energy in different climates using both passive and active measures. Students use their own home or building as a case. In the advanced course they also validate their design using energy performance calculations. The course *Circular Building Products for a Sustainable Built Environment* aims more towards professionals and shows opportunities that circular building products can offer, investigating various perspectives (design, technology, business case, implementation methods).

The course *Sustainable Urban Development*—embedded in the Amsterdam Institute for Advanced Metropolitan Solutions (AMS), a cooperation of TU Delft and Wageningen University—introduces first steps regarding urban challenges using a holistic, systemic and transdisciplinary approach, and integrating knowledge from urban planning, urban design, urban engineering, systems analysis, policy making, social sciences and entrepreneurship. Other courses that are offered are for example *Co-creating Sustainable Cities*, which focuses on citizen interaction, and *Rethink the City*, which focuses on new approaches to local urban challenges and sustainable urban transitions in the Global South.

4 Changes in the Educational Context at the University and Faculty Level

As exemplified above, environmental sustainability challenges were gradually bottom-up integrated in academic courses and educational initiatives at the TU Delft, to fit a growing demand to integrate environmental sustainability in construction and real estate management education. The goal of the educational initiatives was to prepare future professionals not only with immediate knowledge, but also with a flexible mind-set to adapt know-how to real applications. With such a methodological capacity students can be prepared to make their impact for achieving sustainable goals in their future professions. Based on these experiences as well as the sharing of experiences from the departmental and educational programming, TU Delft nowadays responds to sustainability challenges with formally stated goals.

The Bachelor Education of the Faculty of Architecture and the Built Environment (TU Delft 2022:1) now states that a student with a Bachelor Degree should be able *to connect human needs and the environment from a perspective of sustainability*, including design, technique and society and have knowledge and understanding about the life cycle of the built environment including initiative and programme, (re)design, (re)development and (re)use. The Bachelor programme also foresees basic knowledge about, amongst other, urban and building physics, climate design, climate systems in buildings and global dimensioning of urban mobility, green and water infrastructures. A Bachelor student is further trained to identify societal urgencies, digital opportunities and to critically and ethically reflect, research and communicate. Scientific and systemic approaches are further elaborated in the Master's programme where students can specialize in Architecture, Urbanism, Architectural Engineering and Technology, Landscape Architecture, and Management in the Built Environment. The education in Management in the Built Environment (MBE 2023) focuses explicitly on working with end users and stakeholders towards a sustainable built environment. Amongst other, Master students explore and assess solutions for the development of management of buildings, portfolios and urban areas.

According to the setting of Masters education (TU Delft 2022:2), an MBE graduate has thorough knowledge of design and construction management (DCM), real estate management (REM), housing management (HM) and urban development management (UDM). An MBE graduate achieves advanced knowledge, intellectual and inquisitive skills and competencies for understanding, researching and analysing societal, legal, financial, economic, commercial, entrepreneurial, policy and informational processes and procedures (TU Delft 2022:2). Combined with design skills and collaborative settings, this supports MBE graduates to design, plan, and decide about appropriate context-sensitive management processes, business models and governance strategies in a particular context (TU Delft, 2022:2).

5 Discussion: Detected Educational Challenges and Opportunities

The research confirms some critical factors that can enable universities to embed environmental sustainability into university areas, such as: a strong policy environment, resourcing of strategies, and encouragement of leaders and environmental sustainability advocates (Ralph and Stubbs 2014). For universities to comprehensively address sustainability, a 'learning for sustainability' approach needs to be embedded across every aspect of institutional operations in a synergistic way (Ralph and Stubbs 2014). In the field of construction and real estate management education it is perceived that environmental and social sustainability are still poorly embedded compared to the economical perspective.

The various initiatives respond to needed changes regarding engineering disciplines as pointed out by (Kamp 2020) as well as the moral and ethical responsibility of universities to develop and facilitate the use of the new technologies to empower society, particularly with regards to social change and social justice (Kamp 2020). Educational initiatives develop social consciousness by getting familiar with the human aspects and ethical issues, and interdisciplinary, systems thinking and interpersonal communication skills. Instructors are regarded as facilitators and collaborators rather than sources of authority (El-adaway et al. 2015).

Our experiences confirm that to achieve more comprehensive education for sustainable development, open communication among all stakeholders should be facilitated (Weiss et al. 2021). Feedback as well as reflection should be encouraged, and strong informal collaboration and communication can compensate for a lack of formalized leadership support from the top (Weiss et al. 2021).

Given the fast progress of sustainable technologies, design solutions and management practices, a constant renewal and flexibility of curriculum design is perceived, which is mainly focused in the development of elective courses and new educational initiatives. The theme of environmental sustainability inherently brings interests from colleagues across departments and faculties and their expertise on sustainable management topics could be more integrated, looking systematically at technological, financial, economical, management, policy, social, design, institutional, behavioural, and other perspectives. The adoption of innovative solutions and sustainable measures by various stakeholders as presented in electives would demands a systemic review of changes in the standard curriculum that ought to be made institutionally and behaviourally.

Combining the strengths of the various bottom-up educational initiatives and aligning their content would certainly benefit developing a more integrated view on implementing education for sustainable construction and real estate management. Next to energy and material sustainability, social sustainability and affordability are being more emphasized in management education. One way to approach these challenges could be to identify the educational needs on various scales (building, portfolio, district), with particular attention for various related core stakeholders interests. Now environmental sustainability is mainly positioned as one of the many goals or context aspects of management in the built environment that are related to other goals, such as safety, reliability, adequateness, affordability, sufficiency and efficiency of the designed, constructed and used buildings, assets and portfolios.

Looking at the perceived instability and wide range of elective courses, a more holistic view on addressing environmental sustainability challenges seems to be needed in the standard curriculum, including for example the development of socially sustainable practices, changing policy developments, bottom-up stakeholder initiatives, process innovations, supply side perspectives and collaborative entrepreneurship, just transition and spatial justice challenges, economical behaviour, design and management of public goods, and so on.

Given the need to use integrated approaches to achieve environmental sustainability, thematic overlap now can occur in various courses, leading to a perceived lesser need for new elective courses on environmental sustainability. On the other hand, new active teaching methods such as those developed in the Summer School, student competitions such as the Solar Decathlon, and MOOC's are popular for students. Moreover, there is a trend to test sustainability principles within a specific local context and to analyse and reconstruct decision-making processes. Enthusiasm for cross-border, inter-disciplinary and intra-disciplinary working was also detected with students, professionals and teachers alike. More emphasis could still be placed on how to effectively incorporate blended and on-line education in management education, although this might increase the needed staff effort.

Moreover, internal reflection groups showed that there is an ongoing debate whether sustainability should be a separate specialisation or integrated in the standard curriculum. Pathway for new specialisations can be challenging as they might be perceived as competing with regular programmes; initiatives have to prove that a sufficient number of students can be reached. In practice, temporarily optional courses can provide a response to the curriculum needs, as new subjects need time to gain their place in the curriculum. Further elaboration of principles and implementation of teaching methods from isolated initiatives within the Master and Bachelor education could lead to optimization of the standard curriculum. This should be strategically planned in co-creation and well-balanced with shared responsibilities. Course revisions can be related to the management of construction phases from initiative to development with a focus on the building scale, or related to the phases from maintenance to programming of operations with a focus on the asset and portfolio level.

6 Conclusion

There is a high need for competencies to manage visions, tactics and operations including energy and sustainability requirements at the level of individual buildings, facilities, building portfolios, districts and energy systems. This requires the development of an adapted academic curriculum for construction and real estate management. This paper explored how construction and real estate management education is integrating environmental sustainability, using as a case the development of education on Management in the Built Environment at the TU Delft Faculty of Architecture and the Built Environment.

Bottom-up initiatives showed that the developed education responds to design, technological, business and urban sustainability challenges, managing stakeholder relations and developing strategies to cope with the energy transition and other policy and governance challenges. Experiences show that facilitating behaviour change, adapting management practices to include stakeholder dialogue, innovation management for sustainable business development, and projected institutional revisions based on policy development can be core issues for the curriculum development. This can be facilitated by students working in multidisciplinary teams and competitive simulation environments, and teachers empowering participative teaching methods, vision formation and strategic thinking, and interdepartmental, cross-faculty and cross-cultural knowledge exchange.

On the other hand, environmental sustainability education is perceived as being still incomplete and scattered in various courses with a certain lack of alignment top-down. The university has a vision on integrating sustainability in its curriculum, but currently bottom-up environmental sustainability initiatives are more likely to be positioned in elective courses and MOOC's, which has the disadvantage that only a limited amount of regular students are reached. Standard curriculum changes are often difficult to achieve as the sizes of the Bachelor and Master programme are fixed. Adding a new theme to the curriculum would imply that other themes get less attention.

In conclusion, this paper showcased that further efforts are still needed for systematically introducing environmental sustainability in education. The research finds that middle-out it could be explored further how the regular curriculum can be improved based on the experiences from the various bottom-up initiatives and top-down developments, exploring the connections between research, education and application of environmental sustainability on various scales and levels.

Acknowledgements The authors would like to thank the staff members and students of the Faculty of Architecture and the Built Environment of the TU Delft for their input and numerous tips and suggestions for curriculum improvements.

References

Altomonte S (2009) Environmental education for sustainable architecture. Rev Eur Stud 1(2):Art.2. https://doi.org/10.5539/res.v1n2p12

Bekebrede G, van Bueren E, Wenzler I (2018) Towards a Joint local energy transition process in urban districts: The GO2Zero simulation game. Sustainability 10(8):2602. https://doi.org/10.3390/su10082602

BOSS (2023) Education management in the built environment. Boss, TU Delft. https://www.bosstudelft.nl/education/mbe/#:~:text=The%20Management%20in%20the%20Built,buildings%2C%20areas%20and%20housing%20stock. Consulted 9 Feb 2023

Brundiers K, Barth M, Cebrián G, Cohen M, Diaz L, Doucette-Remington S, Dripps W, Habron G, Harré N, Jarchow M, Losch K, Michel J, Mochizuki Y, Rieckmann M, Parnell R, Walker P, Zint M (2021) Key competencies in sustainability in higher education—toward an agreed-upon reference framework. Sustain Sci 16(1):13–29. https://doi.org/10.1007/s11625-020-00838-2

Brundtland GH, Khalid M (1987) Our common future. Oxford University Press

EED (2018) Directive (EU) 2018/2002 of the European parliament and of the council of 11 December 2018 amending Directive 2012/27/EU on energy efficiency (Text with EEA relevance). EP, CONSIL, 328 OJ L (2018). http://data.europa.eu/eli/dir/2018/2002/oj/eng.

El-adaway I, Pierrakos O, Truax D (2015) Sustainable construction education using problem-based learning and service learning pedagogies. J Prof Issues Eng Educ Pract 141(1):05014002. https://doi.org/10.1061/(ASCE)EI.1943-5541.0000208

EPBD (2010) Directive 2010/31/EU of the European parliament and of the council of 19 May 2010 on the energy performance of buildings (recast). EP, CONSIL, 153 OJ L. http://data.europa.eu/eli/dir/2010/31/oj/eng

EPBD (2018) Directive (EU) 2018/844 of the European parliament and of the council of 30 May 2018 amending directive 2010/31/EU on the energy performance of buildings and directive 2012/27/EU on energy efficiency (Text with EEA relevance). PE/4/2018/REV/1. https://eur-lex.europa.eu/legal-content/EN/TXT/?uri=celex%3A32018L0844

eu2020.de (2020) The New Leipzig Charter. https://eurocities.eu/latest/what-is-the-new-leipzig-charter/. Consulted 9 Feb 2023

Feyen W, Debackere K, Olivares M, Gfroerer R, Stam E, Mumby-Croft B, Keustermans L (2019) Student entrepreneurship at research-intensive universities - From a peripheral activity towards a new mainstream. League of European Research Universities. https://www.leru.org/publications/student-entrepreneurship-at-research-intensive-universities-from-a-peripheral-activity-towards-a-new-mainstream. Consulted 9 Feb 2023

Green Deal (2019) Communication from the commission to the European parliament, the council, the European economic and social committee and the committee of the regions. The European Green Deal, COM/2019/640 final. European Commission. https://eur-lex.europa.eu/legal-content/EN/TXT/?qid=1576150542719&uri=COM%3A2019%3A640%3AFIN

Gething B, Puckett K (2019) Design for climate change. RIBA Publishing

Grosseck G, Țîru LG, Bran RA (2019) Education for sustainable development: evolution and perspectives: a bibliometric review of research 1992–2018. Sustainability 11(21):6136. https://doi.org/10.3390/su11216136

IEA (2019) Global Status Report for Buildings and Construction 2019. International Energy Agency. https://www.iea.org/reports/global-status-report-for-buildings-and-construction-2019, consulted 9 February 2023.

Kamp A (2020) Navigating the landscape of higher engineering education subtitle: coping with decades of accelerating change ahead. 4TU.Centre for Engineering Education

MBE (2023) Management in the built environment. https://www.tudelft.nl/en/architecture-and-the-built-environment/about-the-faculty/departments/management-in-the-built-environment. Consulted 9 Feb 2023

Morelli J (2011) Environmental sustainability: a definition for environmental professionals. J Environ Sustain 1(1):1–10. https://doi.org/10.14448/jes.01.0002

Oliveira S, Marco E, Gething B (2018) Energy-efficient design and sustainable development. In: Leal Filho W (Red) Encyclopedia of sustainability in higher education. Springer International Publishing, pp 1–10. https://doi.org/10.1007/978-3-319-63951-2_130-1

Paris Agreement (2015) Paris agreement. United Nations. https://unfccc.int/process-and-meetings/the-paris-agreement/the-paris-agreement. Consulted 9 Feb 2023

Ralph M, Stubbs W (2014) Integrating environmental sustainability into universities. High Educ 67(1):71–90. https://doi.org/10.1007/s10734-013-9641-9

RED (2018) Directive (EU) 2018/2001 of the European parliament and of the council of 11 December 2018 on the promotion of the use of energy from renewable sources (recast) (Text with EEA relevance). EP, CONSIL, 328 OJ L. http://data.europa.eu/eli/dir/2018/2001/oj/eng

Renovation Wave (2020) Communication from the commission to the European parliament, the council, the European economic and social committee and the committee of the regions, a renovation wave for Europe—greening our buildings, creating jobs, improving lives. (testimony of Renovation Wave). https://eur-lex.europa.eu/legal-content/EN/TXT/?qid=1603122220757&uri=CELEX:52020DC0662

Sassi P (2006) Strategies for sustainable architecture. Taylor & Francis. https://doi.org/10.4324/9780203480106

SDE (2021) Solar decathlon Europe. TU Delft, The Netherlands. https://www.delftsolardecathlon.com/en/solar-decathlon-europe. Consulted 9 Feb 2023

TU Delft (2023) Zero energy design: an approach to make your building sustainable. https://online-learning.tudelft.nl/courses/zero-energy-design/. Consulted 9 Feb 2023

TU Delft (2022–1) Studentenstatuut Bacheloropleiding Bouwkunde Studiejaar 2022–2023. Faculteit Bouwkunde, TU Delft, The Netherlands, pp 1–49

TU Delft (2022–2) Student charter master of science architecture, urbanism and building sciences, Academic Year 2022–2023. Faculty of Architecture and the Built Environment, TU Delft, The Netherlands, pp 1–52

US DOE (2023) Solar decathlon. U.S. Department of Energy, United States. https://www.solardecathlon.gov/. Consulted 9 Feb 2023

UN (2015) Transforming our world: the 2030 agenda for sustainable development. United Nations. https://sdgs.un.org/publications/transforming-our-world-2030-agenda-sustainable-development-17981. Consulted 9 Feb 2023

UN (2017) New urban agenda. Habitat III, United Nations, A/RES/71/256, ISBN: 978–92–1–132731–1. https://habitat3.org/wp-content/uploads/NUA-English.pdf. Consulted 9 Feb 2023

UNFCCC (1997) Kyoto protocol to the United Nations framework convention on climate change. UNFCCC. https://unfccc.int/documents/2409. Consulted 9 Feb 2023

Vare P, Rieckmann M, Lausselet N (2022) Introduction. In: Vare P, Lausselet, N, Rieckmann M (Red) Competences in education for sustainable development. Springer International Publishing, pp 3–10. https://doi.org/10.1007/978-3-030-91055-6_1

van Bueren E, Daamen T, Chen Y, Franzen A, Heurkens E, Hobma F, Verheul WJ (2016). Urban development management—past, present and future. In: Arkesteijn M, Van der Voordt T, Remoy H, Chen Y (Eds) Dear is durable: liber Amicorum for Hans de Jonge. TU Delft, pp 281–289

Weiss M, Barth M, von Wehrden H (2021) The patterns of curriculum change processes that embed sustainability in higher education institutions. Sustain Sci 16(5):1579–1593. https://doi.org/10.1007/s11625-021-00984-1

Wiek A, Bernstein M, Foley R, Cohen M, Forrest N, Kuzdas C, Kay B, Withycombe Keeler L (2015) Operationalising competencies in higher education for sustainable development. In: Barth M, Michelsen G, Rieckmann M, Thomas I (eds) Handbook of higher education for sustainable development. Routledge, pp 241–260

Wiek A, Redman A (2022) What do key competencies in sustainability offer and how to use them. In: Vare P, Lausselet N, Rieckmann M (Red) Competences in education for sustainable development. Springer International Publishing, pp 27–34. https://doi.org/10.1007/978-3-030-91055-6_4.

Wright TSA (2002) Definitions and frameworks for environmental sustainability in higher education. High Educ Pol 15(2):105–120. https://doi.org/10.1016/S0952-8733(02)00002-8

Meeting Stakeholder Needs Through Authentic Assessment of Carbon Literacy and Digital Skills

Ann Hindley

Abstract This chapter explores the embedding of Carbon Literacy and digital skills into a UK undergraduate tourism module which is designed to meet tourism employer needs and expectations of graduate employability skills. Carbon Literacy is an awareness of the carbon dioxide costs from everyday activities and it scaffolds learning on the complex relationship between climate change and tourism. An authentic assessment of the key learning outcomes is based on the identified needs of a tourism destination and illustrated via easy-to-use and freely available digital software such as an App, narrated video or website builder.

Keywords Climate change · Carbon literacy · Digital skills · Authentic assessment · Mobile app

1 Introduction

Green and digital skills gaps have been identified by a multi-disciplinary partnership of European universities, industry partners and tourism sector representatives within destination management, accommodation, visitor attractions, travel agencies, tour operators, food and beverage operations (Next Tourism Generation Alliance 2022). Employers look to recruit work-ready (Carlisle et al. 2021a, b) tech-savvy graduates who are flexible, creative and effective communicators (Jopp 2020) but who also have sustainability knowledge (Carlisle et al. 2021a, b). Digital fluency is considered necessary to solving sustainability problems, as basic literacy is insufficient in 'a digitally mediated society' (Alexander et al. 2019, p. 14). Therefore, the local and interdisciplinary context of climate change is increasingly important to effective university teaching (Devlin and Samarawickrema 2022), as is the need for developing digitally competent people in a constantly changing digital environment (Oberländer et al. 2020).

A. Hindley (✉)
Liverpool Business School, Liverpool John Moores University, Liverpool, UK
e-mail: a.m.hindley@ljmu.ac.uk

T. Wall et al. (eds.), *Sustainability in Business Education, Research and Practices*, World Sustainability Series, https://doi.org/10.1007/978-3-031-55996-9_12

Higher education will need to evolve and adapt to changes in the global ecosystem by integrating the Sustainable Development Goals into teaching and offering valuable and relevant learning experiences that prepare students for the future (Pelletier et al. 2022). However, Devlin and Samarawickrema (2022) wonder how university educators will engage in the task of integrating the complex or 'wicked' problems of climate change into the curricula, as wicked problems are unstructured, cross-cutting, and relentless (Weber 2008). The need for graduates to acquire systems-thinking cross-cutting skills, in addition to critical-thinking and lateral-thinking skills is also identified, along with a collaborative approach between universities and businesses in developing sustainability training and course materials (Change Agents UK et al. 2020). In addition, Molthan-Hill and Blaj-Ward (2022) make a specific call for climate-relevant assessments that create climate solutions, which suggests the need for authentic assessment.

Digital literacy comprises the skills and competencies required to use digital technology in work, learning and leisure and a digital mindset that develops critical thinking and reflective life-long learning (Martzoukou et al. 2020). However, at many universities "digital literacy is either taken for granted or assumed to be at an adequate level" (Murray and Pérez 2014, p. 85). Therefore, practice and policy change in higher education is needed for digital development that will prepare graduates for the "digitally immersive industry workplace" (Morgan et al. 2022, p. 271). Teaching, learning and assessment needs to include technical and operational functionality of hardware, software, and digital platforms, in addition to the cognitive skills of analysing, evaluating, communicating (Morgan et al. 2022). Nevertheless, there is limited knowledge on the importance and role of new technologies in authentic assessment (Jopp 2020) and a lack of disseminated information on the use of technology-enhanced assessment (Keppell et al. 2015).

2 Authentic Assessment

To provide sustainability and digital employability skills, "teaching and assessment needs to focus on shifting from 'recall-on-paper' to 'do-it-for-real'" (Knight and Drysdale 2020, p. 58). This requires design of meaningful assessments that are authentic and use contemporary digital tools, replicating work-based tasks that have real-world relevance and diverse outcomes (Devlin and Samarawickrema 2022). Although there is no widely accepted definition of authentic assessment, the themes focus on industry-related real-world assessment tasks which have clear links to intended learning outcomes; tasks that motivate, engage, and challenge students while developing critical thinking and problem-solving skills; and tasks that aim for integration of cross-disciplinary knowledge through multiple assessment methods and with multiple solutions (Jopp 2020, p. 945). These are reflected in the continuum by (Mueller 2018) which defines the attributes of traditional and authentic assessment. In essence, authentic assessment can positively influence student engagement (Jopp 2020) when it is work-related or real-world in nature (Porcaro et al. 2016) and encourages iterative development-work or learning opportunities (Boud et al. 2018) that result in worthwhile and genuine outputs (Herrington and Herrington 2006).

Nevertheless, Jopp (2020) also suggests more clarity is needed on why higher education should implement authentic assessment. Herbert et al. (2020) usefully review the long-running debate on responsibility for work-readiness and whether this lays with employers or higher education. Further reasons for not engaging with authentic assessment include lack of clarity on how learning outcomes would develop employability skills (Jopp 2020); institutionalised approaches to teaching (Villarroel et al. 2018); institutional systems and standardisation (Mimirinis 2019); resource constraints and the need for efficient marking processes (Hart et al. 2011); and the lack of connectedness to industry (Villarroel et al. 2018). In addition, a fear or reluctance of staff to innovate through digital practice (ibid.) can be due to a variation in digital skills (Hamalainen et al. 2021), digital competence (Hilli 2020), age (Hamalainen et al. 2021), genders (Alanazi 2019) and value placed on digital (Lu and Han 2022).

Students can also find authentic assessment using digital technologies challenging. Innovative summative assessment design, such as blogs enhanced by multimedia, should not be constrained by traditional submission methods that require conversion from digital to essay format (Cartner and Hallas 2020; Knight and Drysdale 2020). This is not a pedagogy-first approach where the appropriate technology has been selected to fit the teaching and learning approach (Matthews 2021) and is likely related to institutional processes designed to limit plagiarism and contract cheating in traditional assessment formats (Bretag et al. 2020; Knight and Drysdale 2020). Although formative assessments provide opportunities to be used 'as' and 'for' learning (National Forum for the Enhancement of Teaching and Learning in Higher Education 2017), from the student perspective they are poorly motivated and find no incentive in doing extra work for formative tasks that are not graded (Meedya 2019). Further, a lack of digital skills is a disincentive to engage in formative assessment tasks where more time can be spent on learning the technology than on the task itself (ibid).

The transformative use of digital technologies is challenging (Blundell et al. 2020). It impacts the comfort and passive engagement with traditional learning and assessment that can result in first-order, surface, or cognitive learning (Meedya et al., 2019). In active learning the intrinsic motivation, self-belief, self-determination, and autonomy of students is augmented with supportive peer learning and guidance by tutors (Zepke and Leach 2010). This active engagement leads to second-order, transformative, or meta learning that generates original work (Meedya et al. 2019; Sterling 2011) while developing confidence, competences, and resilience (Zepke and Leach 2010). Students' ability to adapt and persevere to an unstructured learning environment (Schrum et al. 2021) can assist with a more effective transition to the workplace (Herbert et al. 2020).

3 Digital and Climate Change in Assessment

The importance and role of new technologies in authentic assessment and the use of technology-enhanced assessment is under-investigated (Jopp 2020; Keppell et al. 2015) and Table 1 shows only nine returns from a Web of Science (WoS) search. Digital is often used to improve digital skills (Nieminen et al. 2022) and it could be argued that this is achieved when an employability assessment uses a digital viva (Scott and Unsworth 2018) or personal and professional reflection uses video (Sargent and Lynch 2021). However, authentic assessments can provide scope for needed digital and creativity skills through the presentation of highly engaging visuals for marketing (Spanjaard et al. 2022) and financial data (van Rensburg et al. 2021). Nevertheless, there is recognition that sector specific digital software and skills, such as in construction, is lacking in the curriculum due to funding and upskilling issues (Balogun et al. 2021).

Authentic assessment and climate change is also under-investigated, and Table 1 shows zero returns in a Web of Science search. This corresponds with findings by Molthan-Hill and Blaj-Ward (2022) that identified limited literature on climate change learning assessment in higher education. Despite examples of authentic tools being used within teaching and learning (Emblen-Perry 2019) there is little consideration of assessments that explore solutions (Cross and Congreve 2021) or the effectiveness of different assessment tools (Cebrián et al. 2019). This seems at odds with a system seeking to build student agency and enhance employability skills through authentic assessment. It nonetheless reflects institutional and societal contradictions that can place students somewhere between worry and hope, and fragility and resilience (Hiser and Lynch 2021). Molthan-Hill and Blaj-Ward (2022) therefore call for climate-relevant assessments that create climate solutions, while Chankseliani et al. (2021) identify the need for research that gauges the impact of activities on society.

Table 1 Web of science search

Search criteria Abstract + English + Article + 2018–2022	Total returns	Exclusions	Reviewed or scanned
"authentic assessment" and "digital" and "climate change"	0	0	0
"authentic assessment" and "climate change"	0	0	0
"authentic assessment" and "digital"	12	3	9
"assessment" and "digital" and "climate change"	116	116*	0
"teaching" and "learning" and "assessment" and "climate change"	22	19*	3
"teaching" and "learning" and "assessment" and "digital"	367	45**	322

* 'assessment' here does not relate to or does not focus on 'student assessment'

**duplicates

A further gap in the literature relates to student assessment that combines digital and climate change (Table 1). Despite the under-investigation of climate change assessment, digital is used in the broader teaching, learning and assessment landscape. The 322 abstracts cover the Web of Science categories education, ICT, business, engineering, environment, medical and healthcare. The top four journals and numbers of articles are Education Sciences (14), International Journal of Emerging Technologies in Learning (7), Sustainability (7) and BMC Medical Education (6). A scan of the abstracts indicates articles tend towards assessment of digital tools in teaching and learning, rather than digital is used in assessment.

4 Digital Tools in Teaching, Learning and Assessment

Covid19 accelerated the digital transition in some institutions, others were already using Learning Management Systems (LMS), Virtual Learning Environments (VLE), video conferencing and collaborative platforms and blended learning to some extent (Table 2). The LMS/VLE reduces memory requirements and promotes learning through the use of integrated activities (Sastre-Merino 2021). As a repository the LMS/VLE can be used for asynchronous recordings and uploaded tasks (Kanetaki 2022) or the recording of weekly formative group work (Tang 2022). Blended learning draws together online and face-2-face components, which allowed medical students on their internship to receive online competence-oriented seminars for their electives (Streitlein-Bohme et al. 2021).

A range of decentralised technologies now enhance student engagement (Mimirinis 2019). Social media platforms and communications tools are used within teaching delivery. Assessment tools include multi-media, audio and e-assessment tools (Table 3). Podcasts have been used for active participation in learning within clinical chemistry (Baecker 2022) and as an option for a formative assessment in nursing (Meedya et al. 2019). Story Maps (digital storytelling technology) which build narrative around digitized dynamic maps, images and videos have been used for a group summative assessment on qualitative research in geography (Dickinson and Telford 2020). e-Assessment has been used in modules such as accounting (Helfaya 2019).

Simulations can train students in real-life situations (Cutumisu et al. 2020) or use professional tools such as GIS (Zhang 2018). Gamification can enhance engagement through collection of points, badges and rewards (Campillo-Ferrer et al. 2020), with serious games building skills and knowledge (Elsherbiny and Al Maamari 2021) through Apps such as Kahoot, Wordwall and Mentimeter (Zhyhadlo 2022). However, although student self-assessment often suggests good digital skills (Cho et al. 2021; De Rossi and Restiglian 2019; Freire and Rodriguez 2022; Sailer et al. 2021; Vereshchahina et al. 2018), digital competences show a lack of "digital creation skills…designing apps, digital games, virtual environments, and interfaces" (Martzoukou et al. 2020, p. 1433). This suggests a need for change that allows "for the creation of new tasks, previously inconceivable" (Puentedura 2006).

Table 2 Systems, conferencing and blended learning

Learning management systems (LMSs) and virtual learning environments (VLEs)	***LMS/VLE*** (Drugova 2021; Hilli 2020; Kliziene et al. 2021; Knivsberg et al. 2022; Masek et al. 2021; Pinilla et al. 2021; Schramm et al. 2022; Smolyaninova and Bezyzvestnykh 2019); ***Canvas*** (Membrillo-Hernandez et al. 2021), ***Moodle*** (Freire and Rodriguez 2022), ***Blackboard*** (Alanzi and Al-Habib 2020); ***M-learning tools*** (Egilsdottir et al. 2021); ***e-labs*** (Duszenko et al. 2022)
Video conferencing platforms & collaborative platforms	***Video conferencing*** (Drijvers et al. 2021; Pokrovskaia et al. 2021); *Teams* (Ruzic-Baf et al. 2021; Sobaih et al. 2021); ***e-Lumen*** (Membrillo-Hernandez et al. 2021); ***Zoom*** (Guiter et al. 2021; Membrillo-Hernandez et al. 2021; Ruzic-Baf et al. 2021; Schramm et al. 2022; Sobaih et al. 2021); ***Merlin*** (Ruzic-Baf et al. 2021); ***Skype*** (Ruzic-Baf et al. 2021);
Blended learning	***Blended Learning*** (Alvarez-Nieto et al. 2017; Fannakhosrow and Nourabadi 2020; Streitlein-Bohme et al. 2021; Volejnikova-Wenger et al. 2021)

Table 3 Social and multi-media tools

Social media platforms and communications tools	***Social media*** (Pokrovskaia et al. 2021); ***Facebook*** (Huang et al. 2018; Sobaih et al. 2021); ***We-Chat*** (Tong, et al. 2020); ***What's App*** (Sobaih et al. 2021); ***YouTube*** (Park et al. 2021); ***Messengers*** (Pokrovskaia et al. 2021); ***REMIND*** (Membrillo-Hernandez et al. 2021); ***Chats*** (Krishnamurthy 2021)
Multi-media tools	***Multimedia*** (Ajeesh and Rukmini 2022; Dinmore 2019); ***video*** (Volejnikova-Wenger et al. 2021); ***videography*** (Dinmore 2019; Jorm et al. 2019); ***slowmation*** (Meedya et al. 2019); ***Story Maps*** (Dickinson and Telford 2020); ***digital storytelling*** (Annacontini and Paiano 2019; Cornillie 2021; Limone et al. 2021; Sunderland et al. 2021); ***Apps*** (Ang et al. 2018);
Audio, interactive and e-assessment tools	***Audio*** (Donkin et al. 2019); ***Audacity*** (Gonzalez et al. 2019); ***PantoDict*** (Bock et al. 2021); ***podcasts*** (Cartner and Hallas 2020; Gudmanian and Sydorenko 2020; Schobel et al. 2021); ***Interactive*** (Mu et al. 2019); ***digital interactive notebook*** (Galanti et al. 2021); ***e-assessment*** (Alvarez-Nieto et al. 2018; Bashir et al. 2021; Helfaya 2019; Rivera-Mata 2021; Viitaharju et al. 2021); ***e-portfolio*** (Eliyahu-Levi and Chen 2022; Fu et al. 2022; Hegarty and Thompson 2019)

Table 4 Games, simulators and realities

Game-based learning	***Game-based learning*** (Elsherbiny and Al Maamari 2021; Liu, Shaikh, and Gazizova 2020), ***gamification*** (Ang et al. 2018; Campillo-Ferrer et al. 2020), ***gamified class (G-Class)*** (Malahito and Quimbo 2020), ***serious games*** (Bardelli et al. 2022; S. Mueller et al. 2020); ***Games & quizzes digital quiz*** (Pereira et al. 2022), ***Kahoot*** (Toma et al. 2021; Zhyhadlo 2022), ***Wordwall and Mentimeter*** (Zhyhadlo 2022)
Learning through simulators, 3S-technology, AR/VR	***Simulations and simulators*** (Almeida and Bastos 2018; Cutumisu et al. 2020; Egilsdottir et al. 2021; Langewitz et al. 2021; Lenes et al. 2020); ***digital anatomy table*** (da Silveira et al. 2022); ***3D*** (Rabattu et al. 2023); ***3S technology - GeoGebra*** (Diaz-Nunja et al. 2018); ***GIS, SPS, remote sensing*** (Zhang 2018); ***augmented reality*** (Moore et al. 2020; Muangpoon et al. 2020; Tosto et al. 2021); ***virtual reality*** (Martin et al. 2021; Perron et al. 2021)

5 Case Study: Carbon Literacy and Digital Skills Authentic Assessment

Green and digital skills shortages have been identified by the tourism industry, most recently by Next Tourism Generation Alliance https://nexttourismgeneration.eu/ Therefore, a medium-sized university in North-West England included, in the final year of an undergraduate tourism degree, a compulsory 20-credit module that was designed to build climate change knowledge and digital skills. The content cuts across the six transformation areas of sustainable development: climate change, decarbonisation, and energy; smart cities; food biosphere and water; consumption and production; digital revolution; human capacity and demography. Scaffolding moved students toward stronger understanding and skill acquisition that resulted in the conceptualisation of Mobile Apps designed to improve visitor experience while simultaneously increasing awareness of carbon literacy.

A range of challenges and solutions in the accommodation, transport and visitor attractions sectors were explored. These introduced students to new concepts ranging from the circular economy to net zero and Smart Cities to Virtual Reality. Carbon Literacy delivery used the NTU Carbon Literacy for Universities Course training toolkit https://carbonliteracy.com/toolkits/universities-colleges/ consisting of four sessions: the science of climate change; climate justice and future scenarios; carbon footprint calculations; individual and community actions. The sustainability audit training introduced the work of Students Organising for Sustainability (SoS) and the role of the Institute of Environmental Management and Assessment (IEMA). The university Sustainability Team provided data on the university carbon footprint and facilitated a practical audit with results discussed and solutions suggested, so students could appreciate businesses at the operational level. This led to the first assessment, a written report: "Critically evaluate the steps that our society (tourism businesses, government, and tourists) needs to take to become carbon neutral by 2038."[1]

Building on the Carbon Literacy knowledge, an authentic assessment followed. "Your local Destination Management Organisation (DMO) is aiming to improve the visitor experience and simultaneously increase awareness of carbon literacy through an App, which you will conceptualise and present using one of the recommended formats." For illustrative purposes, one of these formats was a "10-min narrated video (using ClipChamp or similar) of the created App (using Wix.com App designer or similar)". Students were introduced to the free Wix website builder which automatically provides a mobile-optimized version. The easy drag and drop functionality meant no coding was required. The free ClipChamp video allowed audio capture and demonstration of the App design on a desktop.

Guidance for the authentic assessment needed the student to represent a business and demonstrate their App concept to a DMO. The narrative to support the visual demonstration needed to be based on research but delivered in an authoritative voice (references were in a separate word document accompanying submission). This required an understanding of the visitor statistics; an evaluation of the current visitor experience; sensitivity towards any challenges; development of an innovative concept that would enhance existing offerings or be a new offering; reference to example Apps that were successful in other destinations or sectors; justification of design and content decisions including user-interface and user-experience; benefits of the main offering, paying particular attention to draw out carbon literacy features.

The authentic assessment was designed against the learning outcomes and the rubric was designed against the learning outcomes and authentic assessment (Table 5). Guidance to students was used as a checklist and to build feedback during the marking process. All students using the video and App demonstrated digital skills and carbon literacy understanding. The outstanding grades not only submitted high quality concepts but presented detailed and informed professional outputs.

6 Assessing the Case Study Through Andre's Five Practice Model

Molthan-Hill and Blaj-Ward (2022) recommend use of Rae Andre's Five Practice Model (André 2020) to assess the teaching of climate change and development of climate leaders. Building on the climate change element of the case study the Five Practice Model is applied.

Get the truth The tourism and climate change context helped explain how climate change impacts tourism, and vice versa. As non-scientists tend to rely on the media for information, source scrutiny is required. Therefore, in addition to assessing the

[1] As a Manchester based charity, the Carbon Literacy Project which accredits Carbon Literacy training uses the date 2038 which is in line with Greater Manchester's plans to be carbon neutral by that point.

Table 5 Learning outcomes mapped to authentic assessment and rubric

Learning outcomes	Authentic assessment	Rubric
To evaluate the use of **technology** to discover and **assess trends** and their **impact on the tourism system**	Your local DMO is aiming to improve the **visitor experience** and simultaneously increase **awareness of carbon literacy** through **an App**, which you will conceptualise and **present using one of the recommended formats**	Demonstrate knowledge and critical understanding of **tourism related technologies**, the concept of **smart cities/smart destinations**
To critically examine **potential solutions** and provide recommendations to **tourism stakeholders**	Your local DMO is aiming to **improve the visitor experience** and simultaneously **increase awareness of carbon literacy** through an App, which you will **conceptualise and present** using one of the recommended formats	Use of a **critical and evaluative approach that uses theories and cases to analyse** the use of technology for a **specific city destination and a set of stakeholders**
To demonstrate a variety of **presentation skills (written, visual and/or verbal)**	Your local DMO is aiming to improve the visitor experience and simultaneously increase awareness of carbon literacy through an App, which you will conceptualise and **present using one of the recommended formats**	**Innovative and creative presentation idea**, well suited for stakeholders, a destination and recommended application(s)

currency and source of news items, materials also included case studies, sector examples, academic commentary, policy documents and legislation. Rational thinking is required to understand the climate science, which is more challenging than quick, easy emotional responses. Designed to build knowledge and confidence, the NTU Carbon Literacy training, selected for its international content and appeal to Business Schools, contains a range of resources from a scientific simulator to a personal footprint calculator. It was important for students to engage in independent activities and group discussions, as the personal and social process of truth come from observations, emotions, experience, and beliefs (André 2020).

Assess the risks Those believing that climate change risk has global consequences and is not easily reduced can feel fearful, but the students own 'understanding of the truth' allowed them to reduce fear and apply learning. Lay explanations assisted with understanding the science behind the carbon model and how greenhouse gases affect natural systems. The business-as-usual approach to risk that makes climate uncertainties appear more manageable, was contextualised in local and international tourism contexts through assessment of risks to life and livelihoods caused by more extreme weather conditions. Although feelings of discomfort could arise from inconsistency between attitudes and behaviour, the students recognised they had agency to change their own behaviours.

Weigh the stakes Interactive tools from the Carbon Literacy training materials assisted students with understanding the wider energy industry. Using the Global Carbon Atlas, they could determine the main producers of carbon emissions by country and fossil fuel type. The En-ROADS simulator allowed exploration and testing of solutions to keep warming down to 1.5-degrees; it also provided an understanding of the complexity and difficulties with achieving this given the impact of national government policies. By example, contextualised tourism examples considered Sustainable Aircraft Fuel (SAF) and conflicting UK government policies such as Air Passenger Duty reductions on domestic flights.

Define the Business of Business Tourism has long been aware of its positive and negative impacts on planet and people. The finite natural capital is essential to tourism and the quality of life of locals and visitors. The rich vein of academic ethics and sustainability resources are supplemented with data from international bodies. Within business school settings, the curriculum has been influenced by UN Global Compact and the PRME initiative for Business Schools. The 2030 Agenda has created greater focus on the SDGs, while the UK/Europe tourism skills council have identified a 'green' skills gap. It is important, therefore, to provide future graduates with the knowledge and skills needed by industry and to protect the tourism resource.

Engage Global Leadership At the global level, students made recommendations on reducing carbon through financial incentives and disincentives. They appreciated the challenges of achieving agreement between nations at the Conference for the Parties (COP) and then maintaining that. They further appreciated how self-interest and behaviour change is a challenge for national governments, businesses, and themselves as consumers. Students were able to suggest ways to motivate consumer behaviour change through their App concepts, while advocating Carbon Literacy training for businesses.

7 Conclusions

The interactive content created curiosity and improved awareness of previously unrecognised personal consumption patterns and their related carbon impacts. Combined with the design and content decision-making needs of the authentic assessment, this gave the students agency. Meeting stakeholder needs of improved visitor experience and climate-conscious consumers, their App concepts were demonstrated and justified using video presentations. The knowledge and experience gained by the students provided employability skills required in industry, while giving them confidence to make recommendations on carbon reduction initiatives and use their digital skills.

Practical implications in designing the module content, the relationship between the subject area, Carbon Literacy training and digital skills needs to be clear, with consideration given to scaffolding. The competencies, attributes and literacy gained by students through this tourism module were made possible by tutor upskilling. The Carbon Literacy training is often delivered free to Business School academics via the PRME network, although there are paid alternatives. The tutor can deliver training using available toolkits from the Carbon Literacy Project. The App/website creator software was free (for design purposes only), with an easy to use drag and drop functionality making this suitable for students and tutors.

Future research is recommended that takes a longitudinal approach to the acquisition of climate change knowledge and digital skills. How have digital skills and Carbon Literacy influenced career choice? Is Carbon Literacy evident in the work and home environment after graduation?

References

Ajeesh AK, Rukmini S (2022) Integrating hyperreal literature with CALL in English language curriculum for engineering studies in India: an empirical study of the impact on students' learning. World J Eng 19(2):254–262. https://doi.org/10.1108/wje-07-2021-0393

Alanazi AM (2019) E-assessment through Tablets and Smart phones: an attitudinal assessment o teachers. Ijeri-Intern J Educ Res Innov (11):1–17

Alanzi T, Al-Habib DK (2020) The use of social media by healthcare quality personnel in Saudi Arabia. J Environ Public Health 2020. https://doi.org/10.1155/2020/1417478

Alexander B, Ashford-Rowe K, Barajas-Murph N, Dobbin G, Knott J, McCormack M, … Weber N (2019). *Horizon report 2019 higher education edition.* Retrieved from Louisville, CO. https://library.educause.edu/resources/2019/4/2019-horizon-report

Almeida GD, Bastos CAR (2018) Application of the cognitive walkthrough method to evaluate the usability of PhET simulations package to teach physics. Int J Inf Commun Technol Educ 14(4):34–48. https://doi.org/10.4018/ijicte.2018100103

Alvarez-Nieto C, Parra-Anguita G, Linares-Abad M, Grande-Gascon ML, Lopez-Medina IM (2017) Developing digital educational materials for nursing and sustainability: the results of an observational study. Paper presented at the 9th International Conference on Education and New Learning Technologies (EDULEARN), Barcelona, SPAIN

Alvarez-Nieto C, Richardson J, Parra-Anguita G, Linares-Abad M, Huss N, Grande-Gascon ML, … Lopez-Medina IM (2018) Developing digital educational materials for nursing and sustainability: the results of an observational study. Nurse Educ Today 60:139-146. https://doi.org/10.1016/j.nedt.2017.10.008

André R (2020) Lead for the planet: Five practices for confronting climate change. University of Toronto Press

Ang ET, Chan JM, Gopal V, Shia NL (2018) Gamifying anatomy education. Clin Anat 31(7):997–1005. https://doi.org/10.1002/ca.23249

Annacontini G, Paiano AP (2019) How to assess A meaningful technology-integrated tool for the reflective educators. Dig Educ Rev (35):36–61

Baecker D (2022) Introducing audio podcasts into a practical laboratory course for pharmacy students as a novel tool for performance assessment. Pharmacy 10(2):11. https://doi.org/10.3390/pharmacy10020040

Balogun TB, Awonuga OO, Abowen-Dake R (2021) Investigating digital technological competencies amongst black Asian minority ethnic construction students in the UK. J Eng Des Techn ahead-of-print(ahead-of-print). https://doi.org/10.1108/jedt-08-2021-0449

Bardelli S, Del Corso G, Ciantelli M, Del Pistoia M, Lorenzoni F, Fossati N, … Cuttano A (2022) Improving pediatric/neonatology residents' newborn resuscitation skills with a digital serious game: DIANA. Front Pediat 10:17. https://doi.org/10.3389/fped.2022.842302

Bashir A, Bashir S, Rana K, Lambert P, Vernallis A (2021) Post-COVID-19 adaptations; the shifts towards online learning, hybrid course delivery and the implications for biosciences courses in the higher education setting. Front Educ 6:13. https://doi.org/10.3389/feduc.2021.711619

Blundell Ç, Lee K-T, Nykvist S (2020) Moving beyond enhancing pedagogies with digital technologies: frames of reference, habits of mind and transformative learning. J Res Technol Educ 52(2):178–196

Bock A, Elvers D, Goloborodko E, Kramer C, Kniha K, Holzle F, … Modabber A (2021) An innovative PantoDict program for reporting panoramic radiographs using automatic speech recognition in dental education: a randomized observer-blinded study. Oral Surg Oral Med Oral Pathol Oral Radiol 132(1):104-111. https://doi.org/10.1016/j.oooo.2020.10.002

Boud D, Dawson P, Bearman M, Bennett S, Joughin G, Molloy E (2018) Reframing assessment research: through a practice perspective. Stud High Educ 43(7):1107–1118. https://doi.org/10.1080/03075079.2016.1202913

Bretag T, Harper R, Rundle K, Newton PM, Ellis C, Saddiqui S, van Haeringen K (2020) Contract cheating in Australian higher education: a comparison of non-university higher education providers and universities. Assess Eval High Educ 45(1):125–139. https://doi.org/10.1080/02602938.2019.1614146

Campillo-Ferrer JM, Miralles-Martinez P, Sanchez-Ibanez R (2020) Gamification in higher education: impact on student motivation and the acquisition of social and civic key competencies. Sustainability 12(12).https://doi.org/10.3390/su12124822

Carlisle S, Zaki K, Ahmed M, Dixey L, McLoughlin E (2021b) The imperative to address sustainability skills gaps in tourism in Wales. Sustainability 13(3):1161

Carlisle S, Ivanov S, Dijkmans C (2021) The digital skills divide: evidence from the European tourism industry. J Tour Fut ahead-of-print(ahead-of-print). https://doi.org/10.1108/JTF-07-2020-0114

Cartner H, Hallas J (2020) Aligning assessment, technology, and multi-literacies. E-Learn Dig Media 17(2):131–147. https://doi.org/10.1177/2042753019899732

Cebrián G, Segalàs J, Hernández À (2019) Assessment of sustainability competencies: a literature review and pathways for future research and practice. Central Eur Rev Econ Manag 3(3):19–44

Change Agents UK, EAUC, & Cook, I. (2020) Future Graduate Skills: A Scoping Study Retrieved from https://www.changeagents.org.uk/skills-for-the-future

Chankseliani M, Qoraboyev I, Gimranova D (2021) Higher education contributing to local, national, and global development: new empirical and conceptual insights. High Educ 81(1):109–127. https://doi.org/10.1007/s10734-020-00565-8

Cho WC, Gill P, Aung PP, Gu J, Nagarajan P, Ivan D, … Torres-Cabala CA (2021) The utility of digital pathology in improving the diagnostic skills of pathology trainees in commonly encountered pigmented cutaneous lesions during the COVID-19 pandemic: a single academic institution experience. Annals Diagn Pathol 54:7. https://doi.org/10.1016/j.anndiagpath.2021.151807

Cornillie F, Buendgens-Kosten J, Sauro S, Van der Veken J (2021) There's always an option: collaborative writing of multilingual interactive fanfiction in a foreign language class. CALICO J 38(1):17–42. https://doi.org/10.1558/cj.41119

Cross ID, Congreve A (2021) Teaching (super) wicked problems: authentic learning about climate change. J Geogr High Educ 45(4):491–516. https://doi.org/10.1080/03098265.2020.1849066

Cutumisu M, Ghoman SK, Lu C, Patel SD, Garcia-Hidalgo C, Fray C, … Schmolzer GM (2020) Health care providers' performance, mindset, and attitudes toward a neonatal resuscitation computer-based simulator: empirical study. Jmir Seri Games 8(4):14. https://doi.org/10.2196/21855

da Silveira CR, Dias PEM, Oenning AC, de Brito RB, Turssi CP, Oliveira LB (2022) Digital anatomy table in teaching-learning process of the temporomandibular joint anatomy. Eur J Dent Educ 26(1):131–137. https://doi.org/10.1111/eje.12680

De Rossi M, Restiglian E (2019) Hybrid solutions for didactics in higher education: an interdisciplinary workshop of 'Visual Storytelling' to develop documentation competences. Tuning J High Educ 6(2):391–419. https://doi.org/10.18543/tjhe-6(2)-2019pp175-203

Devlin M, Samarawickrema G (2022) A commentary on the criteria of effective teaching in post-COVID higher education. High Educ Res Dev 41(1):21–32. https://doi.org/10.1080/07294360.2021.2002828

Diaz-Nunja L, Rodriguez-Sosa J, Lingan SK (2018) Teaching of geometry with GeoGebra software in high school students of an educational institution in Lima. Propositos Y Representaciones 6(2):235–251. https://doi.org/10.20511/pyr2018.v6n2.251

Dickinson S, Telford A (2020) The visualities of digital story mapping: teaching the 'messiness' of qualitative methods through story mapping technologies. J Geogr High Educ 44(3):441–457. https://doi.org/10.1080/03098265.2020.1712686

Dinmore S (2019) Beyond lecture capture: Creating digital video content for online learning - a case study. J Univ Teach Learn Pract 16(1):11

Donkin R, Askew E, Stevenson H (2019) Video feedback and e-Learning enhances laboratory skills and engagement in medical laboratory science students. BMC Med Educ 19(1):12. https://doi.org/10.1186/s12909-019-1745-1

Drijvers P, Thurm D, Vandervieren E, Klinger M, Moons F, van der Ree H, ... Doorman M (2021) Distance mathematics teaching in Flanders, Germany, and the Netherlands during COVID-19 lockdown. Educ Stud Mathem 108(1-2):35-64. https://doi.org/10.1007/s10649-021-10094-5

Drugova E, Zhuravleva I, Aiusheeva M, Grits D (2021) Toward a model of learning innovation integration: TPACK-SAMR based analysis of the introduction of a digital learning environment in three Russian universities. Educ Inform Technol 26(4):4925–4942. https://doi.org/10.1007/s10639-021-10514-2

Duszenko M, Frohlich N, Kaupp A, Garaschuk O (2022) All-digital training course in neurophysiology: lessons learned from the COVID-19 pandemic. BMC Med Educ 22(1):14. https://doi.org/10.1186/s12909-021-03062-3

Egilsdottir HO, Heyn LG, Brembo EA, Byermoen KR, Moen A, Eide H (2021) Configuration of mobile learning tools to support basic physical assessment in nursing education: longitudinal participatory design approach. JMIR Mhealth Uhealth 9(1):16. https://doi.org/10.2196/22633

Eliyahu-Levi D, Chen S (2022) Hebrew in Tel Aviv, Beijing, Denver, and Melbourne: assessing peer learning in a multicultural environment. Interact Learn Environ 12. https://doi.org/10.1080/10494820.2022.2059519

Elsherbiny MMK, Al Maamari RHH (2021) Game-based learning through mobile phone apps: effectively enhancing learning for social work students. Soc Work Educ 40(3):315–332. https://doi.org/10.1080/02615479.2020.1737665

Emblen-Perry K (2019) Can sustainability audits provide effective, hands-on business sustainability learning, teaching and assessment for business management undergraduates? Int J Sustain High Educ 20(7):1191–1219. https://doi.org/10.1108/ijshe-10-2018-0181

Fannakhosrow M, Nourabadi S (2020) Blended learning, innovative pattern in teachers professional development in Educatuibak system of Iran. Revista Praxis Educ 16(38):306–322

National Forum for the Enhancement of Teaching and Learning in Higher Education (2017) Expanding our understanding of assessment and feedback in irIsh higher education. Forum Insights. Retrieved from https://hub.teachingandlearning.ie/wp-content/uploads/2021/06/95.-NF-2017-Expanding-our-Understanding-of-Assessment-and-Feedback-in-Irish-Higher-Education.pdf

Freire T, Rodriguez C (2022) The transformation to an online course in higher education results in better student academic performance. Ried-Revista Iberoamericana De Educacion a Distancia 25(1):299–322. https://doi.org/10.5944/ried.25.1.31465

Fu H, Hopper T, Sanford K, Monk D (2022) Learning with digital portfolios: teacher candidates forming an assessment identity. Canad J Schol Teach Learn 13(1):22. https://doi.org/10.5206/cjsotlrcacea.2022.1.11108

Galanti TM, Baker CK, Morrow-Leong K, Kraft T (2021) Enriching TPACK in mathematics education: using digital interactive notebooks in synchronous online learning environments. Interac Technol Smart Educ 18(3):345–361. https://doi.org/10.1108/itse-08-2020-0175

Gonzalez JMM, Gea EMV, Ariz MDH (2019) Learning of audacity for the edition and production of digital didactic contents. Pixel-Bit- Revista De Medios Y Educacion (55):7–21. https://doi.org/10.12795/pixelbit.2019.i55.01

Gudmanian A, Sydorenko S (2020) A web resource map for a university course in the history of the English language. Inform Technol Learn Tools 76(2):126–136

Guiter GE, Sapia S, Wright AI, Hutchins GGA, Arayssi T (2021) Development of a remote online collaborative medical school pathology curriculum with clinical correlations, across several international sites, through the Covid-19 pandemic. Med Sci Educ 31(2):549–556. https://doi.org/10.1007/s40670-021-01212-2

Hamalainen R, Nissinen K, Mannonen J, Lamsa J, Leino K, Taajamo M (2021) Understanding teaching professionals' digital competence: What do PIAAC and TALIS reveal about technology-related skills, attitudes, and knowledge? Comput Hum Behav 117:15. https://doi.org/10.1016/j.chb.2020.106672

Hart C, Hammer S, Collins P, Chardon T (2011) The real deal: using authentic assessment to promote student engagement in the first and second years of a regional law program. Leg Educ Rev 21(1/2):97–121

Hegarty B, Thompson M (2019) A teacher's influence on student engagement: using smartphones for creating vocational assessment e-portfolios. J Inform Technol Educ Res 18:113–159. https://doi.org/10.28945/4244

Helfaya A (2019) Assessing the use of computer-based assessment-feedback in teaching digital accountants. Acc Educ 28(1):69–99. https://doi.org/10.1080/09639284.2018.1501716

Herbert IP, Rothwell AT, Glover JL, Lambert SA (2020) Graduate employability, employment prospects and work-readiness in the changing field of professional work. Intern J Manag Educ 18(2):13. https://doi.org/10.1016/j.ijme.2020.100378

Herrington J, Herrington A (2006) Authentic conditions for authentic assessment: aligning task and assessment. Paper presented at the Critical Visions. Thinking, learning and researching in higher education. 29th HERDSA Annual Conference., Perth, Western Australia

Hilli C (2020) Extending classrooms through teacher collaboration in virtual learning environments. Educat Action Res 28(4):700–715. https://doi.org/10.1080/09650792.2019.1654901

Hiser KK, Lynch MK (2021) Worry and hope: what college students know, think, feel, and do about climate change. J Comm Engag Schol 13(3):7

Jopp R (2020) A case study of a technology enhanced learning initiative that supports authentic assessment. Teach High Educ 25(8):942–958

Jorm C, Roberts C, Gordon C, Nisbet G, Roper L (2019) Time for university educators to embrace student videography. Camb J Educ 49(6):673–693. https://doi.org/10.1080/0305764x.2019.1590528

Kanetaki Z, Stergiou C, Bekas G, Troussas C, Sgouropoulou C (2022) Evaluating remote task assignment of an online engineering module through data mining in a virtual communication platform environment. Electronics 11(1):14. https://doi.org/10.3390/electronics11010158

Keppell M, Suddaby G, Hard N (2015) Assuring best practice in technology-enhanced learning environments. Res Learn Technol 23

Kliziene I, Taujanskiene G, Augustiniene A, Simonaitiene B, Cibulskas G (2021) The impact of the virtual learning platform EDUKA on the academic performance of primary school children. Sustainability, 13(4). https://doi.org/10.3390/su13042268

Knight GL, Drysdale TD (2020) The future of higher education (HE) hangs on innovating our assessment - but are we ready, willing and able? Higher Educ Pedag 5(1):57–60. https://doi.org/10.1080/23752696.2020.1771610

Knivsberg IC, Kopperud SE, Bjork MB, Torgersen G, Skramstad K, Kvaal SI (2022) Digitalised exercise material in forensic odontology. Int J Legal Med 136(1):381–390. https://doi.org/10.1007/s00414-021-02740-7

Krishnamurthy N (2021) Teaching freshmen chemistry in India during the COVID-19 pandemic: student perspectives and challenges. J Chem Educ 98(12):3884–3891. https://doi.org/10.1021/acs.jchemed.1c00813

Langewitz W, Seixas UPD, Hunziker S, Becker C, Fischer MR, Benz A, Otto B (2021) Doctor-patient communication during the Corona crisis web-based interactions and structured feedback from standardized patients at the University of Basel and the LMU Munich. Gms J Med Educ 38(4):19. https://doi.org/10.3205/zma001477

Lenes A, Klasen M, Adele A, Garetz U, Proch-Trodler C, Schenkat H, Sopka S (2020) Crisis as a chance. A digital training of social competencies with simulated persons at the Medical Faculty of RWTH Aachen, due to the lack of attendance teaching in the SARS-Cov-2 pandemic. Gms J Med Educ 37(7):6. https://doi.org/10.3205/zma001375

Limone P, Toto GA, Cafarelli B (2021) The decision-making process and the construction of online sociality through the digital storytelling methodology. Electronics 10(20):18. https://doi.org/10.3390/electronics10202465

Liu ZY, Shaikh ZA, Gazizova F (2020) Using the concept of game-based learning in education. Int J Emerg Technol Learn 15(14):53–64. https://doi.org/10.3991/ijet.v15i14.14675

Lu JJ, Han FF (2022) Challenges of using online technology in the pandemic: voices from English language and literacy primary school teachers in central China. Education 3–13:1–12. https://doi.org/10.1080/03004279.2022.2048678

Malahito JAI, Quimbo MAT (2020) Creating G-Class: a gamified learning environment for freshman students. E-Learn Dig Media 17(2):94–110. https://doi.org/10.1177/2042753019899805

Martin D, Touron J, Asencio EN (2021) Flipped training in a virtual 3D environment to foster teaching competences. Revista De Educ (391):89–117. https://doi.org/10.4438/1988-592x-re-2021-391-472

Martzoukou K, Fulton C, Kostagiolas P, Lavranos C (2020) A study of higher education students' self-perceived digital competences for learning and everyday life online participation. J Document 76(6):1413–1458. https://doi.org/10.1108/jd-03-2020-0041

Masek A, Ana A, Muda W, Paimin ANB, Ab Halim FB, Barliana MS, … Sari AR (2021) The digital pedagogical tools for effective learning during pandemic: facilitating TVET students' learning process in Malaysia and Indonesia. J Eng Sci Technol 16:34-41

Matthews A (2021) Sociotechnical imaginaries in the present and future university: a corpus-assisted discourse analysis of UK higher education texts. Learn Media Technol 46(2):204–217. https://doi.org/10.1080/17439884.2021.1864398

Meedya S, Moroney T, Nielsen W, Bokat IN (2019) Digital explanations and nursing students' perception of learning science. Nurse Educ Pract 41:6. https://doi.org/10.1016/j.nepr.2019.102636

Membrillo-Hernandez J, Ramirez-Cadena MD, Ramirez-Medrano A, Garcia-Castelan RMG, Garcia-Garcia R (2021) Implementation of the challenge-based learning approach in Academic Engineering Programs. Intern J Inter Des Manufact Ijidem 15(2–3):287–298. https://doi.org/10.1007/s12008-021-00755-3

Mimirinis M (2019) Qualitative differences in academics' conceptions of e-assessment. Assess Eval High Educ 44(2):233–248. https://doi.org/10.1080/02602938.2018.1493087

Molthan-Hill P, Blaj-Ward L (2022) Assessing climate solutions and taking climate leadership: how can universities prepare their students for challenging times? Teach High Educ. https://doi.org/10.1080/13562517.2022.2034782

Moore A, Daniel B, Leonard G, Regenbrecht H, Rodda J, Baker L, … Mills S (2020) Comparative usability of an augmented reality sandtable and 3D GIS for education. Intern J Geograp Inform Sci 34(2):229-250. https://doi.org/10.1080/13658816.2019.1656810

Morgan A, Sibson R, Jackson D (2022) Digital demand and digital deficit: conceptualising digital literacy and gauging proficiency among higher education students. J High Educ Policy Manag 44(3):258–275

Mu EW, Pusic M, Coneybeare M, Meehan SA (2019) Learning analytics to enhance dermatopathology education among dermatology residents. J Drugs Dermatol 18(12):1231–1236

Muangpoon T, Osgouei RH, Escobar-Castillejos D, Kontovounisios C, Bello F (2020) Augmented reality system for digital rectal examination training and assessment: system validation. J Med Internet Res 22(8):13. https://doi.org/10.2196/18637

Mueller S, Soriano D, Boscor A, Saville NM, Arjyal A, Baral S, … Kostkova P (2020) Mantra: improving knowledge of maternal health, neonatal health, and Geohazards in women in rural Nepal using a mobile serious game. Front Public Health 8:14. https://doi.org/10.3389/fpubh.2020.584375

Mueller J (2018) Authentic assessment toolbox. http://jfmueller.faculty.noctrl.edu/toolbox/whatisit.htm

Murray MC, Pérez J (2014) Unraveling the digital literacy paradox: how higher education fails at the fourth literacy. Iss Inf Sci Inform Technol 11:85

Next Tourism Generation Alliance (2022) Plan for sectoral cooperation to address skills needs in the tourism sector. Blueprint for Tourism Sectoral Skills. Retrieved from https://nexttourismgeneration.eu/wp-content/uploads/2022/06/Final-Next-Tourism-Generation-Blueprint-Report.pdf

Nieminen JH, Bearman M, Ajjawi R (2022) Designing the digital in authentic assessment: is it fit for purpose? Assess Eval High Educ. https://doi.org/10.1080/02602938.2022.2089627

Oberländer M, Beinicke A, Bipp T (2020) Digital competencies: a review of the literature and applications in the workplace. Comput Educ 146:103752. https://doi.org/10.1016/j.compedu.2019.103752

Park JC, Kwon HJE, Chung CW (2021) Innovative digital tools for new trends in teaching and assessment methods in medical and dental education. J Educ Evalu Health Prof 18:1–7. https://doi.org/10.3352/jeehp.2021.18.13

Pelletier K, McCormack M, Reeves J, Robert J, Arbino N, Dickson-Deane C, … Bessette LS (2022) 2022 EDUCAUSE horizon report teaching and learning edition. Retrieved from Boulder, CO. https://library.educause.edu/resources/2022/4/2022-educause-horizon-report-teaching-and-learning-edition

Pereira AC, da Silva MAD, Patel US, Tanday A, Hill KB, Walmsley AD (2022) Using quizzes to provide an effective and more enjoyable dental education: a pilot study. Eur J Dent Educ 26(2):404–408. https://doi.org/10.1111/eje.12716

Perron JE, Coffey MJ, Lovell-Simons A, Dominguez L, King ME, Ooi CY (2021) Resuscitating cardiopulmonary resuscitation training in a virtual reality: prospective interventional study. J Med Internet Res 23(7):10. https://doi.org/10.2196/22920

Pinilla S, Cantisani A, Kloppel S, Strik W, Nissen C, Huwendiek S (2021) Curriculum development with the implementation of an open-source learning management system for training early clinical students: an educational design research study. Adv Med Educ Pract 12:53–61. https://doi.org/10.2147/amer.s284974

Pokrovskaia NN, Leontyeva VL, Ababkova MY, Cappelli L, D'Ascenzo F (2021) Digital communication tools and knowledge creation processes for enriched intellectual outcome-experience of short-term e-learning courses during pandemic. Future Internet 13(2):22. https://doi.org/10.3390/fi13020043

Porcaro PA, Jackson DE, McLaughlin PM, O'Malley CJ (2016) Curriculum design of a flipped classroom to enhance haematology learning. J Sci Educ Technol 25(3):345–357

Puentedura R (2006) Transformation, technology, and education. Strengthening your district through technology. http://hippasus.com/resources/tte/

Rabattu PY, Debarnot U, Hoyek N (2023) Exploring the impact of interactive movement-based anatomy learning in real classroom setting among kinesiology students. Anat Sci Educ 9. https://doi.org/10.1002/ase.2172

Rivera-Mata J (2021) How to teach online? Recommendations for the assessment of online exams with University students in the USA in times of pandemic. Ijeri-Intern J Educ Res Innov (15):188–202. https://doi.org/10.46661/ijeri.5003

Ruzic-Baf M, Kadum S, Kvaranta K (2021) Online teaching and grading during the Covid-19 pandemic - attitudes of Croatian students. J Educ Cult Soc 12(2):399–411. https://doi.org/10.15503/jecs2021.2.399.411

Sailer M, Stadler M, Schultz-Pernice F, Franke U, Schoffmann C, Paniotova N, … Fischer F (2021) Technology-related teaching skills and attitudes: validation of a scenario-based self-assessment instrument for teachers. Comput Human Beh 115:12. https://doi.org/10.1016/j.chb.2020.106625

Sargent J, Lynch S (2021) None of my other teachers know my face/emotions/thoughts': digital technology and democratic assessment practices in higher education physical education. Technol Pedagog Educ 30(5):693–705. https://doi.org/10.1080/1475939x.2021.1942972

Sastre-Merino S, Nunez-Del-Rio MC, Caravantes A, Bravo-Ramos JL (2021) Perceptions of engineering faculty members of online teaching due to COVID-19. Int J Eng Educ 37(6):1567–1581

Schobel T, Zajonz D, Melcher P, Lange J, Fischer B, Heyde CE, … Ghanem M (2021) Podcasts as a teaching tool in orthopaedic surgery Is it beneficial or more an exemption card from attending lectures? Orthopade 50(6):455-463. https://doi.org/10.1007/s00132-020-03956-y

Schramm L, Friedrich P, Schuttler J, Lutcke B (2022) Interdisciplinary interactive blended learning concept in times of a pandemic - pain medicine "totally digital." GMS J Med Educ 39(1):8. https://doi.org/10.3205/zma001527

Schrum K, Majury N, Simonelli AL (2021) Authentic learning across disciplines and borders with scholarly digital storytelling. Teach Learn Inquiry-the Issotl J 9(2):16. https://doi.org/10.20343/teachlearninqu.9.2.8

Scott M, Unsworth J (2018) Matching final assessment to employability: developing a digital viva as an end of programme assessment. Higher Educ Pedag 3(1):373–384. https://doi.org/10.1080/23752696.2018.1510294

Smolyaninova O, Bezyzvestnykh E (2019) Implementing teachers' training technologies at a federal university: E-portfolio, digital laboratory, PROLog module system. Intern J Online Biomed Eng 15(4):69–87. https://doi.org/10.3991/ijoe.v15i04.9288

Sobaih AE, Salem AE, Hasanein AM, Abu Elnasr AE (2021) Responses to COVID-19 in higher education: students' learning experience using microsoft teams versus social network sites. Sustainability 13(18):12. https://doi.org/10.3390/su131810036

Spanjaard D, Garlin F, Mohammed H (2022) Tell me a story! Blending digital storytelling into marketing higher education for student engagement. J Mark Educ 1–16. https://doi.org/10.1177/02734753221090419

Sterling S (2011) Transformative learning and sustainability: Sketching the conceptual ground. Learn Teach Higher Educ 5(11):17–33

Streitlein-Bohme I, Woestmann B, Vollmar HC, Bohme K (2021) We can also do online - evaluation of the accompanying digital seminar of the elective subject "General Practice" during intership (PJ) at Ruhr-University Bochum. Gms J Med Educ 38(4):6. https://doi.org/10.3205/zma001469

Sunderland N, Robinson K, Burgess A (2021) Overcoming future professionals' fear of digital storytelling. Aust Soc Work 74(1):13–28. https://doi.org/10.1080/0312407x.2020.1762098

Tang JT (2022) A practical action research of portfolio assessment on building the learning community for graduate students in Taiwan. Syst Pract Action Res 24. https://doi.org/10.1007/s11213-021-09583-8

Toma F, Diaconu DC, Popescu CM (2021) The use of the Kahoot! learning platform as a type of formative assessment in the context of pre-university education during the COVID-19 pandemic period. Educ Sci 11(10):18. https://doi.org/10.3390/educsci11100649

Tong PR, An ISD, Zhou YJ (2020) Holistic and dynamic: teacher-researcher reflections on operating mobile-assisted learning tasks supported by WeChat for Chinese as a foreign language. Instr Sci 48(6):729–763. https://doi.org/10.1007/s11251-020-09524-3

Tosto C, Hasegawa T, Mangina E, Chifari A, Treacy R, Merlo G, Chiazzese G (2021) Exploring the effect of an augmented reality literacy programme for reading and spelling difficulties for children diagnosed with ADHD. Virt Real 25(3):879–894. https://doi.org/10.1007/s10055-020-00485-z

van Rensburg CJ, Coetzee SA, Schmulian A (2021) Developing digital creativity through authentic assessment. Assess Eval High Educ 47(6):857–877. https://doi.org/10.1080/02602938.2021.1968791

Vereshchahina T, Liashchenko O, Babiy S (2018) English language teachers' perceptions of hybrid learning at university level. Adv Educ (10):88–97. https://doi.org/10.20535/2410-8286.130734

Viitaharju P, Yliniemi K, Nieminen M, Karttunen AJ (2021) Learning experiences from digital laboratory safety training. Educ Chem Eng 34:87–93. https://doi.org/10.1016/j.ece.2020.11.009

Villarroel V, Bloxham S, Bruna D, Bruna C, Herrera-Seda C (2018) Authentic assessment: creating a blueprint for course design. Assess Eval High Educ 43(5):840–854. https://doi.org/10.1080/02602938.2017.1412396

Volejnikova-Wenger S, Andersen P, Clarke KA (2021) Student nurses' experience using a serious game to learn environmental hazard and safety assessment. Nurse Educ Today 98:7. https://doi.org/10.1016/j.nedt.2020.104739

Zepke N, Leach L (2010) Improving student engagement: ten proposals for action. Act Learn High Educ 11(3):167–177

Zhang XL (2018) Practice teaching of landscape survey course based on ecognition remote sensing image interpretation technology. Educat Sci-Theory Pract 18(5):1411–1423. https://doi.org/10.12738/estp.2018.5.038

Zhyhadlo OY (2022) Application of digital game-based tools for formative assessment at foreign language lessons. Inform Technol Learn Tools 87(1):139–150. https://doi.org/10.33407/itlt.v87i1.4703

Sustainable Internationalisation Through Collaborative Online Intercultural Learning

Amanda Mason, Farrah Robert, Adriana Webb, Augusto Mendes Siega, Ain Kamalia Mazlin, and Lee Mei Chee Jamie

Abstract This chapter explores the potential of collaborative online intercultural learning as a vehicle for sustainable intercultural learning through the reflections of practitioners and participants. As universities strive to produce graduates as global citizens, intercultural skills are seen as key to healthy citizenship. Student mobility is often viewed as the most effective route to this, but many students are unable to participate in long-term mobility, and short-term mobility has a high carbon footprint. The Covid-19 pandemic provided an opportunity to explore the potential of a more inclusive and sustainable approach to internationalisation using experiential learning around global business issues. This paper adopts a reflective methodology, drawing from the evaluative reflections of tutors and students engaged in three short virtual exchanges which took place over a two-year period. The exchange participants were students enrolled on undergraduate business courses at a UK university and a partner college in Malaysia. The findings demonstrate that a short online exchange can open students' minds to different world views and initiate the processes involved in critical intercultural learning but cannot replace the experiential learning associated with physical mobility. The paper concludes by exploring ways to enhance the experience and extend the opportunity more widely.

A. Mason (✉) · A. Webb · A. M. Siega
Liverpool Business School, Liverpool John Moores University, Liverpool, UK
e-mail: a.mason@ljmu.ac.uk

A. Webb
e-mail: adrianawebb@hotmail.co.uk

A. M. Siega
e-mail: augustosiega@gmail.com

F. Robert · A. K. Mazlin · L. M. C. Jamie
YPC International College, Kuala Lumpur, Malaysia
e-mail: farrah@ypccollege.edu.my

A. K. Mazlin
e-mail: ainkamalia.m@ypccollege.edu.my

L. M. C. Jamie
e-mail: meicl@ypccollege.edu.my

T. Wall et al. (eds.), *Sustainability in Business Education, Research and Practices*, World Sustainability Series, https://doi.org/10.1007/978-3-031-55996-9_13

Keywords Intercultural learning · Sustainable internationalisation · Virtual exchange · Experiential learning · COIL

1 Introduction

There is a broad consensus that sustainability themes need to be embedded in university curricula, and as climate change is a serious threat not only to sustainable development but to life on earth itself, universities and business schools have a broader responsibility to address climate change in their operations as well as their curricula (Molthan-Hill and Blaj-Ward 2022). At the same time, global issues such as climate change require cooperation and collaboration across cultures and geographical boundaries. Graduates, therefore, need not only subject-specific knowledge but also intercultural and other key transferable skills. International mobility has long been considered a way to develop intercultural skills, but a very small proportion of students in Europe, and particularly the UK, engage in long-term mobility (UUKi 2020). To increase opportunities for students to gain '*international experiences*', a growing number of UK (UUKi 2020) institutions are promoting short-term mobility. There is, however, increasing awareness in the internationalisation community about the high carbon footprint associated with mobility (Shields 2019) and its apparent conflict with the sustainability agenda (e.g. Proctor 2022).

Collaborative online intercultural learning (COIL) or virtual exchange (VE) are approaches to providing intercultural learning which can be more environmentally sustainable (Leal Filho et al. 2023) and inclusive (Wimpenny et al. 2022). This chapter seeks to explore the potential of COIL in the development of intercultural awareness and other key transferable skills of students on business-related undergraduate courses in Liverpool, UK and Kuala Lumpur, Malaysia. This case study draws on the evaluative reflections of practitioners and participants involved in one of three short exchanges which took place over a two-year period.

2 Literature Review

2.1 *Internationalisation of Higher Education*

Growth in the internationalization of higher education (HE) is driving the expansion of tertiary systems and institutions globally. Universities recognise the need for international cooperation and collaboration with respect to both research and teaching. Global competencies are essential for graduates to examine local, global and intercultural issues, to comprehend and appreciate the perspectives and world views of others, to participate in open, appropriate and effective collaborations with people from different cultures, and to act for collective well-being and sustainable development (OECD 2018).

In the UK, the internationalisation agenda has often been driven by the recruitment of international students to pay for the increasing costs of educating its home students. In 2020–21, international students accounted for 22% of the total student population in the UK (Universities UK 2022) which includes more than 12,000 Malaysian students. This has been accompanied by steady growth in transnational education, the UK being a 'world leader' in this area, and Malaysia coming second only to China as a host country in terms of the numbers of UK TNE students (Universities UK 2022).

Significantly less attention is given to the internationalisation of the curriculum (IoC) (Leask 2015) and approaches where students can develop intercultural skills by drawing on their own cultural backgrounds (Caruana 2014) and the diversity of the campus (Sawir 2013). Despite over a decade of research in this area, there are limited examples of good practice and, as noted by Leask et al. (2018), we are still far away from any form of internationalisation which is both inclusive and accessible rather than elitist and exclusive.

2.2 *Student Mobility*

Internationalisation of the curriculum is often misinterpreted as providing opportunities for its home students to travel abroad, and promoting HE student mobility continues to be an EU priority, albeit with a focus on inclusion and diversity (European Commission 2022). There is general agreement that student mobility has multiple benefits for students. A systematic review by Roy et al. (2019) investigated outcomes of both short- and long-term mobility programmes and identified a wide range of benefits to students from cultural awareness and cultural sensitivity to personal outcomes such as higher levels of self-efficacy. Students perceive these benefits too: a study of 3557 students from across Europe who had engaged in Erasmus programmes revealed three important factors for a successful Erasmus experience: '*becoming more independent, gaining another perspective on the way things are at home, and interacting with people from different origins*' (Asodar et al. 2017). Student mobility within Asia has been driven and encouraged due to economic growth, national competitiveness, and regional development in the early twenty-first century. Similarly, more Asian students expect to be mobile internationally in the upcoming decades, with students searching for better education opportunities and experience (Chan 2012).

There are several issues surrounding student mobility. Firstly, the proportion of students participating in outward mobility remains low in Europe generally and particularly in the UK, with the most recent pre-pandemic figures showing that only 7.4% of 2018–19 graduates in the UK had overseas experience (UUKI 2020). This figure is even lower for the most disadvantaged groups such as those from minority ethnic groups and low socioeconomic backgrounds (UUKI 2017) hence one of the key priorities for Erasmus + 2021–2027 is to be more inclusive (European Commission 2022).

There is also the challenge to the widely held assumption that merely interacting with people from another culture during a period of mobility will lead to intercultural learning. For example, studies have shown that intercultural interactions need careful management for learning to take place (Pederson 2010; Nolan et al. 2022).

Finally, there is the issue of sustainability. A recent study (Shields 2019) estimated that the total greenhouse gas emissions associated with international student mobility globally in 2014 were equivalent to the total annual emissions of an entire nation (for example, Croatia or Tunisia). Thus, there is a growing awareness in the international community about the tension between internationalisation and sustainability goals (Proctor 2022; Leal Filho et al. 2023) with virtual mobility or COIL promoted as a more sustainable and inclusive approach to internationalisation.

2.3 Collaborative Online International Learning (COIL)

COIL which emerged from the SUNY COIL project (Rubin 2016), is now widely used as an '*umbrella*' term (Wimpenny et al. 2022) to describe any initiative which enables students to interact across geographical boundaries using digital technology. Opportunities for intercultural interactions without the need for travel have a long history in language teaching and learning, from low-technology letter pen-friend communication to more complex technology-assisted '*telecollaboration*', and more recently '*virtual exchange*' (Lewis and O'Dowd 2017; O'Dowd 2021). There are additional benefits to COIL as well as the reduction in environmental costs. A systematic review exploring the contribution of digital transformation to the sustainability of HEIs found that collaborative online learning can foster intercultural awareness as well as other transferable skills (Viera Trevisan et al. 2023). A small study of engineering students showed that those who participated in a COIL initiative performed better in their engineering assignment than those that had not (Appiah-Kubi and Annan. 2020). It is argued that COIL can offer more inclusive accessible approaches that oppose Western hegemony (Wimpenny et al. 2022) and challenge the colonial view of internationalisation (Heleta and Chasi 2022). A study by Naicker et al. (2022) suggested that the key barriers to successful COIL included technological challenges, language, and partner dynamics while the key enablers were a structured course outline, student's receptiveness to COIL and increased academic self-efficacy.

Despite the reported benefits of COIL and the digital transformation opportunities afforded by the pandemic (Chang and Gomes 2022), many of the pre-pandemic resistance and institutional barriers to virtual mobility remain (Garcés and O'Dowd 2021; Jørgensen et al. 2022). This chapter, therefore, seeks to explore the potential of COIL through the reflections of practitioners and undergraduate student participants in a series of trial projects. To facilitate the comparison of student's experience of COIL and physical mobility, COIL experiences will be referred to as virtual exchange in the following sections.

3 Methods

3.1 Background and Context of the COIL Project

Liverpool John Moores University (LJMU) in the UK, and YPC International College in Malaysia, have a long-standing successful partnership established in 2005. Students in Malaysia study four undergraduate degree programmes, two validated programmes from the Faculty of Engineering and Technology and two franchised programmes from the School of Business and Management. Before the pandemic, several short-term mobility trips were organised which enabled ten students per trip from LJMU to visit YPC during reading week, a period when no normal classes are scheduled. However, although potentially transformative to those students involved, these trips were highly selective and available to a tiny proportion of students (‘*elitist and exclusive*’), and are high cost in both monetary terms and in their carbon footprint. For example, the total greenhouse gas emissions associated with air travel alone to Kuala Lumpur for ten students and a member of staff is approximately 33 tonnes. This is equivalent to the total annual emissions of five people from the UK or 2.5 Malaysians.[1] To extend the provision of an international experience more widely, we decided to try a virtual exchange project. The pandemic provided an opportunity to do this, and an extra-curricular non-credit-bearing activity was deemed to be a low-risk approach. This chapter provides a case study outlining our approach and draws on qualitative reflections to evaluate the learning and experience.

3.2 The Exchange Programme

The first exchange was scheduled during reading week in November 2020. Students were recruited three weeks before the start date. Students across all levels on three undergraduate business programmes (two of which are delivered at YPC) were sent an email by their programme leader and invited to complete an online application demonstrating their interest and commitment. All LJMU applicants were accepted, and once the numbers had been confirmed, YPC promoted the exchange via a poster to attract an equal number of students.

A site was created on the university’s virtual learning environment, Canvas, and students were enrolled so that they could access the Zoom sessions and course materials. The programme is summarised in Table 1. The task was to develop a presentation comparing how businesses in Liverpool and Kuala Lumpur were dealing with the impact of their teams. The guided reflection included sections on student learning in relation to culture and identity, personal growth and the challenges experienced during the exchange.

---

[1] This equivalence is calculated using Climatewatch figures of annual emissions per capita in 2019 in the UK (6.42 tonnes per capita) and Malaysia (12.4 tonnes per capita).

Table 1 Summary of exchange activities

Schedule	Activities	Session type	Platform
Day 1	Separate briefings to students at each institution focusing on cultural identity and intercultural communication	Facilitated session	Zoom
Day 2	Ice-breaking, team activities and task brief	Facilitated session	Zoom
Day 3–7	Students worked within their teams on an assigned group task	Independent (with optional facilitator support)	Students' own choice
Day 8	Submission of presentation Peer and staff evaluation via an online voting system	Independent	Canvas
Day 10	Virtual prize-giving ceremony	Facilitated	Zoom
End of semester	Submission of individual guided reflection (800–1,000 words) Audio feedback provided Certificate of participation awarded	Independent	Canvas

The student reflections included evaluations and suggestions which we embedded in the following iterations of the exchange in March 2021 and 2022. This included more ice-breaking sessions. In total, 57 students participated in the projects.

3.3 *The Study*

To evaluate the exchanges, a reflective approach was adopted (Schön 1983). The student reflections were reviewed by the first and second authors. The most detailed and insightful reflections of current students were identified (n = 6) and the students were invited to co-author the paper. Four students responded positively. The students shared their experiences, focussing on the benefits and challenges of participating, and agreed to write their own individual reflections. These reflections were shared with the practitioners, who critically reflected on the overall project.

4 Findings

4.1 *Reflections from Practitioners*

From the perspective of LJMU, and the first author, the exchanges provided an opportunity for students across several undergraduate programmes to engage with students from very different cultures. Due to the short duration of the programme, it required very little input in terms of resources, yet for participants, it opened a window into a different culture and initiated the first steps of a journey into intercultural learning. By learning from their Malaysian peers, they expanded their horizons and changed their view of their place in the world, but perhaps more importantly they realised how much they had in common with their Malaysian peers.

The greatest challenge was recruiting participants from LJMU when students already have a full curriculum as well as work and other responsibilities to juggle. This meant we had to tread a delicate balance between providing sufficient teacher-led sessions and engaging tasks without overloading and overwhelming students. By demonstrating that these exchanges can work, I have been able to encourage colleagues to embed similar initiatives within their courses.

From the perspective of YPC International College and the second author, the international virtual exchange initiative improved accessibility and participation in cross-cultural learning as well as fostered exchange opportunities among underrepresented student groups. These opportunities helped develop our future leaders for an ever-globalizing and interconnected world, especially in relation to digital transformation. The international virtual exchange initiative is unique as it provides an innovative international networking platform, and engagement and cultural exchange opportunities without any financial barriers. This then creates a pathway for students from a myriad of institutions to connect and actively engage with each other.

4.2 *Reflections from Participants*

Augusto

> It was interesting to work with the Malaysian students and listen to their perspectives on the West. The main thing that I learnt from the experience was how to behave in a more formal international corporate setting as I had no previous experience of this, particularly in respect to understanding cultural differences. The exchange program helped me to understand the impact of Covid-19 on businesses as well as how it was felt on a personal level by all students. This helped me accept that the extraordinary conditions I was living in during lockdown were also faced by other people, and it reduced my levels of anxiety.
>
> I also learnt a little about Malaysia, its different cultures and how the people from those different cultures live together. The exchange program extended my cultural horizons so I could experience working in an international setting. I would feel comfortable working with Malaysians in the future. We discovered that working remotely is easy. By writing ‘handover’

comments about our own contributions to the project, we only had calls when necessary, which overcame the constraints of the time difference. Overall, the final project worked out well and my group managed to deliver a good presentation. This experience expanded my academic curriculum and has enhanced my CV.

The exchange program took part during the reopening phase of the pandemic in the UK. However, in Malaysia, they were still facing strict measures meaning that students were keener to engage in activities in the virtual medium, and I struggled to keep some of the UK students engaged. The motivation levels dropped significantly after the first meeting and some students gave up halfway through the exchange program.

I have good memories of my peers in Malaysia, but we did not keep in touch. The relationship felt virtual, and so the friendships did not last. As far as long-distance/virtual relationships go, I feel this was inevitable. Although I learnt about Malaysian culture, I did not live those experiences. I feel that this virtual exchange program could be improved by introducing some in-person aspects that have a low carbon footprint. For instance, students in the UK could go to a Malaysian restaurant in the Liverpool city area, visit a local temple of one of the religions in Malaysia, or perhaps visit the Malaysian Embassy in London.

In the summer of 2022, I spent two weeks on a physical exchange programme in India. Compared to this experience of a physical exchange, the virtual one felt somewhat formal and did not provide in-depth personal value. I did feel that we learned about a culture by listening to the students on the other end about their experiences. In a physical exchange program, those differences would be learnt tacitly.

Adriana

I took part in the virtual exchange in November 2020. It was a fun way to interact with other students and get insights into their culture and hear new perspectives. We discussed how our general days in university would differ and the types of things we enjoyed in our spare time. We found some similarities in our student culture, but that each country had many different traditions. It was also interesting for the Malaysian students to share their thoughts on the UK and for us to give them a perspective of our day-to-day life away from typical British stereotypes, particularly as none of the JMU students was from London which is what the Malaysian students were most familiar with. Learning how to connect over Zoom and finding ways to collaborate over a long distance was something new. It enhanced my digital skills and gave me more confidence. I am sure this will be useful in the future as the world adapts to new ways of working post-pandemic and is evidence I could use to show future employers.

I completed a study-abroad year in the USA in 2021/2022. After completing the virtual exchange, I felt well-equipped for my pre-departure to the USA. During pre-departure sessions for the exchange to the USA, we would speak to other international students who would be attending. After completing the virtual exchange, I felt much more well-equipped and confident participating in these online meetings. I found that the main difference between completing a university exchange in-person to online was the level of connection you can make with people and the way in which you can understand the new culture by experiencing it. Inevitably, the longer time makes it easier to create long-lasting friendships and connections. However, even when participating in online classes and sessions when abroad, the majority of the time we would prefer meeting up in groups to discuss work and complete tasks rather than online. I think it can be difficult sometimes to replicate the same level of engagement and cohesion virtually.

Jamie

Personally, I have experienced both the physical student exchange programme which took place before the pandemic occurred and the virtual one which happened during the pandemic.

Based on my personal experience, I feel that we can't compare the two as both sessions presented a different level of engagement.

I personally enjoyed and preferred the physical one when I was a host to UK students. This is because we were able to interact face-to-face which I feel created a stronger connection and bond throughout. I have maintained those friendships to this day. During the programme held at YPC International College, we participated in ice-breaking, and games and lastly presented a report at the end of the programme. It was truly an eye-opening experience as we got to experience and share each other's opinions and ideas coming from different regions and cultural perspectives. Other than that, we get to spend time with the students during lunch and outings which was nice. My overall experience was a positive one and I believe all the students enjoyed it too.

My view on the virtual student exchange programme is slightly different, as there are a few challenges that we faced during the programme. I think overall it was a good learning experience for everyone, however, it lacked engagement and participation from some of the students. I realised that during the virtual programme, many students were too shy to make conversation or share their thoughts. In addition, the difference in time zones made it hard for the students to set up meetings for discussions which brought a lesser quality outcome at the end of the day. I believe that when given time, experience, and better planning, the virtual programme would thrive and provide a better experience more like the physical programme.

Ain

I experienced both virtual exchanges in May 2022 and was also a host to ten LJMU students who visited Kuala Lumpur in July 2022. From both experiences, I have gained a new mindset of 'abundance', in the sense that I believe that opportunities are endless for me. In addition, I have formed new friendships & connections.

During both exchanges, I found that the UK students I met and talked to tended to be more comfortable in voicing their opinion even if their opinion differs from the majority, albeit toned down during the virtual exchange. They also have this sense of curiosity about the world and different cultures around them, which could be mistaken as ignorant if one is not used to it. I found it rather endearing. Compared to the UK students, Malaysian students were shy to talk, particularly in the virtual setting. However, given the opportunity and time to express themselves, the ideas they give out are amazing.

Obviously, there are some challenges that come with the programmes I've participated in. When it comes to the physical student exchange, one of the limitations was the cost as some students were fully funded whereas additional participants who were interested in participating had to cover the costs themselves. Certain trips were too costly, for example, the trip to the Royal Selangor Pewter factory was RM 100 (approximately £20), which was expensive for us host students.

For me, the main limitations of both experiences would be the length of the programs. It would be amazing if both programmes were longer than a few weeks. This would give us enough time to adjust to other students' communication styles and personalities. The biggest challenge of the virtual exchange was the time difference. During that time, I had a problem converting UK time to Malaysian time, which led to our meeting being held at midnight! It was my mistake, but I should've known better.

Overall, I enjoyed my time during both student exchange programmes. However, the physical one was much more enjoyable for several reasons. The interactions I had with the UK students were much more meaningful and we were able to experience trips to different locations in Kuala Lumpur together in the real world. The human aspect of this is what makes it more special in my heart and a joy to be a part of.

I hope in the future, more students, both Malaysian and UK students will have the opportunity to gain more insight into each other's culture. In the Malay language, we have this saying 'rambut sama hitam, hati lain-lain' which means we might look the same but we also have different opinions on numerous matters. This also works in reverse, in a way that we might have a different appearance from each other, but we are not so different from each other anyway. I think that's the beauty of humanity.

5 Discussion

All the student authors in this paper had experience with a both virtual and physical exchange. It is clear from their reflections that a short virtual relationship cannot create the deeper connections that are developed when they can '*see 'the other' as a real being who exists in a physical world*' (Magnier 2022). However, there were clear benefits in terms of confidence building, enhanced digital communication skills and a general broadening of horizons. To be considered a true 'virtual exchange' in the definition of O'Dowd (2021), these opportunities must be embedded in the curriculum. This would enable extended periods of engagement to enhance the student experience and improve learning outcomes. In such a short period of time, students were not able to build relationships, and staff were not able to include sufficient activities to foster more critical intercultural learning.

Issues highlighted by students included the time difference, communication barriers and lack of engagement. Intercultural communication strategies including linguistic accommodation were introduced to students at the start of each exchange, but an extended programme would have allowed more ongoing reflection and practice to ensure all students felt confident to participate fully. Students were alerted to the eight-hour time difference during the recruitment phase, and the majority adopted appropriate strategies to overcome this challenge including the use of text-communication tools. Allowing students the flexibility to select their own technology for communication enabled this, but any international project needs to consider such practical challenges in the development stage.

The lived experience of a physical exchange cannot be replicated online but using a variety of media such as students' own videos and photographs or a cultural experience, such as a visit to a Malaysian restaurant, or even virtual reality (Akdere et al. 2021), could help bring the culture to life and incentivise engagement.

COIL and virtual exchange can provide an inclusive and sustainable way to foster students' intercultural awareness and enhance their digital and other transferable skills, but there are still institutional and policy barriers to widespread uptake (Garcés and O'Dowd 2021). By trialling these extra-curricula exchanges, we can begin to dismantle some of those barriers and begin the process of embedding opportunities within our programmes. A summary of lessons learned and suggested strategies that could be adopted by other HE institutions are shown in Table 2.

Despite the climate impact of student mobility, few within HE would argue that we should completely abandon it. But institutions must ensure that they integrate their internationalisation policies with those on equality and diversity as well as sustainability. One study has shown that experience of virtual mobility can lead to uptake of physical mobility (Lee et al. 2021) so if virtual exchange opportunities are integrated

Table 2 Lessons learned and suggested strategies

Lessons learned	Suggested strategies
Extra-curricular activities are low risk, can stimulate student curiosity and can break down institutional barriers but are not inclusive	Embed opportunities early in the curriculum to foster curiosity and ensure inclusivity, use extracurricular activities to pilot ideas and provide incentives for engagement
Practical challenges such as time differences and digital access need addressing	Consider these first and allow flexibility in terms of choice of technology
Online meetings appear more formal and relationship building is, therefore, more time-intensive	Provide sufficient time and opportunities for students to build relationships
Students may lack confidence due to personal, language or cultural reasons	Provide guidance and practice on intercultural communication strategies including linguistic accommodation and technology use (captions/text chat)
Experiential learning is limited online compared to physical exchange	Use a variety of media (pictures/video/cultural experiences) to bring culture alive

at the start of undergraduate programmes, students from disadvantaged groups might be more likely to take part in physical mobility. Short-term mobility is usually selective because of financial and other practical constraints, so if these initiatives continue to grow, institutions should ensure that selection processes facilitate the inclusion of those that have the least cultural capital rather than the most. Rigorous evaluation of these programmes must ensure that short-term mobility has the outcomes that are intended and are not simply pleasure jaunts. For all mobility programmes, replacing air travel with water, train and road options should be considered where possible. As argued by a growing number of voices, if universities integrate the concepts of internationalization and sustainability, they will be better placed to tackle global challenges and create a better future for all (Gregersen-Hermans 2021; Leask and Gayardon 2021; Ramaswamy et al. 2021; Leal Filho et al. 2023).

6 Conclusion

This pilot study has shown that a short extra-curricula COIL programme can broaden students' horizons, build confidence and open their minds to further opportunities. Provided basic practical issues are sufficiently addressed, such programmes are financially and environmentally sustainable so that they can be scaled up to facilitate more inclusive internationalisation at home. COIL cannot replicate completely the benefits of physical mobility but as technology continues to develop, further research and exploratory practice should be conducted to find new ways to enhance the online experience. The students in this pilot study all had sufficient access to the internet and appropriate technologies, but this is not the case in HE globally so more effort is required to provide equal access to international education. As time is running out to save the planet, HE institutions need to move quickly to integrate education for sustainable development with internationalisation goals as well as their diversity and inclusion policies, not only for the benefit of students but for the good of society.

References

Akdere M, Acheson K, Jiang Y-L (2021) An examination of the effectiveness of virtual reality technology for intercultural competence development. Int J Intercult Relat 82:109–120. https://doi.org/10.1016/j.ijintrel.2021.03.009

Appiah-Kubi P, Annan E (2020) A review of collaborative online international learning. Int J Eng Pedagog Manag Syst Fac Publ 10(1):109–124. https://doi.org/10.3991/ijep.v10i1.11678

Asodar M, Atai M, Baten L (2017) Successful Erasmus experience: analysing perceptions before, during and after Erasmus. J Res Int Educ 16(1):80–97. https://doi.org/10.1177/147524091770433

Caruana V (2014) Re-thinking global citizenship in higher education: from cosmopolitanism and international mobility to cosmopolitanisation, resilience and resilient thinking. High Educ Q 68(1):85–104. https://doi.org/10.1111/hequ.12030

Chan SJ (2012) Shifting patterns of student mobility in Asia. High Educ Pol 25:207–224. https://doi.org/10.1057/hep.2012.3

Chang S, Gomes C (2022) Why the digitalization of international education matters. J Stud Int Educ 26(2):119–127. https://doi.org/10.1177/10283153221095163

Climate Watch Historical GHG Emissions (2022) World Resources Institute, Washington, DC. https://www.climatewatchdata.org/ghg-emissions

European Commission (2022) Erasmus+ priorities 2021–2027 for higher education mobility. https://erasmus-plus.ec.europa.eu/programme-guide/part-a/priorities-of-the-erasmus-programme/objectives-features

Garcés P, O'Dowd R (2021) Upscaling virtual exchange in university education: moving from innovative classroom practice to regional governmental policy. J Stud Int Educ 25(3):283–300. https://doi.org/10.1177/1028315320932323

Gregersen-Hermans J (2021) Toward a curriculum for the future: synthesizing education for sustainable development and internationalization of the curriculum. J Stud Int Educ 25(4):461–548. https://doi.org/10.1177/10283153211031033

Heleta S, Chasi S (2022) Rethinking and redefining internationalisation of higher education in South Africa using a decolonial lens. J High Educ Policy Manag. Published online 13 November 2022. https://doi.org/10.1080/1360080X.2022.2146566

Higher Education Statistics Agency (HESA) (2019) Higher education student statistics: UK 2018/19. Higher Education Statistics Agency (HESA). https://www.hesa.ac.uk/news/16-01-2020/sb255-higher-educationstudent-statistics

Jørgensen M, Mason A, Pedersen R, Harrison R (2022) The transformative learning potential in the hybrid space between technology and intercultural encounters. J Stud Int Educ 26(3):318–333. https://doi.org/10.1177/1028315320976030

Leal Filho W, Viera Trevisan L, Dinis MAP, Sivapalan S, Wahaj Z, Liakh O (2023) Ensuring sustainability in internationalisation efforts at higher education institutions. Int J Sustain High Educ vol. ahead-of-print No. ahead-of-print. https://doi.org/10.1108/IJSHE-10-2022-0333

Leask B (2015) Internationalizing the curriculum. Routledge, London

Leask B, de Gayardon A (2021) Reimagining internationalization for society. J Stud Int Educ 25(4):323–329. https://doi.org/10.1177/10283153211033667

Leask B, Jones E, de Wit H (2018) Towards inclusive intercultural learning for all. University World News. 7th December. https://www.universityworldnews.com/post.php?story=20181205093157690

Lee J, Leibowitz J, Rezek J (2021) The impact of international virtual exchange on participation in education abroad. J Stud Int Educ 26(2):202–221. https://doi.org/10.1177/10283153211052777
Lewis T, O'Dowd R (2017) Introduction to online cultural exchange and this volume. In O'Dowd R, Lewis T (eds) Online intercultural exchange. Routledge, pp 3–20
Magnier A-A (2022) 'A mountain to climb: climate-proofing our programmes. Our changing climate: member forum magazine. European Association for International Education, Spring 2022. https://www.eaie.org/our-resources/library/publication/Forum-Magazine/2022-spring-forum.html
Molthan-Hill P, Blaj-Ward L (2022) Assessing climate solutions and taking climate leadership: how can universities prepare their students for challenging times? Teach High Educ 27(7):943–952. https://doi.org/10.1080/13562517.2022.2034782
Naicker A, Singh E, van Genugten T (2022) Collaborative online international learning (COIL): preparedness and experiences of South African students. Innov Educ Teach Int 59(5):499–510. https://doi.org/10.1080/14703297.2021.1895867
Nolan E, Héliot YF, Rienties B (2022) Encouraging intercultural interaction by cultural specific learning design. J Stud Int Educ. https://doi.org/10.1177/10283153221145083
O'Dowd R (2021) Virtual exchange: MOVING forward into the next decade. Comput Assist Lang Learn 34(3):209–224. https://doi.org/10.1080/09588221.2021.1902201
OECD (2018) Preparing our youth for an inclusive and sustainable world. The OECD PISA global competence framework. https://www.oecd.org/pisa/Handbook-PISA-2018-Global-Competence.pdf
Pederson P (2010) Assessing intercultural effectiveness outcomes in a year-long study abroad. Int J Intercult Relat 34(1):70–80. https://doi.org/10.1016/j.ijintrel.2009.09.003
Proctor D (2022) Editorial. In our changing climate: member forum magazine. European Association for International Education, Spring 2022. https://www.eaie.org/our-resources/library/publication/Forum-Magazine/2022-spring-forum.html
Ramaswamy M, Marciniuk DD, Csonka V, Colò L, Saso L (2021) Reimagining internationalization in higher education through the United Nations sustainable development goals for the betterment of society. J Stud Int Educ 25(4):388–406
Roy A, Newman A, Ellenberger T, Pyman A (2019) Outcomes of international student mobility programs: a systematic review and agenda for future research. Stud High Educ 44(9):1630–1644. https://doi.org/10.1080/03075079.2018.1458222
Rubin J (2016) The collaborative online international learning network: online intercultural exchange in the State University of New York Network of Universities. In O'Dowd R, Lewis T (eds) Online intercultural exchange. Routledge, pp 263–270
Sawir E (2013) Internationalisation of higher education curriculum: the contribution of international students. Glob Soc Educ 11(3):359–378. https://doi.org/10.1080/14767724.2012.750477
Schön DA (1983) The reflective practitioner: how professionals think in action. Temple Smith
Shields R (2019) The sustainability of international higher education: student mobility and global climate change. J Clean Prod 217:594–602. https://doi.org/10.1016/j.jclepro.2019.01.291
Universities UK International (2017) Widening participation in UK outward student mobility: a picture of participation. https://www.universitiesuk.ac.uk/topics/international/widening-participation-uk-outward
Universities UK International (2020) International facts and figures 2020. https://www.universitiesuk.ac.uk/universities-uk-international/insights-and-publications/uuki-publications/international-facts-and-figures-2020

Universities UK International (2022) International facts and figures 2022. https://www.universitiesuk.ac.uk/universities-uk-international/insights-and-publications/uuki-publications/international-facts-and-figures-2022

Viera Trevisan L, Eustachio JHPP, Dias BG, Leal Filho W (2023) Digital transformation towards sustainability in higher education: state-of-the-art and future research insights. Environ Dev Sustain. https://doi.org/10.1007/s10668-022-02874-7

Wimpenny K, Finardi KR, Orsini-Jones M, Jacobs L (2022) Knowing, being, relating and expressing through third space global south-north coil: digital inclusion and equity in international higher education. J Stud Int Educ 26(2):279–296. https://doi.org/10.1177/10283153221094085

Sustainability Competence Assessment Validation

Alice Annelin and Gert-Olof Boström

Abstract Education for sustainability (EfS) is recognised by institutions such as UNESCO, national education acts and higher education as of key importance. Business schools have been implementing the UN's sustainable development goals, creating organisations such as the PRME. Support for implementing an advanced EfS is on demand. To plan education for sustainability, teachers need to know the students' level of competencies for sustainability before they come to class to incorporate a learning progression approach. Thus, a formative assessment about student competence for sustainability is needed. This study provides a validation of a Sustainability Competence Assessment tool. Questionnaire scales that measure sustainability competence have been tested in prior research, but much work is required to validate the scales. We have designed an assessment tool that includes all eight key sustainability competencies (KSC). Several statistical tests (Hair et al. Eur Bus Rev 31(1):2–24, 2019; Lans et al. J Clean Prod 62:37–47 2014) have been conducted to validate the sustainability competencies scales with participants at a university business school, comparing results across modules and programs. Results show there is some validity in using all eight KSC, but more investigation is needed to improve the tool's validity.

Keywords Key sustainability competencies · Assessment · Validate · Education for sustainability · Sustainable development

1 Introduction

It is well understood that education plays a vital role in the development of student competencies for sustainability (Wiek et al. 2011; UNESCO 2017; Shephard et al. 2019). Since the higher education act law in Sweden introduced a mandate that

A. Annelin (✉) · G.-O. Boström
USBE Umeå University, Biblioteksgränd 6, 90187 Umeå, Sweden
e-mail: alice.annelin@umu.se

G.-O. Boström
e-mail: gert-olof.bostrom@umu.se

T. Wall et al. (eds.), *Sustainability in Business Education, Research and Practices*, World Sustainability Series, https://doi.org/10.1007/978-3-031-55996-9_14

education for sustainability is applied in higher education (The Swedish Council for Higher Education 2020), universities in Sweden have developed policies, visions and missions that include sustainable development, Agenda 2030 and the United Nation's (UNs) sustainable development goals (SDGs). However, advancing student sustainability competencies in practice by the lecture, in the classroom, has been difficult for individual teachers (O'Byrne et al. 2015) since support mechanisms have not been readily available.

While many lecturers are engaged in the effort to ensure that future leaders can work towards achieving the goals (Lozano et al. 2015; Leal Filho et al. 2021), it has been recognised that student competency for sustainability can vary from year to year (Annelin and Boström 2022). Also, many teachers feel a need to better provide learning progression, rather than disjointed add-on lectures or seminars (Wals 2014). To advance the education for sustainability (EfS) in higher education and provide a learning progression approach to EfS teachers need to know what level of competence for sustainability each student group has before they come to class.

A formative self-assessment of the sustainability competencies (Redman and Wiek 2021) can help support teachers with information about the student cohort's attitudes towards sustainability and their abilities and skills to have an integrated problem-solving approach to sustainability. While the knowledge about sustainability has already been taught across disciplines in higher education throughout many countries (Wals 2014; Bamber et al. 2016; Leal Filho et al. 2021), working with the sustainability competencies can help the cohort to develop a mindset for sustainability by addressing the whole student. This student-centred approach allows the teacher to adapt the class to the students' needs each semester, providing insight into the cohort's sustainability competence at each stage.

A sustainability competence assessment tool (Annelin and Boström 2022) can be used to follow each cohort throughout their educational programs and set goals to advance the sustainability competency of the cohort in line with the expected learning outcomes of each module in the program. As the teacher groups of each semester work together to use the information gained from the sustainability competence assessment tool, the teachers can integrate sustainability into the curriculum with a learning progression approach (Wilson and Lehrer 2021). The tool can be used alongside other assessment tools (Redman et al. 2021) that measure the students' knowledge about sustainability and other skills that enable students to work towards sustainability transformations.

Several frameworks describe the key sustainability competencies (Wiek et al. 2011; Brundiers et al. 2020; Redman and Wiek 2021) as a set of skills and abilities that can help transform mindsets to be able to work towards a sustainable future. The key sustainability competencies (KSC) include systems thinking, strategic thinking, futures thinking, values thinking, interpersonal thinking, intrapersonal thinking and implementation thinking can together lead to an integrated problem-solving competency for sustainability. We have designed an assessment tool with these KSC scales that has been tested across student cohorts at different levels in the programs to provide teachers with valid and reliable information about their students' sustainability competency. With a valid and reliable (Hair et al. 2020) sustainability competence assessment tool, teachers can advance their goals for sustainability education throughout their curriculum and improve their lesson plans for sustainability.

In the next section, the key sustainability competencies assessment will be discussed. In Sect. 3, the method used to validate the assessment scales will be presented and in Sect. 4 results of validation tests will be produced. Finally, a discussion about how to advance this work will conclude this paper.

2 Literature Review: The Key Sustainability Competencies

There are several insights that discuss research about the key sustainability competencies (KSC) and have developed frameworks that describe the competencies (Barth et al. 2007; Wiek et al. 2011; Brundiers et al. 2020; Redman and Wiek 2021). Although this research has given different labels to the types of sustainability competence, there are eight different sustainability competencies recognised as important for student education (Brundiers et al. 2020; Redman and Wiek 2021). The eight KSC include systems thinking, strategic thinking, future/anticipatory thinking, values/norms thinking, interpersonal/collaborative thinking, intrapersonal (self-awareness/mind-set) thinking, implementation/action thinking, which together create an integrated problem-solving competency for sustainability. While there might be some slight variation in the way researchers understand the competencies for sustainability, they all agree that students need to develop sustainability competencies to be able to transform society towards a sustainable future.

There is also some literature reviewed (Lozano et al. 2017; Backman et al. 2019; Birdman et al. 2021) about pedagogical approaches that teachers may want to use to teach for sustainability. Evidently, the activities we choose to use in the classroom to teach for sustainability can help to encourage engagement with sustainability problems to provide solutions and a positive mindset (Shephard 2008), rather than anxiety and pessimistic denial. For teachers to better plan classroom activities for sustainability education the teacher also needs to know about the student group's sustainability competency before they come to class (Annelin and Boström 2022). There have been a wide variety of different qualitative and quantitative assessments that have measured student competencies for sustainability (Redman et al. 2021). While there have been advantages and disadvantages to many of these approaches to sustainability competence assessment, the quantitative questionnaire survey has had its benefits in reaching larger samples, efficiently and with actionable results.

While qualitative assessment tools, such as journal reflective writing, focus group interviews and observation, have made some progress in assessing student sustainability competence (Redman et al. 2021), quantitative sustainability competence assessment tools have not been robustly tested. The most frequently used quantitative assessment tool by teachers and action researchers is the questionnaire survey that provides evidence about student perceptions of their competence for sustainability (Lans et al. 2014; Azeiteiro et al. 2015; Savage et al. 2015; Ateskan and Lane

2018; Meza et al. 2018; Molderez and Fonseca 2018; Ploum et al. 2018; Cabral and Lochan Dhar 2019). The self-assessment questionnaire is an effective tool for formative assessment (Andrade 2019) and can help students to reflect about sustainability in a manner that can develop their intrapersonal competency for sustainability (Galt et al. 2013). Thus, a sustainability competence assessment questionnaire tool (Annelin and Boström 2022) can provide a student learning experience as well as assessment for the teacher to use as a planning tool for classroom activity and curriculum development.

Quantitative self-assessment questionnaires have been criticised for their reliability and validity of measurements (Holdsworth et al. 2018; Cebrián et al. 2019), but with robust statistical tests and concrete use of the questionnaire, the tool can become quite advantageous as well (Hair et al. 2020; Redman et al. 2021). Although much work has been done to assess student competencies with help from the self-assessment tool, there is still a need for a validation study that can test the KSC across educational programs and between the different student cohorts. Such a study will require statistical testing and validation development over several years. Thus, this research questions

RQ1. Do the sustainability competence scales validate across higher education programs?

RQ2. Do the sustainability competence scales validate across higher education between student cohorts on the same program?

3 Method

3.1 Questionnaire Design

The questionnaire was designed to capture the student's perception of competence for sustainability. The participants are first asked some background questions to help control for demographic differences between the students. The main sustainability competence scales are then presented in different question formats using plain language, to help decrease common method bias (Podsakoff et al. 2003). Although the scales are taken from previously tested research to reduce context-induced bias, many of the scales are not well validated (Savage et al. 2015; Meza et al. 2018); except the scale for intrapersonal thinking competence that is taken from the New Environmental Paradigm (Dunlap 2008; Hawcroft and Milfont 2010; Zhu and Lu 2017).

To reduce any desirability bias, the participants responses are completely anonymous to the researchers. The research is considered exempt from an Internal Review Board inspection, since it does not provide sensitive information defined in Sweden under the Act (2003: 460). We followed ethical guidelines for conducting the surveys. This includes that participation is voluntary, consent was given to using the data for research purposes, and anonymity and confidentiality is guaranteed.

3.2 Variable Measurements

The KSC scales are taken from a literature review (Annelin and Boström 2022) conducted about questionnaire assessment scales used to measure the KSC. The scales designed in the Savage et al. (2015) and Meza et al. (2018) studies have been replicated for systems, strategic, futures, values and interpersonal thinking competency. The question they put to participants asks to rank on a 5-point Likert-type scale from strongly disagree to strongly agree, *how much do you agree with the following statements?* The participants answer three items per sustainability competence scale with the prefix, *I feel confident and competent to...*After several pilot tests using these scales, we found that the validity and reliability could be too low, so we have worked on the face validity of the scales, which includes an adaption of the language of the scales and the number of items on the scales. These changes will be discussed further in the results and conclusion sections.

The implementation thinking competency for sustainability has been designed by the authors and based on the discussion about the competence in the Brundiers et al. (2020) paper as well as the definition described in the Redman and Wiek (2021) study. Four items were used to measure the construct for implementation thinking. The question asks *How well are you able to...* to rank their ability to act for sustainability on a 5-point Likert-type scale, from Not at all well to Extremely well.

The intrapersonal thinking competency for sustainability is taken from a well-tested scale (Dunlap 2008; Zhu and Lu 2017) called the New Environmental Paradigm. The scale asks participants *How much do you agree with the following statements?* and to rank on a 5-point Likert-type scale, from strongly disagree to strongly agree, their response to 15 items. The items are varied in format, such that some are positively stated, and some are negatively stated, which adds to the validity of the statements. The 15 items are said (Zhu and Lu 2017) to be valid across different contexts in different countries, which indicates that the scale could be the most robust scale of all the variable measurements of this study.

3.3 Data Collection

We conducted this study at Umeå School of Business, Economics and Statistics (USBE) at the social science faculty of Umeå University in the north of Sweden. There are about 3000 students that study at USBE, which are recruited from across Sweden and internationally. Data was first collected in the beginning of the Autumn term of 2020, where we asked every student at the business school to participate in a pilot study. After removing those with missing values, there were 62 full responses to the sustainability scales that gives a response rate of 2%. This was, of course, less than ideal, so we decided to test the questionnaire with one module in each semester. We asked the teachers to introduce the questionnaire at the beginning of a lecture

Table 1 Response rates

Program	Module	Semester	2021 (%)	2022 (%)
Business-english	Management accounting	1	68	59
	Service marketing	4	55	63
	Financial accounting	4	89	
	Accounting and finance master	7	48	63
Business and economics-swedish	Introduction to business administration	1		54
	Thesis in business or economics	8		43
Civil engineering	Industrial economics	1	64	93
	Financial and management accounting	5		62

and give the students 10–15 min to respond. This helped to improve the response rate tremendously.

Our second pilot test was in a module at the end of the same semester in 2020, and we had a response rate of 80%. The method of introducing the project and giving the students time to take the questionnaire during lecture time has helped response rates ever since. The questionnaire link is now provided to teachers with a three-minute video attached to the questionnaire, so that the students can listen to information from the researchers about the project before they take the questionnaire.

The questionnaire is provided in both English and Swedish. After the pilot tests, we adjusted some of the background questions and wording in the Swedish language, to improve the function of the questionnaire. The original questionnaire by Savage et al. (2015) and the questionnaire we designed with all eight KSC were then tested in 2021 and 2022, in several modules at the beginning and end of each program in three different programs. The programs involved with this study include two business programs, one is taught in English, and one taught in Swedish, and one civil engineering program that is taught in Swedish and in cooperation with the Department of Mathematics and Mathematical Statistics at the science and technology faculty. The response rates of each cohort can be seen in Table 1.

3.4 Validity and Reliability Tests

The tests are first conducted between two modules (see Panel A, Table 2) at the same time by using the original sustainability competence questionnaire by Savage et al. (2015). Then all the sustainability competence scales are tested with data from different modules in the same year. Validity and reliability tests were conducted for all student cohort data following Hair et al. (2019)'s research on tests for ordinal data.

Table 2 Descriptive statistics-sustainability competence constructs

Panel A: Savage et al.'s (2015) sustainability competence scales

	Mean	sd	Spearman correlation				
			Value	Inter	Strategy	Systems	Future
FA 2021							
Value	3.53	0.72	1				
Inter	4.18	0.52	0.4023*	1			
Strategy	3.46	0.78	0.4873*	0.2202	1		
Systems	3.57	0.77	0.5754*	0.1845	0.6431*	1	
Future	3.76	0.61	0.3964*	0.5347*	0.4088*	0.3541*	1
N = 40							
SM 2021							
Value	3.62	0.56	1				
Inter	4.11	0.49	0.5333*	1			
Strategy	3.41	0.69	0.6512*	0.5883*	1		
Systems	3.57	0.75	0.4804*	0.4323*	0.6781*	1	
Future	3.88	0.63	0.5882*	0.7200*	0.7005*	0.5560*	1
N = 36							

Panel B: All Sustainability Scales tested across modules in 2022

	Mean	sd	Spearman correlation						
			Act	Intra	Value	Inter	Strategy	Systems	Future
IBA 2022									
Act	3.47	0.47	1						
Intra	3.37	0.29	0.0335	1					
Value	3.21	0.63	0.2021*	0.182	1				
Inter	3.75	0.56	0.1850	0.0075	0.2482*	1			
Strategy	3.25	0.74	0.2230*	0.0811	0.5268*	0.1571	1		
Systems	3.42	0.68	0.3184*	0.0998	0.4521*	0.2090*	0.6027*	1	
Future	3.51	0.62	0.1094	0.1828	0.5248*	0.4104*	0.3906*	0.5119*	1
N = 106									
TBE 2022									
Act	3.23	0.57	1						
Intra	3.33	0.24	−0.1252	1					
Value	3.40	0.56	0.2273	0.3372*	1				
Inter	3.92	0.43	0.5841*	−0.1589	0.4456*	1			
Strategy	3.25	0.80	0.2218	0.4766*	**0.7971***	0.3402*	1		
Systems	3.90	0.44	−0.0473	0.3247*	0.5722*	0.2319	0.4362*	1	
Future	3.79	0.35	0.1644	0.0349	0.3784*	0.2643	0.1082	0.4766*	1

(continued)

Table 2 (continued)

Panel B: All Sustainability Scales tested across modules in 2022									
	Mean	sd	Spearman correlation						
			Act	Intra	Value	Inter	Strategy	Systems	Future
N = 42									
FMA 2022									
Act	3.50	0.66	1						
Intra	3.23	0.22	0.1350*	1					
Value	3.33	0.57	0.6107*	0.0205	1				
Inter	3.96	0.52	**0.7176***	0.4007*	0.3536*	1			
Strategy	3.17	0.69	0.3086*	−0.3214*	0.0628	0.0871	1		
Systems	3.57	0.46	0.3349*	0.2435*	-0.1071	0.5934*	−0.0147	1	
Future	3.44	0.38	0.2385*	0.3107*	0.1070	0.4957*	−0.1542*	0.1621*	1
N = 31									
SM 2022									
Act	3.34	0.56	1						
Intra	3.32	0.28	0.0779	1					
Value	3.39	0.61	0.5085*	0.1063	1				
Inter	4.05	0.58	0.3284*	−0.0109	0.4971*	1			
Strategy	3.36	0.63	0.2387	0.1277	0.5903*	0.3278*	1		
Systems	3.57	0.62	0.3547*	−0.0556	0.6861*	0.4757*	0.5118*	1	
Future	3.66	0.53	0.4272*	0.0107	0.6107*	0.3956*	0.4594*	0.7062*	1
N = 41									

Notes FA = Financial Accounting, SM = Service Marketing
Notes IBA = Intro. Business Administration, TBE = Thesis in business or economics, FMA = Finance and Management Accounting, SM = Service Marketing

Firstly, the indicator loadings can be checked to see if they are above 0.50 and preferably at 0.78, but not too high either and that all weights of indicators of latent variables are statistically significant (p-value = 0.001). Secondly, internal consistency is tested by a reliability test for how the items measure the construct. Cronbach's coefficient alpha test result is considered acceptable when the level is greater than 0.70 (Cronbach 1951), or when the composite reliability test is at a value between 0.60 and 0.70 (Hair et al. 2019).

Thirdly, convergent validity can be tested by checking that the average variance extracted (AVE) value is above 0.50 (Hair et al. 2019). Fourthly, discriminant validity can be tested with the heterotrait-monotrait (HTMT) ratio, which is accepted at a value not significantly different from 1 or above 0.90 (Hair et al. 2019). Also, collinearity of the formative measure that can be tested by the variance inflation factor (VIF), which is acceptable at values above 1 and below 3 (Hair et al. 2019).

4 Results

4.1 Descriptive Statistics

The mean scores, standard deviation and spearman correlation coefficients are presented for the sustainability competence constructs in Table 2, Panels A and B, to check for potential discriminant validity problems. Since most of the correlated coefficients are below $r = 0.70$, it is likely that there are low discriminant validity problems with the sustainability competence scales. Therefore, there is some evidence that the constructs are measuring different aspects of sustainability competence.

However, the civil engineering program (see Panel B) there are some correlated coefficients slightly higher than 0.70 between the strategy, systems and values sustainability competence constructs. This result indicates further tests are required using the HTMT ratio. There are also some non-statistically significant coefficients, mostly with intrapersonal thinking competence. This result indicates that the NEP scale used for self-awareness or attitudes towards sustainability may not be valid in a business school context. This result is contrary to what we were to expect when using the NEP and deserves further investigation.

4.2 Reliability and Validity Tests of the Original Sustainability Competence Scales in 2021

Firstly, we checked the factor indicators for a level of 0.50 and above, and they were all acceptable. Secondly, we checked for internal consistency reliability with the Cronbach alpha test and the Composite reliability test (see Tables 3, 4 and 5). Results of the reliability tests for the original five sustainability scales (see Table 3) taken from Savage et al. (2015) and tested with two different modules in 2021 show that the Cronbach alpha and composite reliability are mostly acceptable. However, the values thinking competence for the service marketing module show slightly lower

Table 3 Reliability tests of Savage et al.'s (2015) sustainability competence scales

2021	Cronbach's alpha		Composite reliability (rho_a)		Average variance extracted (AVE)	
	SM	FA	SM	FA	SM	FA
Future	0.602	0.545	0.642	0.579	0.353	0.290
Inter	0.540	0.578	0.554	0.607	0.257	0.327
Strat	0.718	0.760	0.823	0.985	0.503	0.560
Sys	0.657	0.682	0.880	0.725	0.499	0.443
Value	0.373	0.842	0.360	0.847	0.156	0.638

Notes SM = Service Marketing, FA = Financial Accounting

Table 4 HTMT ratio for discriminant validity of original sustainability competence scales in 2021

2021	Future		Inter		Strat		Sys	
	FA	SM	FA	SM	FA	SM	FA	SM
Future								
Inter	1.022	1.075						
Strat	0.680	0.922	0.292	0.840				
Sys	0.679	0.959	0.359	0.717	0.824	1.024		
Value	0.547	1.213	0.447	1.142	0.554	1.308	0.711	1.008

Notes SM = Service Marketing, FA = Financial Accounting

Table 5 Variance inflation factor collinearity test of original sustainability competence scales in 2021

2021	SM	FA
Future1	1.388	1.094
Future2	1.370	1.302
Future3	1.090	1.216
Inter1	1.110	1.092
Inter2	1.354	1.333
Inter3	1.234	1.424
Strat1	1.226	1.304
Strat2	1.650	2.347
Strat3	1.772	2.043
Sys1	1.885	1.320
Sys2	2.122	1.784
Sys3	1.174	1.426
Value1	1.052	1.707
Value2	1.050	2.232
Values3	1.101	2.502

Notes SM = Service Marketing, FA = Financial Accounting

than acceptable value for internal consistency. Therefore, these results indicated that the scales could have weak reliability and need further development. For example, adapting the language or creating new statements might be required.

The average variance expected (AVE) test for convergent validity scores are also quite low for these scales in both samples (see Table 3). The Strategy thinking competence scale show higher than 0.50 variance in the scales, while systems thinking competence is a little lower and values thinking competence scale has a higher than 0.50 variance in the financial accounting module. Thus, the original scales show that they can have some convergent validity problems that need to be investigated. There is also evidence that the scales have some problems with discriminant validity (see Table 4). While the service marketing module sample showed acceptable validity between the systems, strategy and interpersonal thinking competence for sustainability, the financial accounting module sample had very good scores for discriminant validity across the scales.

The variance inflation factor (VIF) is then used to evaluate the collinearity between the item indicators for each structure. Results of the two samples for the original sustainability competence scales show acceptable scores between 1 and 3 (see Table 5). Thus, evidence indicates that there is no critical collinearity problems between the indicators of the original sustainability competence constructs.

4.3 Reliability and Validity Tests of All Sustainability Scales Between the Modules in 2022

The factor loadings test of the indicators for all sustainability competency scales showed that there were low factor loadings for the intrapersonal thinking competency scale (measured by the NEP), so several indicators were deleted for each sample. Once checked, each construct remained with at least three item indicators to measure the construct, to ensure that the construct measurement is not too narrow (Clark and Watson 2019). The same tests for samples within the three programs in the first semester and seventh semester in 2022, show that most sustainability competence scales have an acceptable reliability score (see Table 6). Again, the intrapersonal thinking competence (the NEP) scale indicates further possible problems with this construct, and the futures and systems thinking competence for sustainability scales show some weakness in the civil engineering program sample. Systems thinking competence scale also showed some weak reliability with the business in Swedish program sample. Thus, results show that further development of the sustainability competence assessment tool should be conducted in future research.

The AVE (see Table 6) results show weaker scores in the samples between modules in 2022. While the systems, strategy and implementation (act) thinking competency for sustainability scales show acceptable scores at 0.50 and above, this is not consistent across all samples. Also, the other sustainability competency scales show some problems with lower than 0.50 AVE scores. The intrapersonal thinking competence for sustainability (NEP) scale shows the most problems with low AVE scores.

However, the discriminant validity test HTMT ratio (see Table 7) indicate acceptable scores between the implementation competence (Act) scale and all other scales across all modules on the business in English program, but only on one of the modules in the other two programs in 2022 samples. There are some issues with discriminant validity with most sustainability competence scales except the values and strategy thinking competency for sustainability scales. Thus, the problematic scales may have problems that they are not measuring different constructs. Future research could investigate the items that define and measure the constructs further.

The VIF test for collinearity between the item indicators for each construct in the between programs samples revealed acceptable scores between 1 and 3 (see

Table 6 Reliability tests for samples between modules and programs in 2022

2022	Cronbach's alpha		Composite reliability (rho_a)		Average variance extracted (AVE)	
Civil engineering	IE	FMA	IE	FMA	IE	FMA
Act	0.7707	0.8197	0.8009	0.8231	0.4697	0.5322
Future	0.6686	−0.5232	0.6910	0.6054	0.4156	0.2207
Inter	0.4428	0.6207	0.4426	0.6365	0.2019	0.3488
Intra	0.3502	−0.0038	0.3524	0.2815	0.1535	0.1013
Strat	0.7757	0.7948	0.7825	0.8305	0.5358	0.5790
Sys	0.6831	0.3009	0.7034	0.3526	0.4220	0.1424
Value	0.7395	0.5639	0.7590	0.5740	0.4967	0.3034
Business in Swedish	IBA	TBE	IBA	TBE	IBA	TBE
Act	0.6843	0.7476	0.8342	0.7864	0.3764	0.4484
Future	0.6030	0.4414	0.6141	0.4105	0.3418	0.1862
Inter	0.6999	0.4357	0.7346	0.4565	0.4449	0.2169
Intra	0.5417	0.0474	0.5487	0.0684	0.2723	0.0239
Strat	0.7659	0.7987	0.8120	0.9180	0.5194	0.5588
Sys	0.6948	0.3642	0.7083	0.4342	0.4340	0.1901
Value	0.5516	0.5527	0.5559	0.5659	0.2898	0.2933
Business in English	MA	SM	MA	SM	MA	SM
Act	0.8148	0.8313	0.8294	0.8461	0.5294	0.5581
Future	0.6684	0.5918	0.6693	0.6083	0.3764	0.3359
Inter	0.5766	0.7621	0.6407	0.8281	0.3400	0.5282
Intra	0.6774	0.4668	0.6866	0.5070	0.4091	0.0605
Strat	0.8064	0.7281	0.8093	0.7603	0.5832	0.4844
Sys	0.7606	0.7360	0.7829	0.7957	0.5166	0.5137
Value	0.7028	0.6613	0.7099	0.6682	0.4373	0.3954

Notes IE = Industrial Economics, FMA = Financial and Management Accounting, IBA = Intro. Business Administration, TBE = Thesis in Business or Economics, MA = management Accounting, SM = Service Marketing

Table 8). The remaining indicators used to measure each construct after factor loadings were tested show to have little critical collinearity problems between the indicators of all sustainability competence construct scales. However, again, the intrapersonal thinking competency for sustainability scales (NEP) show some inconsistency between indicators in most of the samples that had low factor loadings and so need to be deleted. Conversely, the service marketing sample had no low factor loadings and showed no critical collinearity issues when using the VIF test (see Table 8, and the SM column).

A comparison of results between programs and with Table 8 show that items 1, 2, 3, 5, 10 and 15 should be deleted, since there is little support for their validity in

Table 7 HTMT ratio of all sustainability competence scales between programs in 2022

2022	Act		Future		Inter		Intra		Strat		Sys	
Business program in english	SM	MA	SM	MA	SM	MA	SM	MA	SM	MA	SM	MA
Future	0.700	0.569										
Inter	0.406	0.506	0.632	0.759								
Intra	0.429	0.487	0.435	0.281	0.407	0.329						
Strat	0.346	0.408	0.748	0.959	0.418	0.836	0.391	0.194				
Sys	0.403	0.459	0.982	0.874	0.565	0.797	0.405	0.191	0.721	0.919		
Value	0.700	0.527	0.895	0.949	0.702	0.812	0.437	0.272	0.811	0.755	0.825	0.655
Business program in swedish	IBA	TBE	IBA	TBE	IBA	TBE	IBA	TBE	IBA	TBE	IBA	TBE
Future	0.254	0.510										
Inter	0.312	1.143	0.711	0.961								
Intra	0.295	1.518	0.575	1.673	0.231	1.289						
Strat	0.288	0.765	0.635	0.569	0.422	0.642	0.207	0.914				
Sys	0.451	0.726	0.853	1.738	0.392	1.326	0.229	1.436	0.876	0.897		
Value	0.340	0.663	0.938	1.161	0.472	1.178	0.487	0.919	0.881	0.918	0.818	1.588
Civil engineering	IE	FMA	IE	FMA	IE	FMA	IE	FMA	IE	FMA	IE	FMA
Future	0.558	1.006										
Inter	0.858	0.941	0.835	1.117								
Intra	0.731	1.355	0.657	2.898	1.046	1.736						
Strat	0.334	0.346	0.665	0.807	0.669	0.309	0.563	0.651				
Sys	0.640	1.107	0.952	2.153	0.845	1.498	0.764	2.434	0.869	0.831		
Value	0.710	0.718	0.737	1.251	0.620	0.757	0.615	2.213	0.562	0.705	0.814	0.854

Notes IE = Industrial Economics, FMA = Financial and Management Accounting, IBA = Intro. Business Administration, TBE = Thesis in Business or Economics, MA = management Accounting, SM = Service Marketing

Table 8 Variance inflation factor test for collinearity of all sustainability competence scales between programs in 2022

2022	Business in english		Business in swedish		Civil engineering	
VIF	MA	SM	IBA	TBE	IE	FMA
Act1	1.612	2.796	1.217	2.789	1.731	3.251
Act2	1.723	2.246	1.462	1.634	1.688	3.618
Act3	1.896	1.675	1.321	2.038	1.290	2.136
Act4	1.591	1.733	1.277	1.249	1.483	2.297
Future1	1.294	2.059	1.144	1.040	1.558	1.051
Future2	1.284	2.340	1.267	1.304	1.257	1.030
Future3	1.540	1.319	1.247	1.263	1.331	1.045
Inter1	1.126	1.611	1.487	1.120	1.040	1.160
Inter2	1.256	1.465	1.322	1.079	1.166	1.289
Inter3	1.201	1.583	1.345	1.203	1.132	1.278
Strat1	1.863	1.564	1.323	1.481	1.739	1.708
Strat2	1.802	1.591	2.188	3.117	1.738	1.837
Strat3	1.631	1.294	1.938	2.481	1.441	1.580
Sys1	1.563	1.288	1.362	1.175	1.569	1.001
Sys2	1.472	1.643	1.485	1.172	1.379	1.116
Sys3	1.600	1.815	1.288	1.004	1.252	1.116
Value1	1.490	1.152	1.145	1.131	1.592	1.236
Value2	1.431	1.508	1.132	1.147	1.575	1.155
Values3	1.273	1.460	1.198	1.226	1.339	1.141
Intra1	1.847	1.877				
Intra2		1.942				
Intra3		2.297				
Intra4		1.671		1.011	1.045	
Intra5		1.496				
Intra6		1.628			1.044	
Intra7	1.251	2.179	1.083			
Intra8		1.685		1.011	1.089	
Intra9		1.907				1.009
Intra10		1.788				
Intra11	1.549	1.807	1.210			
Intra12		2.162				1.006
Intra13		2.039	1.201	1.016		

(continued)

Table 8 (continued)

2022	Business in english		Business in swedish		Civil engineering	
VIF	MA	SM	IBA	TBE	IE	FMA
Intra14		2.139				1.015
Intra15		2.695				

Notes IE = Industrial Economics, FMA = Financial and Management Accounting, IBA = Intro. Business Administration, TBE = Thesis in Business or Economics, MA = management Accounting, SM = Service Marketing

these samples. Items 4 and 8 show the most promising results with the factor loadings and collinearity tests. All other items require further investigation to determine the validity of their measurement for this construct.

5 Discussion and Conclusion

As we have discussed in the introduction, it is recognised that there is a need for a valid and reliable assessment tool to help teachers to learn more about their student cohort's sustainability competence before they come to class. We have conducted a study to test scales that have been provided by earlier research and scales that have been developed based on the discussion about the KSC. Results of this study at a business school have shown that the KSC scales available can help to provide a valid and reliable information about the participant's sustainability competence, but more work is needed to strengthen this tool.

Unexpectedly, we found that the NEP scale that has been validated many times in rigorous prior studies (Zhu and Lu 2017) gave the weakest results when tested for validity and reliability with our business school students. The scale is quite extensive, with 15 items that address strong feelings and attitudes towards sustainability problems. It is possible that students don't feel comfortable considering the statements or revealing their feelings towards the statements. It is also possible that the length of the scale (15 items) is unappealing, and students may be reluctant to answer properly. Reducing the scale to the items that are most valid will have to be established carefully, since the items may only depict some aspects of the attitudes the scale was designed to capture.

The face validity of the scales could be strengthened by adapting some items to be reversed, since all the items from the Savage et al. (2015) study are positively worded. Also, alternative methods of measuring the scale could be beneficial to the sustainability competence assessment tool. Pictural tools (Leutner et al. 2017) or facial scales (Buchanan and Niven 2002; Yon et al. 2020) can help to increase the validity of a questionnaire and varying the format that the questions are presented can help to decrease bias (Podsakoff et al. 2003). Future research could investigate the items that define and measure the constructs further.

Despite some weaknesses, the tests show that the changes we have made to the Savage et al. (2015) sustainability competence scales, including developing the face validity and adding implementation competency, has strengthened the validity of the scales. However, this study concentrates on a business school context and the assessment tool could be tested across disciplines at a whole university. While there is more investigation required to improve the validity and reliability of this tool, mostly regarding the intrapersonal competency (NEP scale), the Sustainability Competence Assessment can be a useful tool to help teachers learn about their student cohort's sustainability competency before they come to class. It is also recognised that adapting the questionnaire to provide a student-centred approach so that participants can easily understand what is asked and can easily navigate through the assessment quickly, and in an engaging manner, is an important next step in this research.

References

Andrade HL (2019) A critical review of research on student self-assessment. Front Educ 4(87):1–13. https://doi.org/10.3389/feduc.2019.00087

Annelin A, Boström G-O (2022) An assessment of key sustainability competencies: a review of scales and propositions for validation. Int J Sustain High Educ.https://doi.org/10.1108/IJSHE-05-2022-0166

Ateskan A, Lane JF (2018) Assessing teachers' systems thinking skills during a professional development program in Turkey. J Clean Produc 172:4348–4356. https://doi.org/10.1016/j.jclepro.2017.05.094

Azeiteiro UM et al (2015) Education for sustainable development through e-learning in higher education: experiences from Portugal. J Clean Produc 106:308–319. https://doi.org/10.1016/j.jclepro.2014.11.056

Backman M et al (2019) Experiential approaches to sustainability education: towards learning landscapes. Int J Sustain High Educ 20(1):139–156. https://doi.org/10.1108/IJSHE-06-2018-0109

Bamber P et al (2016) A comparative review of policy and practice for education for sustainable development/education for global citizenship (ESD/GC) in teacher education across the four nations of the UK. Manag Educ 30(3):112–120. https://doi.org/10.1177/0892020616653179

Barth M et al (2007) 'Developing key competencies for sustainable development in higher education. Int J Sustain High Educ 8(4):416–430. In Adomssent M, Godemann J, Michelsen G (eds). https://doi.org/10.1108/14676370710823582

Birdman J, Wiek A, Lang DJ (2021) Developing key competencies in sustainability through project-based learning in graduate sustainability programs. Int J Sustain High Educ ahead-of-print(ahead-of-print)https://doi.org/10.1108/IJSHE-12-2020-0506

Brundiers K et al (2020) Key competencies in sustainability in higher education—toward an agreed-upon reference framework. Sustain Sci 16:13–29. https://doi.org/10.1007/s11625-020-00838-2

Buchanan H, Niven N (2002) Validation of a facial image scale to assess child dental anxiety. Int J Paediatr Dent 12(1):47–52. https://doi.org/10.1046/j.0960-7439.2001.00322.x

Cabral C, Lochan Dhar R (2019) Green competencies: construct development and measurement validation. J Clean Prod 235:887–900. https://doi.org/10.1016/j.jclepro.2019.07.014

Cebrián G, Hernández À, Segalas J (2019) Assessment of sustainability competencies: a literature review and future pathways for ESD research and practice. Cent Eur Rev Econ Manag 3(3):19–44

Clark LA, Watson D (2019) Constructing validity: new developments in creating objective measuring instruments. Psychol Assess 31(12):1412–1427. https://doi.org/10.1037/pas0000626

Dunlap RE (2008) The new environmental paradigm scale: From marginality to worldwide use. J Environ Educ 40(1):3–18. https://doi.org/10.3200/JOEE.40.1.3-18
Galt RE, Parr D, Jagannath J (2013) Facilitating competency development in sustainable agriculture and food systems education: a self-assessment approach'. Int J Agric Sustain 11(1):69–88. https://doi.org/10.1080/14735903.2012.683569
Hair JF, Howard MC, Nitzl C (2020) Assessing measurement model quality in PLS-SEM using confirmatory composite analysis. J Bus Res 109:101–110. https://doi.org/10.1016/j.jbusres.2019.11.069
Hair JF et al (2019) When to use and how to report the results of PLS-SEM. Eur Bus Rev 31(1):2–24. https://doi.org/10.1108/EBR-11-2018-0203
Hawcroft LJ, Milfont TL (2010) The use (and abuse) of the new environmental paradigm scale over the last 30 years: a meta-analysis. J Environ Psychol 30(2):143–158. https://doi.org/10.1016/j.jenvp.2009.10.003
Holdsworth S, Thomas I, Sandri O (2018) Assessing graduate sustainability attributes using a Vignette/scenario approach. J Educ Sustain Dev 12(2):120–139. https://doi.org/10.1177/0973408218792127
Lans T, Blok V, Wesselink R (2014) Learning apart and together: towards an integrated competence framework for sustainable entrepreneurship in higher education. J Clean Prod 62:37–47. https://doi.org/10.1016/j.jclepro.2013.03.036
Leal Filho W et al. (2021) A framework for the implementation of the sustainable development goals in university programmes. J Clean Prod 299:126915. https://doi.org/10.1016/j.jclepro.2021.126915
Leutner F et al (2017) From Likert scales to images: validating a novel creativity measure with image-based response scales. Personal Individ Differ 106:36–40. https://doi.org/10.1016/j.paid.2016.10.007
Lozano R et al (2015) A review of commitment and implementation of sustainable development in higher education: results from a worldwide survey. J Clean Prod 108:1–18. https://doi.org/10.1016/j.jclepro.2014.09.048
Lozano R et al. (2017) Connecting competences and pedagogical approaches for sustainable development in higher education: a literature review and framework proposal. Sustainability 9(10):1889. https://doi.org/10.3390/su9101889
Meza RMM et al (2018) Strengthening sustainability leadership competencies through university internships. Int J Sustain High Educ 19(4):739–755. https://doi.org/10.1108/IJSHE-06-2017-0097
Molderez I, Fonseca E (2018) The efficacy of real-world experiences and service learning for fostering competences for sustainable development in higher education. J Clean Prod 172:4397–4410. https://doi.org/10.1016/j.jclepro.2017.04.062
O'Byrne D, Dripps W, Nicholas KA (2015) Teaching and learning sustainability: an assessment of the curriculum content and structure of sustainability degree programs in higher education. Sustain Sci 17
Ploum L et al. (2018) Exploring the relation between individual moral antecedents and entrepreneurial opportunity recognition for sustainable development. J Clean Produc 172:1582–1591. https://doi.org/10.1016/j.jclepro.2017.10.296
Podsakoff PM et al (2003) Common method biases in behavioral research: a critical review of the literature and recommended remedies. J Appl Psychol 88(5):879–903. https://doi.org/10.1037/0021-9010.88.5.879
Redman A, Wiek A (2021) Competencies for advancing transformations towards sustainability. Front Educ 6:785163. https://doi.org/10.3389/feduc.2021.785163
Redman A, Wiek A, Barth M (2021) Current practice of assessing students' sustainability competencies: a review of tools. Sustain Sci 16(1):117–135. https://doi.org/10.1007/s11625-020-00855-1
Savage E et al (2015) Experiential learning for sustainability leadership in higher education. Int J Sustain High Educ 16(5):692–705. https://doi.org/10.1108/IJSHE-10-2013-0132

Shephard K, Rieckmann M, Barth M (2019) 'Seeking sustainability competence and capability in the ESD and HESD literature: an international philosophical hermeneutic analysis. Environ Educ Res 25(4):532–547. https://doi.org/10.1080/13504622.2018.1490947
Shephard K (2008) Higher education for sustainability: seeking affective learning outcomes. Int J Sustain High Educ 9(1):87–98. https://doi.org/10.1108/14676370810842201
The Swedish Council for Higher Education (2020) The Swedish Higher Education Act (1992:1434), UHR.se. https://www.uhr.se/en/start/laws-and-regulations/Laws-and-regulations/The-Swedish-Higher-Education-Act/. Accessed 15 Oct 2020
UNESCO (2017) Education for sustainable development goals: learning objectives. UNESCO, Paris, France
Wals AEJ (2014) Sustainability in higher education in the context of the UN DESD: a review of learning and institutionalization processes. J Clean Produc 62:8–15. https://doi.org/10.1016/j.jclepro.2013.06.007
Wiek A, Withycombe L, Redman CL (2011) Key competencies in sustainability: a reference framework for academic program development. Sustain Sci 6(2):203–218. https://doi.org/10.1007/s11625-011-0132-6
Wilson M, Lehrer R (2021) Improving learning: using a learning progression to coordinate instruction and assessment. Front Educ 6:654212. https://doi.org/10.3389/feduc.2021.654212
Yon MJY et al (2020) An introduction to assessing dental fear and anxiety in children. Healthcare 8(2):86. https://doi.org/10.3390/healthcare8020086
Zhu X, Lu C (2017) Re-evaluation of the New Ecological Paradigm scale using item response theory. J Environ Psychol 54:79–90. https://doi.org/10.1016/j.jenvp.2017.10.005

Developing a Pathway to Sustainable Cities and Communities Through a Leadership Education Programme

Fiona Armstrong-Gibbs and Lisa Knight

Abstract This paper explores the design and delivery of a leadership training programme through the lens of the SDGs with a focus on cities, communities, and partnerships for sustainable living (SDGs 11 and 17). The programme was co-designed for an English city's Regional Combined Authority (a devolved authority) addressing the four pillars of leading self, others, organisation, and place. Examining delivery, we identified several diverse and creative approaches to leadership that may or may not contribute to the development of strong, sustainable, and inclusive places. However, we also found that how this leads to enhanced communities with a strong identity and socio-economic base that enriches the wellbeing of its citizens needs further exploration. Given the richness of different geographies, cultures, and heritages the cross-boundary and cross-sector nature of place, leadership in this context is complex. This work is located at the interface of political, public sector, community, and education to explore the function and outcomes of for sustainable place leadership across the city region. In other words, 'place leadership' from a devolved authority perspective needs further investigation to enable leaders to make significant impacts in these settings.

Keywords Sustainable communities · Place · Leadership · Training

1 Introduction

This chapter explores the design and delivery of a senior leader training programme through the lens of the SDGs with a focus on cities, communities, and partnerships for sustainable living (SDGs 11 and 17). The programme was co-designed by

F. Armstrong-Gibbs (✉) · L. Knight
Liverpool Business School, Liverpool John Moores University, Liverpool L3 5UG, England
e-mail: F.D.ArmstrongGibbs@ljmu.ac.uk

L. Knight
e-mail: L.Knight@ljmu.ac.uk

T. Wall et al. (eds.), *Sustainability in Business Education, Research and Practices*, World Sustainability Series, https://doi.org/10.1007/978-3-031-55996-9_15

an academic team in a Business School and a regional Combined Authority (CA) addressing the four pillars of leading self, others, organisation, and place.

There are ten regions in England that have a devolved governance structure. The CAs primary role is to convene their agreed local authorities and build a collective influence on local economic development whilst manging funds from central government (Roberts 2020; Jeffery 2022). They now play a crucial role in central government's place and economic strategy during a period of rising inflation, increased inequalities, and uncertainty (Ayres et al. 2018).

The Combined Authority (CA) consists of six local authorities which were formally constituted in 2017 and represented by a directly elected Metro Mayor. It covers a population of 1.6 million across a diverse landscape of cities, sea, suburban and rural areas with huge disparities in health, social, environmental, and economic outlooks. The diversity of the region has a significant influence on life expectancy, wellbeing, mobility, and opportunities, therefore the role of CA employees and their relationship with partners is a critical area of leadership development. Therefore, exploring how a leadership development programme may support and impact the SDGs three dimensions of sustainable development: economic growth, social inclusion and environmental protection is timely.

The effectiveness and impact of leader and leadership development programmes are also considered, (Day and Harrison 2007; Fitzsimmons and Callan 2020) and we make the distinction that the fourth pillar of place leadership focusses on *leadership* development as it is concerned with the social capital needed to lead at the collective or relational level as opposed to *leader* development that focusses on human capital. Therefore, this programme was designed to encompass the unique challenges for a regional CA and takes a different approach to other types of training and development.

Through a co-production design process, the programme was bound into the context of both the function of the CA and the geography of the region it operates in (Gibney 2014), therefore the delivery was differentiated to meet the needs of the region and allow for place leadership to develop to support these needs (Nicholds et al. 2017; Beer 2014) and support the organisation and key senior leaders working within the organisation to shape the future of place both internally within the CA and externally with a range of collaborating partners (Ayres et al. 2018). Higher Education Institutes (HEI) have a significant role to play in the achievement of SDGs (Serafini et al. 2022), one that was recognised by both the CA and the HEI. In addition to supporting the development of key competencies in sustainable development (Garcia-Feijoo et al. 2020), HEIs have a significant role to play in their regional context working with the broader community on issues of sustainable development (Sassen et al. 2022).

The purpose of the chapter is to.

- Explain the approach to design, delivery and engagement through knowledge and skills training, rather than the pedagogy of executive leadership programmes.
- Explore the emerging topics of discussion in 'place leadership' as UK (United Kingdom) public organisations (post Brexit) focus on leadership development programmes.

- To build on concepts of collaborative place leadership through the data gained from this training programme.

What follows is an outline of the programme design and delivery, the topics covered in part one and part two of the workshop and the emerging discussion synthesised from the sessions.

2 Programme Design and Delivery

Co-design with the CA was an integral element of the overall programme design. This was important to ensure (i) context-specific knowledge was woven through the material in case study selections (ii) learning objectives were tailored to the CA needs and (iii) the programme was aligned to the overall strategy and strategic aims of the CA. Focus groups were held within the organisation to ensure relevancy and currency of content so that skill acquisition was as close to skill application as possible (Moldoveanu and Narayandas 2019), but more significantly to delineate specific challenges and opportunities open to the CA to enhance participant engagement and foster a sense of cohesiveness and community amongst the participants whose roles may not be connected across the organisation. This focus on the main principles of knowledge co-production: context-based, pluralistic, goal orientated, and interactive allowed us to actively build on concepts of place leadership (Norström et al. 2020). Accordingly, initial learning outcomes were drafted as follows:

1. Understand and apply key partnership principles to develop, shape and create sustainable partnerships
2. Synthesise and apply current leadership and management theory to lead partnerships effectively (emphasis on relationship building and communication)
3. *Using a social network lens, identify and explore levels of social capital with internal and external partners*
4. Recognise and prioritise the needs of regional stakeholders
5. Effectively manage collaborative partnerships
6. *Develop collective decision-making processes across the region.*

These learning outcomes were further refined as part of the co-design phase and the decision was taken not to include Learning Outcome 3, due to information overload and time constraints and Learning Outcome 6 was excluded as it was determined to be an emerging outcome rather than achievable in its entirety given the time constraints of the workshops.

A convening event was utilised to set the scene and introduce the programme to all participants, followed by delivery of the four workshops (Leading Self, Leading Others, Leading Organisations, Leading Place) to four cohorts of senior leaders. The ‘Place’ workshop followed a similar delivery format to the preceding workshops; pedagogically, the programme took an authentic approach to learning to reflect the emerging nature of the concept of place leadership and in consideration of the co-production ethos. In line with Roach et al. (2018) authentic learning was understood in its broadest sense whereby participants previous and current experience and resources were utilised to align learning outcomes with real-world context and content. The first three workshops of the programme allowed the participants the opportunity to reflect on their own leadership styles (leading self), leading teams (leading others) and internal relationships in the organisation (leading the organisation), the final 'leading place' workshop focused on external concepts of place and how one contributes to the leadership of their place.

An important, and differentiating, element of the leading place workshop was that at the beginning of the session, expectations were set by the Executive Leadership team who introduced the theme of the session and the emerging view of place leadership within the organisation. In line with all the workshops, participants were briefed at the beginning of each session on the expectations to actively contribute to the session and show respect and consideration to colleagues during discussions and group tasks. Learners explored five areas through pre-reading, paired discussion, group activity and individual reflection. The morning session focussed on exploring individual and group understanding of the meaning of 'Place' and consider leadership theories that underpin some of the approaches organisations and individuals can take when leading place. The afternoon considered how collaborative relationships and trust are built in the places we live and work.

3 Programme Content

3.1 Part 1 Exploring Theories of Place

Individual and group reflection on the concept of place framed the session, with participants responding to definitions from the literature; considering how 'space' becomes 'place.' Participants conception of place was linked to their experience, underpinned by history, geography, politics, and emotional connections (both positive and negative) which was aligned to Tuan's concept of 'Topophilia' (1990). Figure 1 provides a summary of participants' responses to the introductory session: '*what does place mean to you*?'.

People, local, regional, geography, cultures, micro cultures and communities, sense of identity, history, sense of belonging, neighbourhoods, shared experiences, emotional connections,
Infrastructure, built environment, natural resources, housing, assets, unique identity, coast, port, diversity, rural, urban, jobs, employment, transport, collaborative, pride, enjoyment, guardians, conveners, ambassadors, service, procurement, sense of place, tourism, innovation.

Fig. 1 Participants view of place

Table 1 Potential advantages and disadvantages of a place-based approach

Potential advantages	Potential disadvantages
Interventions targeted based on local intelligence	Duplication of effort
Community involvement in issue identification and resolution	Derailment due to poor coordination between central, regional, and local governments
Enhanced innovation	Additional layers of governance
Increased social cohesion	Reverse 'nimby'

(Adapted from Tiernan 2016; Hambleton and Howard 2013; Leech and Worrall 2018; Taylor and Buckly 2016; Wilks et al. 2016).

Further discussion embraced the notion of place as simultaneously a driving force for change and sustainability which was more closely aligned to Nilsen's (2005: 27) definition: "Viewed as process, place is something that can be nurtured and reshaped through individual or collective effort". Furthermore, it was noted that this dual notion of place could foster both a desire to enhance and steward but also encourage prejudice and territoriality (Jackson 2019). Broader links to the available academic literature were then able to be made utilising the potential advantages and disadvantages outlined in Table 1

Most significantly, these contextual conceptions of place led to the posed question: "...if place matters—then what does it mean for leadership?" Collinge and Gibney (2010: 367) with participants asked to consider where and by whom leadership of place was enacted: 'who can enact leadership? How? At what scale and with what consequence?' (Ayres 2014: 22).

3.2 *Place Leadership*

Theories of place leadership were shared, and participants were encouraged to explore their applicability in the context of current leadership practices, both organisationally and individually. Initially, three broad types of leaders of place, identified by Hambleton et al. (2022: 86–87), introduced a discussion about where leadership power is situated: political leaders, public sector leaders, and non-governmental leaders:

- Political Leaders: Leadership based on the role of the elected leader; intended to act as a communicative link between the local community and government.
- Public Sector Leaders: public servants employed within local or central government departments or other public sector organisations whose leadership is based on professional expertise and the management of public sector resources whose role is to plan, manage, and deliver public services.
- Non-governmental Leaders: Community activists, business leaders, social entrepreneurs, trade union leaders, voluntary sector leaders, and religious leaders,

all of whom have a direct influence on the local community by advocating, representing, and providing services to it. (Note: Hambleton et al. use the term 'non-state leader', as our work is UK based, we use the term non-governmental).

Participants identified the potential for both collaboration and conflict in line with Hambleton and Howard's (2013) view that where these leadership positions overlap can often consist of competing interests, values, and an unequal distribution of power. Ayres (2014) notes that the scalability of leadership is an important consideration in determining place and the prevailing ideas and ideals within them, and therefore, utilising examples drawn from the local region, participants were asked to consider the idea that 'micro-level activity from unexpected sources can have a cumulative impact on macro-ideology and place' (Ayres 2014: 22). In response to the idea of scalability in the leadership of place, consideration was given to the proposition that leadership of place is unstable and constantly subject to change (Fairhurst 2007; Hambleton 2019). Jackson's (2019) argument that emergent sources of leadership, or leadership from non-traditional individuals and groups should be given equal access and support to address the problems, and take advantage of the opportunities, presented by place initiated a group discussion on the role of place leadership.

Participants were then presented with the Souza and Jackson's (2019) Leadership Lens Hexad to frame the discussion. The four traditional lens of leadership that might commonly be considered: leadership in relation to the leader as an individual (person), the results achieved (performance), the authority of leadership (position), and how leadership is created and maintained (process) (Grint 2005) were juxtaposed to the two additional, non-traditional, lenses of leadership: where is leadership created (place) and why is leadership created (purpose). Considering leadership through these six lenses facilitated a discussion about their interrelatedness—as Jackson (2019: 13) poses 'how does the place of leadership influence the purpose of leadership? In turn, how does the purpose of leadership influence its place?' Participants were then asked to reflect on, and self-assess, their own leadership practice in the context of place utilising two models of enabling leadership practices (Murphy et al. 2017; Broadhurst et al. 2021a). Individuals were presented with enabling leadership practices as a series of four practice pairs that they were informed can shift the balance of leadership in the context of place to either an administrative or adaptive practice (Uhl-Bien et al. 2007). Building on this, Broadhurst et al's (2021b) model offered examples of enablers of adaptive, transformative leadership versus administrative, transactional leadership that support progress towards a more collaborative approach to leading place. This was an opportunity for participants to start to engage in knowledge transfer and transformation, applying theoretical frameworks to their own current and aspiring practice as a leader of place.

Encouraging participants to reflect more on their individual capacity and capability as a place leader, The Centre for Urban and Regional Studies (2011) five core behaviours of place leaders were utilised as a reflective tool, framing a discussion on effective leadership of place being reliant on this diverse mix of multiple relationships, stakeholders, and their voices. A core consideration in the leadership approach taken by the organisation and individuals to achieve sustainable cities and communities.

3.3 Part 2 from 'Stakeholder Management' to 'Collaborative Relationships'

The afternoon session examined regional relationships from a leadership perspective and offered the opportunity to explore evolution and existing management approach to stakeholder relationships in the public sector. As the outcome was to 'recognise and prioritise the needs of regional stakeholders and help us develop better relationships with partners' we explored routes to move away from the 'stakeholder management' rhetoric to 'collaborative relationships'. Participants were asked in advance to read an article by Schruijer (2020) in either the full text or we created an abridged version to frame the discussion later around collaborative interorganisational relationships.

To give context, we shared Freeman's (1984) stakeholder theory; any group or individual who can affect or is affected by the achievement of the organization's purpose and objectives, and a 'stake' being an interest or a share in an undertaking. This can range from simply an interest in an undertaking at one extreme, to a legal claim of ownership at the other extreme identified by Carroll and Buchholtz (2006). However, we were keen to outline that stakeholder theory, emerged as a response to the shareholder and capitalist activities from 1980s onwards. Furthermore, the wave of 'New Public Management' (Osbourne and Gaebler 1992) thinking and literature introduced business-based ideas to public sector stakeholder management. We suggested this is now an unhelpful and dated approach (Smolović Jones et al. 2015) and one that when applied to a new era of public governance such as English devolution in the 2020s it does not focus on the wider sustainable societal concerns that are highlighted by the SDGs. There are challenges for senior leaders who act exclusively in their organisations best interest as it will likely create tensions with stakeholders. Discussions emerged around the complexity in considering all stakeholders and the challenge when working with and achieving the expectations of others. When asked how one would identify their stakeholders and if those people and groups were aware of what the CA thought of them, this was unclear. Group discussions summarised that the old approach of 'stakeholder management' was not sufficient to support the complexity of the relationships between the CA and others in its operating environment. However, the stakeholder engagement and communication approach of knowing and appreciating others who have *an interest or a share in [the organisations] undertaking* may help to identify with who and where the CA can create a shared purpose.

3.4 Understanding Collaborative Partnerships—A Case Study Approach

To develop an understanding of how collaborative partnership working happens we asked participants to identify in their own practice what they do well and may want to improve on.

We asked, how someone would know if they were a CA stakeholder and once the relationship is established, how is its managed? Where does the power lie in the relationship and how does the CA use its power and to what effect? The purpose of this was to identify the good leadership practice within in the organisation. Sharing the success stories internally (as well as externally) was also seen as important. Learning, reflecting, and listening to others who had navigated complex relationships with good (and sometimes not so good) outcomes was shared. During the design phase, a case study was written by two of the participants, describing a regional investment programme. It explained the approach taken, key partnerships and the results. This was shared with the group for feedback and reflection. Due to the diverse job roles across cohorts, some colleagues were unaware of the project at all, some had limited awareness but found it an informative and engaging update from colleagues. Discussions around respect, power dynamics and social networks were raised—highlighting the importance of social capital. However, an emerging conversation around the role of the CA in communication, governance and convening in place leadership became evident. After this discussion we returned to the pre reading by Schruijer from 2020 who described her experience of facilitating three cases of collaboration in the public sector:

> Interorganizational collaboration refers to the relational process when people from different organizations come together to address a concern or an opportunity and explore their interdependencies and how they could add value to one another, resulting in a joint problem formulation and a joint goal – a joint goal that also serves the interests of the individual organizations (Schruijer 2020: 18)

The case studies were powerful accounts of multi-agency working and gave the opportunity for participants, as senior leaders to reflect on their own interorganisational relationships. They were asked to consider if or how they could move their relationship with stakeholders from management' to 'collaboration' and from 'exchange' to 'long-term relationships'. As senior leaders, many are also responsible for their own team's ability to form good relationships which was acknowledged and adds to the complexity. 'Collaboration' implies a two-way symmetrical model of dialogue and consultation. The need for a better understanding of what the organisation is currently doing well and how colleagues can learn from each other was evident, but how could this be achieved given the internal constraints, complex demands of people and contextual challenges of the region.

3.5 *Exploring Trust*

During the design phase of the programme, it was identified that developing a better understanding of trust should be a key component of the leadership training especially for building strong collaborations. The workshop discussions that followed all centred around commitment, fairness, information sharing, and time spent learning that will contribute to building trust in a relationship. Once gained, trust can help the organisation be resilient, adaptable and improve performance and decision making. Furthermore, as leaders they began to understand their role in creating value for all citizens and stakeholders.

We shared the Stakeholder Value Creation model (Kujala et al. 2019) which identifies trust, reciprocity, and mutual interest as a driver for collaboration. High levels of trust in relationships create joint value for all and better relationships. Kujala et al. (2019) suggest for this to happen a common history, shared experiences and mutual objectives will all support the development of trust. We brough the conversation back to our exercise at the start of the day which identified what Place meant to each participant as part of this trust building activity.

Further complexity was added to the discussion on trust as the organisation is governed by a Metro Mayor and a combined authority of Local Government leaders, which adds a political element to the perception of the organisation by external stakeholders. Trust in political leaders is currently at an all-time low (Quilter-Pinner and Statham 2021; Edelman 2022) so consideration of this on the reputation of the CA is needed. Given this context we introduced the notion of diplomacy for developing relationships and Henisz (2014) Diplomat framework to reflect on.

In concluding the workshop, participants were given the opportunity to reflect on the content and their emerging conceptualisation of place, their roles as a leader of place within a collective and relational network. Furthermore, the significance and potential the CA has as a facilitator with a key role in shaping the future, and supporting sustainable place leadership across the city region.

4 Conclusion

This is the emerging account of our understanding of place leadership and how as a HEI we can support and develop the next generation of leaders in our cities and communities. It is timely that a recently devolved CA is exploring and stating what it means to be a Place Leader for the 21st Century.

4.1 Design, Delivery, and Engagement

A co-designed leadership programme that deliberately asks these challenging questions is both an outward statement of intent and an internal change process in action. It is the beginning of a learning journey where all participants, including the executive team had a clear emotional connection to place, which validated this as a position to learn from. The use of theory, at a high cognitive level—similar to master's level, was applied to personal and profession practice. This allowed a space for building confidence to take this new leadership approach. Participants had the opportunity to develop new areas of knowledge in leadership approaches and building collaborative relationships whilst also building social capital and communication and decision-making skills in a complex environment.

As facilitators we are in a unique and privileged position, working with leaders who practice in the field. English devolution is a new area of governance, politics, and leadership, one that is dynamic and changeable due to the richness of both political and civic leadership. However, our role, could also be viewed as 'a pinch-point, where theorising of public leadership meets the everyday concerns of professionals' (Smolović Jones et al. 2015: 405). Issues around decision making and trust came to the forefront in our sessions.

> 'A shift towards place-based thinking may lead to scholarly research and management practices that deal more effectively – at local levels – with such thorny issues such as social justice, global climate change, alternative energy and economic inequality to name but a few' (Guthey et al. 2014: 62)

Therefore, we suggest that public governance bodies who commission leadership development programmes must take a contextual and collaborative approach in the content design and delivery process. The co design approach builds trust between the HEI and the participants allowing the facilitators to stretch leadership thinking.

4.2 Place Leadership

Our findings suggest that there is an acute awareness and acknowledgement of the natural assets, landscape and environment in the region, its physicality and built environment make the partnerships complex and rewarding to manage. Humanity and heritage are intertwined in places, neighbourhoods, public spaces, and organisations. Disparity, deprivation, and disadvantage have been evident in the region for many years, however they can and must be the drivers for change. There is a strong desire from the CA to be inclusive, equitable, build social cohesion and engaging of all cultures while referencing the heritage and looking forward. This is likely to include plans for a wider collaborative leadership network across the region.

From a leadership perspective there are opportunities, challenges and tensions, the CA acknowledges this complexity. While it might appear that it struggles to keep everyone happy, its people are forging a new path to manage expectations and create value with partners. This is a long-term commitment and will evolve. A focus and desire to change the relationship between the organization and its stakeholders from 'management' to 'collaboration' and from 'exchange' to 'long-term relationships' is evident. Empowering senior leaders in the organisation to explore how to develop more trusting relationships is the path to a sustainable future. Much more is needed but 'collaboration' implies a two-way symmetrical model of dialogue and consultation which a necessary path to achieve SDGs in a community.

4.3 Reflections and Recommendations

Place leadership is a collective, creative endeavour that finds shared values and purpose at its heart. This chapter has allowed us to reflect on the programme and how this creates a pathway to a sustainable leadership of place. We have outlined our early-stage positioning and decision work as leadership development facilitators, presenting the academic literature and theoretical models as well as participants reflections from the delivery. We have yet to measure the longer-term impact of this approach. Further work is also needed on how leadership development programmes can better serve the SDG (Sustainable Development Goals) agenda, this is a small contribution, however sustainability was implicitly spoken about rather than explicitly built into the content and materials used.

There are benefits in combining an understanding of place into leadership programmes for all organisations, especially those that have a role in regional governance and economic development. Embedding a sustainable approach to leadership of places can create sustainable cities, communities, and partnerships. However, the notion of 'place and leadership' from a devolved authority perspective needs further investigation to enable leaders to make significant impacts in these settings. It is our hope that the senior leaders of this relatively recently formed authority have been supported to become better attuned with their place. That they can accept and embrace its complexity and nuances and continue to develop trust and partnerships for sustainable cities and communities. The CA have a role in both facilitating place leadership as well as being place leaders, therefore the notion, practice and adoption of collaborative place leadership needs further investigation to understand its effectiveness.

References

Ayres S (2014) Place-based leadership: reflections on scale, agency, and theory. Reg Stud Reg Sci 1:21–24. https://doi.org/10.1080/21681376.2013.869424

Ayres S, Flinders M, Sandford M (2018) Territory, power and statecraft: understanding English devolution. Reg Stud 52(6):853–864. https://doi.org/10.1080/00343404.2017.1360486

Beer A (2014) Leadership and the governance of rural communities. J Rural Stud 34:254–262. https://doi.org/10.1016/j.jrurstud.2014.01.007

Broadhurst K, Ferreira J, Berkeley N (2021a) Collaborative leadership and place-based development. Local Econ 36(2):149–163. https://doi.org/10.1177/02690942211032553
Broadhurst K, Ferreira J, Berkeley N (2021b) Place leadership: developing a model to guide regional partnerships. Reg Stud 55(3):556–567. https://doi.org/10.1080/00343404.2020.1836343
Carroll A, Buchholtz A (2006) Business and society ethics and stakeholder management, 6th edn. Thomson
Collinge C, Gibney J (2010) Connecting place, policy and leadership. Policy Studies 1(31):379–391
Day DV, Harrison MM (2007) 'A multilevel, identity-based approach to leadership development. Human Resour Manag Rev 17(4):360–373. https://doi.org/10.1016/j.hrmr.2007.08.007
Edelman (2022) Edelman trust barometer 2022. https://www.edelman.com/trust/2022-trust-barometer. Accessed 29 Mar 2023
Fairhurst GT (2007) Discursive leadership. In Conversation with leadership psychology. Sage, Thousand Oaks, CA
Fitzsimmons TW, Callan VJ (2020) The diversity gap in leadership: what are we missing in current theorizing? Leadersh Q 31(4):101347. https://doi.org/10.1016/j.leaqua.2019.101347
Freeman RE (1984) Strategic management: a stakeholder approach. Pitman
García-Feijoo M, Eizaguirre A, Rica-Aspiunza A (2020) Systematic review of sustainable-development-goal deployment in business schools. Sustainability 12(1):1–19. https://doi.org/10.3390/SU12010440
Gibney J (2014) Don't lose sight of context: a commentary on mobilizing cities and regions. Reg Stud Reg Sci 1(1): 25–27. https://doi.org/10.1080/21681376.2013.869430
Grint K (2005) Leadership: limits and possibilities. Palgrave Macmillan, Basingstoke
Guthey GT, Whiteman G, Elmes M (2014) Place and sense of place: implications for organizational studies of sustainability. J Manag Inq 23(3):254–265. https://doi.org/10.1177/1056492613517511
Hambleton R (2019) The new civic leadership: place and the co-creation of public innovation. Public Money Manag 39(4):271–279
Hambleton R, Howard J (2013) Place based leadership and public service innovation. Local Gov Stud 39(1):1–24
Hambleton R, Sweeting D, Oliver T (2022) Place, power and leadership: insights from mayoral governance and leadership innovation in Bristol, UK. Leadership 18(1):81–101. https://doi.org/10.1177/17427150211028122
Henisz W (2014) Corporate diplomacy: building reputations and relationships with external stakeholders. Greenleaf
Jackson B (2019) The power of place in public leadership research and development. Int J Public Leadersh 15(4):209–223. https://doi.org/10.1108/IJPL-09-2019-0059
Jeffery D (2022) Understanding drivers of support for English city-region devolution: a case study of the Liverpool City Region. British Polit. https://doi.org/10.1057/s41293-021-00199-z
Kings Fund (2021) Developing place-based partnerships: the foundation of effective integrated care systems. Kings Fund, London
Kujala J, Lehtimäki H, Freeman RE (2019) A stakeholder approach to value creation and leadership takeaways for leading change. Lead Change Complex World Transdiscipl Perspect 1–21
Leech D, Worrall R (2018) Place-based leadership: both opportunity and challenge. National Health Executive Newsletter, London, March/April
Moldoveanu M, Narayandas D (2019) The future of leadership development. Harv Bus Rev 97(2):40–48. https://doi.org/10.4135/9781446213605.n9
Murphy J, Rhodes ML, Meek JW et al (2017) Managing the entanglement: complexity leadership in public sector systems. Public Adm Rev 77(5):692–704. https://doi.org/10.1111/puar.12698
Nicholds A, Gibney J, Collinge C, Hart D (2017) Making sense of variety in place leadership. Reg Stud 51(2):249–259. https://doi.org/10.1080/00343404.2016.1232482
Nilsen E (2005) Rethinking place in planning: opportunities in Northern and Aboriginal planning in Nunavut, Canada. Can J Urban Res 14(1):22–37

Norström AV, Cvitanovic C, Löf MF, West S, Wyborn C, Balvanera P et al (2020) Principles for knowledge co-production in sustainability research. Nat Sustain 3(1). https://doi.org/10.1038/s41893-019-0448-2

Osbourne D, Gaebler T (1992) Reinventing government: how the entrepreneurial spirit is transforming the public sector. Penguin

Quilter-Pinner H, Statham R (2021) Revealed: trust in politicians at lowest level on record. IPPR Progressive Review. https://www.ippr.org/news-and-media/press-releases/revealed-trust-in-politicians-at-lowest-level-on-record/. Accessed 29 Mar 2023

Roach K, Tilley E, Mitchell J (2018) How authentic does authentic learning have to be? High Educ Pedagog 3(1):495–509. https://doi.org/10.1080/23752696.2018.1462099

Roberts J (2020) The leadership of place and people in the new English combined authorities. Local Gov Stud 46(6):995–1014. https://doi.org/10.1080/03003930.2020.1719076

Sassen R, Hueske A-K, Pontoppidan CA (2022) Guest editorial. Int J Sustain High Educ 23(2):213–217. https://doi.org/10.1108/IJSHE-02-2022-518

Schruijer S (2020) Developing collaborative interorganizational relationships: an action research approach. Full text available at https://www.emerald.com/insight/content/doi/https://doi.org/10.1108/TPM-11-2019-0106/full/html

Serafini PG, de Moura JM, de Almeida MR, de Rezende JFD (2022) Sustainable development goals in higher education institutions: a systematic literature review. J Clean Prod 370:133473

Smolović Jones O, Grint K, Cammock P (2015) Public leadership development facilitation and the crossroads blues. Manag Learn 46(4):391–411. https://doi.org/10.1177/1350507614537020

Souza R, Jackson B (2019) Enacting leadership in the favela: an empirical investigation using a dialogical heuristic instrument. In: Koonce R, van Loon R (eds) The dialogic challenge of leadership development. Information Age Publishing, Charlotte, NZ, pp 99–127

Taylor M, Buckly E (2016) Historical review of place-based approaches. Institute of Voluntary Action Research, London

The Centre for Urban and Regional studies (2011) Leadership of place: transforming communities by changing the way we lead. University of Birmingham, Birmingham

Tiernan A (2016) Collaboration in Place', 2016 Solomon Lecture, Policy Innovation Hub. The Machinery of Government, Brisbane. https:medium.com/themachinery-of-government/collaboration-in-place

Uhl-Bien M, Marion R, McKelvey B (2007) Complexity leadership theory: shifting leadership from the industrial age to the knowledge era. Leadersh Q 18(4):298–318. https://doi.org/10.1016/j.leaqua.2007.04.002

Wilks S, Lahausse J, Edwards B (2016) Commonwealth place-based service delivery initiatives: key learnings project. Research Report No 32, Australian Institute of Family Studies, Melbourne

Embedding Sustainability in Existing Curriculum by Using Authentic Assessments, Rubrics, and Formative Assessments: A Case Study

Lovleen Kushwah, Paulina Navrouzoglou, and Wenya Cheng

Abstract It has been broadly recognised that education is a key enabler in achieving the United Nations Sustainable Development Goals (SDGs) by 2030. In particular, SDG 4 states "*Ensure inclusive and equitable quality education and promote lifelong learning opportunities for all*". We designed authentic assessments in three undergraduate economics courses focussing on sustainable development. The assessment tasks were student-centred providing autonomy to promote curiosity, think critically about real-world sustainable development problems, and come up with practical and creative solutions. Students' competencies for mastering the assessment task were supported by a rubric and formative assessments. We collected students' feedback using an online questionnaire to investigate students' perceptions of the assessment approach in terms of fostering their motivation and autonomy and building their competencies. The results show that authentic assessments combined with rubrics and formative assessments can promote education of sustainable development in an existing curriculum and create a foundation for lifelong learning.

Keywords Education for sustainable development · Higher education · Lifelong learning · Autonomous learning · Authentic assessments · Rubrics · Formative assessments

L. Kushwah · P. Navrouzoglou (✉) · W. Cheng
Adam Smith Business School, University of Glasgow, University Avenue, Glasgow G12 8QQ, UK
e-mail: Paulina.Navrouzoglou@glasgow.ac.uk

L. Kushwah
e-mail: Lovleen.Kushwah@glasgow.ac.uk

W. Cheng
e-mail: Wenya.Cheng@glasgow.ac.uk

T. Wall et al. (eds.), *Sustainability in Business Education, Research and Practices*, World Sustainability Series, https://doi.org/10.1007/978-3-031-55996-9_16

1 Introduction

UNESCO's Education for Sustainable Development (ESD) programmes have underlined the international significance of integrating sustainability into the education of young people. The ESD programme for 2030 emphasises the role of education in achieving the United Nations 2030 Sustainable Development Goals (SDGs). As Orr (2004, p. 27) argues, the problem of sustainability is also the problem *of* education in the sense that education requires shifting emphasis from individual and nation-building to the critical issues of human survival. Therefore, there is a new role for all levels of education to achieve such a transformation.

Human survival issues are complex, challenging and constantly changing. Thus, education should be reoriented to not only equip students with knowledge and competencies related to sustainable development, but also to instil a desire to internalise and continuously develop them. This ties in well with SDG 4 "*Ensure inclusive and equitable quality education and promote lifelong learning opportunities for all*". Lifelong learning "*takes place, occurs at all times in each place, through own's life*" (Laal 2011, p. 473) so education should aim to equip students with competencies which are required to continue their own self-education beyond the end of formal schooling (Candy 1991).

Within this educational context, self-determination theory (SDT) is particularly relevant. According to SDT, it is inherent in human nature to be curious about one's environment and interested in learning and developing one's knowledge (Niemiec and Ryan 2009). This implies that humans have a natural tendency to lifelong learning, and educators can—as the guides of students' learning and development—nurture this inherent tendency with the content, learning outcomes, class environment and pedagogy used in their teaching. An important concept of SDT is intrinsic motivation, which refers to behaviours that are interesting and enjoyable in the absence of external forces (Ryan and Deci 2000). Such motivation can promote psychological health and effective engagement with the world (Deci and Ryan 2015). For instance, a student may enrol in an elective subject related to sustainable development for no reason other than because it is interesting and innately satisfying to learn about it. Similarly, a student may engage in an assessment task because they are curious about understanding the subject at a deep level, rather than simply performing the task to receive a high grade.

Various learning and teaching approaches for promoting ESD in higher education have been suggested. Neus et al.'s (2017) literature review highlights a lack of consistency in the approaches used as well as barriers in embedding sustainability in the curriculum (see also, for example, Moore 2005; Doh and Tashman 2014; Filho et al. 2019). Our School, the University of Glasgow Adam Smith Business School, became an Advanced Signatory of the United Nations Principles of Responsible Management Education in 2020. To contribute to this initiative and overcome common barriers related to time, space and resource constraints, we redesigned the assessment approach of three undergraduate economics courses which focused on sustainable development.

In this paper, we present the new assessment approach which is theoretically based on SDT and principles of authentic assessment. Numerous applications of SDT in the educational context have identified autonomy and competence as necessary conditions for facilitating and maintaining intrinsic motivation (Niemiec and Ryan 2009). If students are autonomous, they are willing to devote time and energy to their studies; and if students are competent, they feel capable and confident to meet the challenges of their studies (Black and Deci 2000). Thus, if curiosity, intrinsic motivation, and an autonomy-supportive teaching environment is promoted, then educators can create the foundations for lifelong learning (Malmquist 2022, p. 3). Such an assessment design is also coherent with the key characteristics of authentic assessment (Villarroel et al. 2018).

An online student survey was conducted to understand students' views about the assessment tasks for developing sustainability and employability skills, use of rubrics for clarity of assessment criteria, level of students' confidence, effectiveness of feedback and their well-being. From the analysis, we found that authentic assessments combined with rubrics and formative assessments can promote ESD in an existing curriculum and create conditions for lifelong learning. Therefore, educators' challenge in promoting ESD could be in realising the need to do things differently than they did in the past.

The paper is structured in four sections. In the first section, the assessment approach is articulated. Second, the methodology is described. Third, the results of the student survey are presented. Finally, conclusions are drawn, and future research lines are discussed.

2 Assessment Design and Description

The three undergraduate economics courses, Environmental Economics, Natural Resource Economics, and Economics of Poverty, Discrimination and Development, which were selected for this case study are elective courses in the third and fourth years that embed SDGs in the curriculum of the programme. Students receive a programme curriculum map that indicates which courses embed SDGs during the enrolment period. Reflecting on SDT, this may be an indication that the students who decided to enrol in these three elective courses (from a list of more than 30 courses) were curious and had intrinsic motivation to learn about topics in sustainability. Hence, we wanted to maintain and facilitate their desire.

The summative assessments of the above courses include a piece of written work on a specific sustainable development problem covered in the class (e.g., climate change, deforestation, poverty). This type of assessment task helps students to better understand the problem at hand and develop problem-solving skills, but it can be modified to foster students' curiosity, autonomy, and competencies, which are the drivers of their inherent desire to learn and develop. We thus modified our assessment approach to create an autonomy-supportive teaching environment that can boost students' desire for lifelong learning (Malmquist 2022).

We promoted curiosity and autonomy by providing students the freedom to explore and choose a real-world sustainable development problem and make recommendations for managers or for relevant practitioners. Below are the descriptions we used to frame the situation-problem:

- Suppose that you are working as an environmental consultant for the Department of Environmental Management in your home country. You should (a) select an urgent environmental problem your home country is currently facing and (b) analyse how to address it.
- Suppose that you are a project leader from a local non-governmental organisation. You need to write a proposal for a pilot poverty alleviation programme for a region or group of people of your choice.

This type of assessment task draws on the knowledge, skills, and competencies that Economics students need to apply in similar real-world situation-problems as future professionals, therefore it is an authentic assessment.(Cumming and Maxwell 1999; Gulikers et al. 2004). Furthermore, the analysis of how to overcome the situation-problem generates processes of problem-solving and application of knowledge and decision-making which correspond to the development of higher-order cognitive skills (Ashford-Rowe et al. 2014). Cognitive challenge is another characteristic of authentic assessment (Villarroel et al. 2018).

The autonomous nature of the assessment task can help students become investigative, resourceful, and responsible but they might not know how to master it. To overcome this barrier, we provided students with relevant information and feedback using a rubric and formative assessments which comprised a series of learning activities, including rubric review, assessment exemplars and self-assessment.

There is a vast educational literature on the usefulness of rubrics (e.g., Jonsson and Svingby 2007; Reddy and Andrade 2010; Brookharta and Chen 2015). An analytic rubric can help students build competence by making the expectations of assessment transparent in advance (Wiggins 1990), albeit evidence on students' perceptions of about rubrics is mixed. For instance, Andrade and Du (2005) found that using rubrics helped some students to focus their efforts, produce a higher quality work, earn a better grade, and feel less anxious, but most students did not fully read the rubric or perceived rubric as a tool for satisfying teacher's demands, rather than a representation of the criteria and standards of the discipline. Along the same lines, Baron and Keller (2003) found that students doubt the usefulness of rubrics for self-assessment and better interpretation of feedback. Research also emphasizes that the rubric design can harm students's innovativeness and creativity by encouraging them to focus on grades rather than on learning (Kohn 2006). Finally, while it is agreed that rubrics can have a positive effect on student learning, it is difficult to identify the direct relationship between the use of rubrics and academic outcomes because of the effects of other teaching practices applied simultaneously (Menéndez-Varela and Gregori-Giralt 2018). Therefore, the impact of rubrics on student performance is not explored in this paper.

There is also evidence of the use of rubrics for assessing students' learning about sustainability in higher education (Mckeown 2011; Tactenberg et al. 2017), as an instrument to operationalise student learning in transdisciplinary sustainable projects (Gulikers and Oonk 2019), and as an indicator of ESD programmes' success (UNESCO Bangkok 2007; Freitas et al. 2022). This study focuses on students' perception of the use of rubrics and formative assessments for developing their motivation, autonomy, and competencies for mastering an authentic assessment task related to sustainability.

The timeline of the assessment is shown in Table 1. The assessment question and rubric were provided to students at the beginning of the semester. The rubric described six performance indicators (as per '*Schedule A of the Code of Assessment*' of the University of Glasgow) combined with definitions for those indicators at different performance levels (Dawson 2017). The definitions were descriptive to highlight the key features of the assessment task to help students become thoughtful and constantly judge the quality of their work, which is a vital skill underpinning the development of self-regulation (Nicol and Macfarlane-Dick 2006). The assessment criteria focused on problem identification, knowledge, analysis, evaluation, recommendation, and communication which are skills for sustainability literacy (Nolet 2009).

Andrade (2005) argues that rubrics are not a replacement for good instruction; students still need models, feedback, and opportunities to ask questions, think, revise, and so on. These skills can be developed by providing direct authentic evaluative experience using formative assessments (Sadler 1987; Villarroel et al. 2018). To have a shared understanding of the rubric with students, we first reviewed the rubric with them in the class. Next, we provided three exemplars to assess using the rubric in groups followed by a discussion facilitated by the instructor. The exemplars were past students' work chosen to be the typical level of designated performances (Sadler 1987). Each group reported the grades they awarded to the three exemplars to the class with a justification involving (at least) two good and bad points as well as two ways to improve. The range of grades and comments assigned to each exemplar were compared until an agreement about an acceptable grade was reached. However, the comments and feedback made in the class could lead to student improvements in learning or performance only if the students compare them with their work and generate internal feedback out of that comparison (Sadler 1987; Nicol 2019). Thus, at the end of the activity, the students completed a self-assessment report to identify (at least) three areas on how they plan to incorporate the comments to improve their draft. The students could also discuss their plan with the instructor during dedicated office hours to incorporate feedback loops (Sadler 1987; Hattie and Timperley 2007).

Table 1 Assessment timeline

Release of assessment question and rubric	Week 1
First formative assessment: rubric review	Week 2
Second formative assessment: exemplars and self-assessment	Week 5
Assessment deadline	Week 10

A summary of the class activity and the three exemplars were made available to the students for future reference points.

Several pieces of evidence from the higher education literature show that providing marking criteria, grade descriptors and exemplars is a useful assessment practice to help students understand the requirements of an unfamiliar and challenging assessment task (Bella et al. 2013), the standards and different methods of approaching the assignment as well as their strengths and weaknesses (Orsmond et al. 2002). Exemplars can also motivate students to emulate the quality and standard (Handley and Williams 2011), and emulation is an almost universal learning method (Sadler 1987).

3 Methodology

Ethics approval for this study was obtained from the College of Social Sciences Ethics Committee at the University of Glasgow (Ref no. 400170120). To understand students' perceptions of authentic assessment and rubric, we collected individual student feedback for the three selected courses via an anonymous online questionnaire. The questionnaire consisted of a mix of multiple-choice, Likert-scale and ranking questions. Students were given one week to complete the online form before the submission deadline. The response rate was approximately 50% in each course with a class size ranging from 70 to 110 students.

Table 2 presents the survey questions. Question 1 explores whether students' receptiveness to teacher's feedback is related to the grade of their assessment. The aim of this question is to gain insights into students' willingness to consider or accept suggestions and constructive criticism of their work, which is an indicator of developing capabilities to assess and regulate their learning and performance. It also gives an insight into how motivated students are in their desire to learn post-assessment. Question 2 uses a list of Likert-scale questions to investigate the effectiveness of rubrics and formative assessments, hence this question focuses on the assessment design. Similar to Question 2, Question 3 asks students to indicate their level of agreement or disagreement with a series of statements to examine students' perceptions of the usefulness of the authentic assessments in fostering sustainability and employability skills, competencies, and autonomy for mastering the task. In Question 4 students rank choices based on what roles of rubric are more important than others in the context of our assessment approach. The results are presented in the next section.

Table 2 Survey questions

1. You are receptive to teacher's feedback on your assignments when the grade is: Low High Always, no matter the grade it Never
2. Please rate each of the following objects on a rating scale of: Strongly agree—Agree—Neutral—Disagree—Strongly disagree. a. I reviewed the coursework rubric before working on the assignment. b. Lecturer must review the rubric with students in advance. c. Rubric helped me understand the expectations and intended learning outcomes of the assignment. d. Rubric will help me improve future work. e. Rubric is easy to understand and follow. f. Rubric should be provided for all assessments. g. The formative assessment helped me better understand course expectations and learning outcomes.
3. In this assignment, you worked on a real-world sustainability management problem. Please rate each of the following objects on a rating scale of: Strongly agree—Agree—Neutral—Disagree—Strongly disagree. a. The assignment improved my awareness about sustainability. b. The assignment improved my literacy about sustainability. c. The assignment improved my problem-solving skills. d. The assignment improved my communication skills. e. The assignment developed my critical and independent thinking skills. f. I have learnt to be investigative from this assignment. g. I have learnt to be resourceful from this assignment.
4. Do you believe that the purpose of the rubric is to ensure (please put the options in order of preference): a. Clear assessment criteria. b. Understand the strengths and weaknesses of my work. c. Reduce anxiety about the assignment. d. Fairness in grading. e. Standard and timely feedback.

4 Results

Overall, across the three courses, 110 students agreed to participate in this survey, out of which 52% were males, 45% were females, and 52% were international students and 47% were domestic students. Hence, it was a well-represented sample (Fig. 1).

The collected data was categorized into four categories as follows: students' intrinsic motivation to accept feedback, promotion of students' competencies and autonomy, students' experience with assessment design, and students' views about the role of the rubric. The results are presented in Figs. 2, 3, 4 and 5. We define agreeability rate as the share of students who agree and strongly agree with survey statements.

As Fig. 2 shows, about 80% of the students indicated that they were receptive to the teacher's feedback regardless of the grade they received. This shows not only students' interest in the topic but also their inherent inclination to learn how to improve their work and perform at a higher level in future. Therefore, our practice

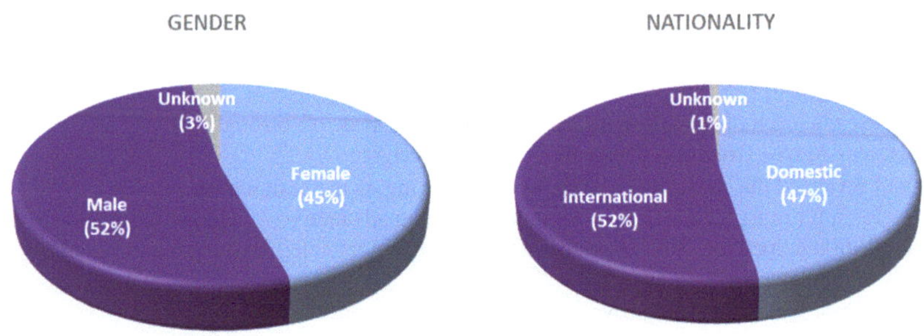

Fig. 1 Demographic characteristics of the sample

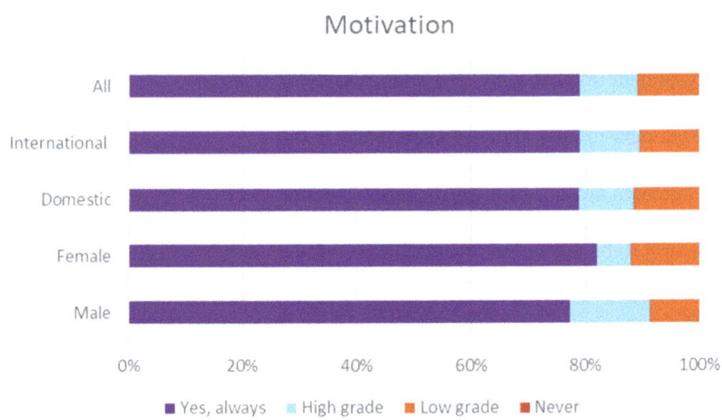

Fig. 2 Students' perceptions about motivation to learn and develop (%)

of using formative assessment and rubric for generating feedback and competencies assisted students with maintaining and increasing their intrinsic motivation. This result is in contrast with some of the papers reporting that students are often driven by high-grade achievement (Romanowski 2004; Hulleman et al. 2008; Van Etten et al. 2008).

Students' experience with authentic assessments supported by rubrics and formative assessments helped them to build competence and autonomy, as it is evident in overall positive students' responses (Fig. 3). In particular, students overwhelmingly agreed that their sustainability literacy (87%) and sustainability awareness (100%) had been promoted. Indeed, none of the students disagreed about these benefits and a high percentage of them voiced strong agreeability on becoming sustainability aware (48%) and sustainability literate (35%).

Students' autonomy was promoted by developing investigative skills (83%), resourcefulness (78%) and critical and independent thinking (74%). There were gains in problem-solving (61%) and communication (52%) skills, however, these were not

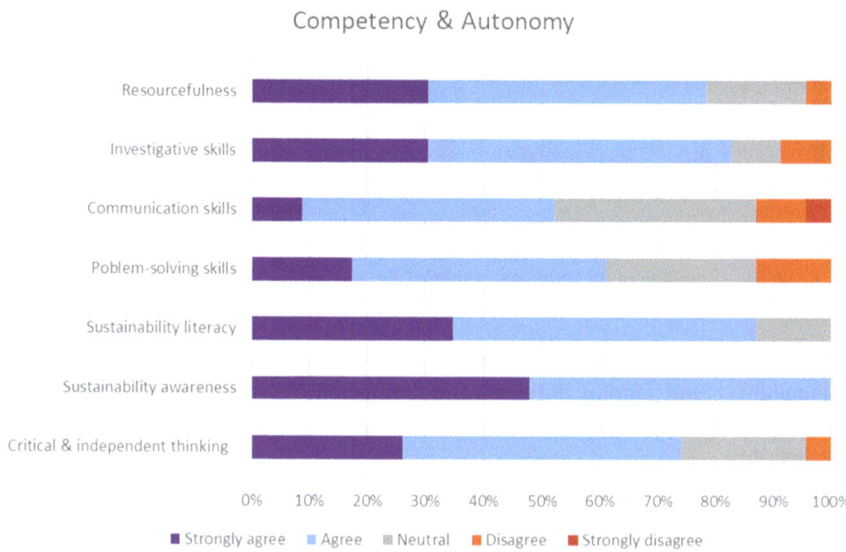

Fig. 3 Students' perceptions about competency and autonomy (%)

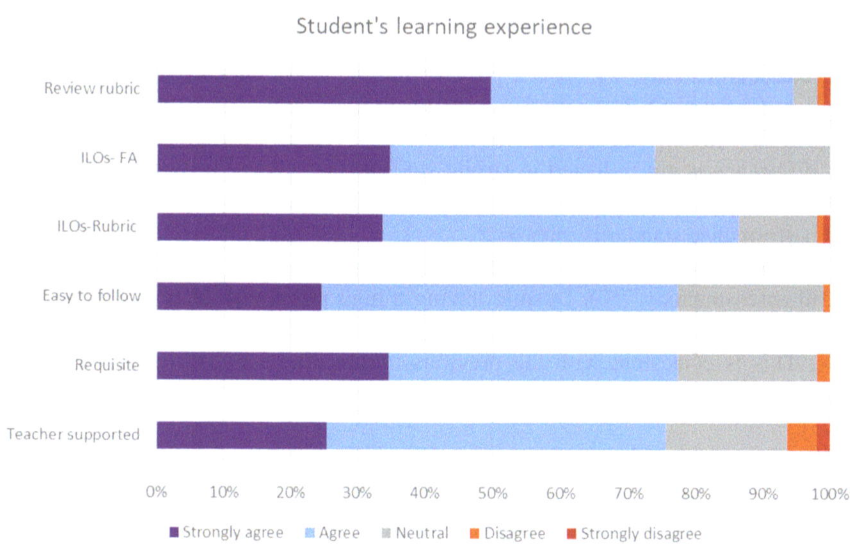

Fig. 4 Students' learning experience with rubric and formative assessments (%)

as high as for other skills. In fact, these two skills had the most disagreement (close to 20%) overall and the highest differences across gender and nationality. This is consistent with Navrouzoglou et al. (2023) findings on diverse students' perspectives in terms of opportunities for developing such skills, and therefore requires further investigation.

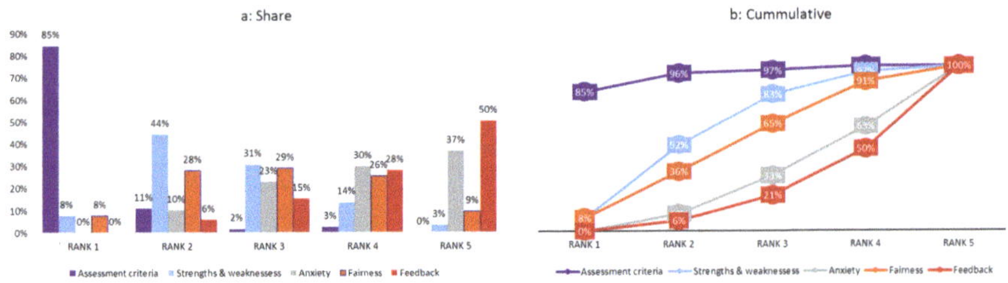

Fig. 5 Students' ranking of the purposes of rubric (%)

In sum, these responses show that the assessment design promoted students' sustainability- and employability-related competences and enabled autonomous learning, both supporting their intrinsic motivation to learn and develop for sustainability.

Considering the students' learning experience with rubrics and formative assessments (FA), the most notable observation is that 94% of students agreed with reviewing rubric before working on summative assessments (Fig. 4). In fact, 50% of the students had a strong positive response to this assessment practice.

Another area that benefitted students was from using rubrics in understanding expectations and intended learning outcomes (ILOs) of the summative assessments which is consistent with previous findings (Andrade 2005). Around 86% of students indicated agreeability, which is a positive finding since students often struggle to translate course ILOs into practice. This result is slightly higher than the students' experience with formative assessments overall (74%), however, it's worth noting this difference is not reflected in strong students' responses. It means that the use of rubrics contributed directly by helping the students understand what course ILOs and performance standards refer to.

Students indicated that rubrics are easy to follow (77%) and they must be provided for all assessments (77%). Despite having a high possibility of reviewing rubrics (94%), the demand for having rubrics for all assessments is surprisingly lower at 77%. This is consistent with the previous finding that some students didn't find rubrics helpful to understand the teachers' expectations and ILOs (Baron and Keller 2003).

When we explored students' views about our overall approach which requires teachers to support students' autonomy using formative assessments and reviewing rubrics, there was a 75% agreement rate with 25% of students voicing their preferences strongly. Around 6% of students indicated disagreement and this could be because these students have stronger preferences for other kinds of feedback, for example, peer-assessment (Nicol and Macfarlane-Dick 2006) or teacher-feedback (Andrade and Du 2005).

In terms of usefulness in improving future work, students were positive (70%), which is less strong compared to other benefits, but it confirms students' motivation to use feedback for improving themselves.

Based on the students' experience with rubric, we used students ranking of the purposes of the rubric to understand what is more and less important to them. In Fig. 5a, the bars in each rank indicate the share of students choosing different purposes. An overall ranking is obtained by looking at the cumulative of students' responses and identifying the sequence in which shares of responses increase in each rank category (Fig. 5b).

Many students indicated clear assessment criteria as one of the top-ranking purposes of the rubric (85%). This was followed by the rubric's ability to help students understand the strengths and weaknesses of their own work (52%), which is important if students were to develop into confident learners. They valued fairness in grading as the third most popular choice (65%), which was followed by the role of rubric in reducing anxiety about the assignment at the fourth place (63%). Receiving standard and timely feedback was ranked the last.

These results confirm that rubrics can help students to better understand the assessment criteria and improve their learning experience of assessment by reducing anxiety about the assessment and its outcome. Such a positive learning environment can make students feel more confident, aligned, and competent. It can change their attitude about assessment, resulting in increasing intrinsic motivation.

5 Conclusions

Education requires transformation to help society to cope with the challenges of sustainability. Humans are born to learn and have a desire to internalise the knowledge, customs, and values that surround them (Niemiec and Ryan 2009, p. 133). Hence, if educators have a mindset of growth, they can enhance students' intrinsic motivation to achieve the required transformation.

This study aimed to show how authentic assessments when combined with rubrics and formative assessments affect learners' autonomous behaviours and intrinsic motivation, build student competencies, improve the student experience, and attitude about assessment, and embed sustainability in an existing curriculum.

The results from the students' experience survey corroborate evidence highlighted in the literature that rubrics can help teachers to better communicate and clarify the assessment criteria, ILOs and their expectations (Mckeown 2011; Andrade 2005). Similar to the literature, this paper highlights the benefits of authentic assessments such as learning, autonomy, motivation, self-regulation, cognition, investigative skills, and resourcefulness by simulating the activities and competencies that they will have to utilize in their future jobs (Villarroel et al. 2018). An authentic assessment on sustainability can further promote sustainability literacy and awareness. Rubrics can also alleviate stress and anxiety from assessments and outcomes for students that can have negative effects on their academic performance. In combination with formative assessments for supporting students to perform an authentic assessment task, rubrics can promote effective academic skills and growth mindset.

A reflection of our experience with the new assessment design is that when educators "*engaged in autonomy-supportive behaviors such as offering choice, providing rationales, focusing on students interests or questions, and other specific autonomy-supportive behaviors, students reported greater interest in the world*" (Ryan and Deci 2020, p. 4). When assessment tasks create curiosity and there is autonomy-supportive learning and teaching environment, the student learning experience can be more positive and effective, resulting in an enhanced intrinsic motivation to learn and engagement with their studies and the world (Malmquist 2022). We believe that this type of assessment approach can create a foundation for lifelong learning, and we plan to explore its potential further.

In the future, we can think about different ways of developing communication skills, for example through peer-to-peer interactions and collaborative learning over a sustained period (Navrouzoglou et al. 2023). For promoting problem-solving skills, we can explore other types of exemplars and use case studies that can serve as practical models of how to explore, understand, and analyse sustainability problems. We want to better understand what types of motivation students have when enrolling in elective courses focused on sustainability and continue finding effective ways to encourage students to develop their inherent potential, deploy positive attributes and behaviors and instil the desire to impact their life and future by having a growth mindset. We also want to explore how students' motivation and perception of our assessment approach differ across different demographic characteristics, such as gender and nationality, and how it contributes to improving students' academic outcomes. Finally, students' perception of our assessment approach could also be explored further by broadening the scope of the survey and incorporating courses that are not focused only on sustainability.

References

Andrade HG (2005) Teaching with rubrics: the good, the bad, and the ugly. College Teaching 53(1):27–30. http://www.jstor.org/stable/27559213

Andrade HG, Du Y (2005) Student perspectives on rubric-referenced assessment. Practical Assess, Res, Eval 10(3). https://doi.org/10.7275/g367-ye94

Ashford-Rowe K, Herrington J, Brown C (2014) Establishing the critical elements that determine authentic assessment. Assess Eval High Educ 39(2):205–222. https://doi.org/10.1080/02602938.2013.819566

Baron J, Keller M (2003) Use of rubrics in online assessment. Evaluation and assessment conference, pp 23–25 November 2003. (2nd: 2003: Adelaide, South Australia)

Bella A, Mladenovicband RP, Price M (2013) Students' perceptions of the usefulness of marking guides, grade descriptors and annotated exemplar. Assess Eval High Educ 38(7):769–788. https://doi.org/10.1080/02602938.2012.714738

Brookhart SM, Chen F (2015) The quality and effectiveness of descriptive rubrics. Educ Rev 67(3):343–368. https://doi.org/10.1080/00131911.2014.929565

Candy P (1991) Self-direction for lifelong learning: a comprehensive guide to theory and practice by Philip C. ISBN 1-55542-303-5

Cumming J, Maxwell G (1999) Contextualising authentic assessment. Assess Educ: Principles, Policies and Practices 6(2):177–194. https://doi.org/10.1080/09695949992865

Dawson P (2017) Assessment rubrics: towards clearer and more replicable design, research and practice. Assess Eval High Educ 42(3):347–360. https://doi.org/10.1080/02602938.2015.1111294

Deci EL, Ryan RM (2015) Self-determination theory. In: Wright JD (ed) International encyclopedia of the social and behavioral sciences, vol 91, pp 486–491. https://doi.org/10.1016/B978-0-08-097086-8.26036-4

Doh JP, Tashman P (2014) Half a world away: the integration and assimilation of corporate social responsibility, sustainability, and sustainable development in business school curricula. Corporate Social Responsibility Environ Manage 21:131–142. https://doi.org/10.1002/csr.1315

Filho WL, Shiel C, Paço A, Mifsud M, Veiga Ávila L et al (2019) Sustainable development goals and sustainability teaching at universities: falling behind or getting ahead of the pack? J Clean Prod 232:85–294. https://doi.org/10.1016/j.jclepro.2019.05.309

Freitas D, Calafell G, Pierson AHC (2022) A rubric to evaluate critical science education for sustainability. Sustainability 14:8289. https://doi.org/10.3390/su14148289

Gulikers J, Oonk C (2019) Towards a rubric for stimulating and evaluating sustainable learning. Sustainability 11(4):969. https://doi.org/10.3390/su11040969

Gulikers J, Bastiaens T, Kirschner P (2004) A five-dimensional framework for authentic assessment. Educ Tech Res Dev 52(3):67–86. https://doi.org/10.1007/BF02504676

Hattie J, Timperley H (2007) The power of feedback. Rev Educ Res 77:81–112. https://doi.org/10.3102/003465430298487

Hulleman CS, Durik AM, Schweigert SA, Harackiewicz JM (2008) Task values, achievement goals, and interest: an integrative analysis. J Educ Psychol 100(2):398. https://doi.org/10.1037/0022-0663.100.2.398

Jonsson A, Svingby G (2007) The use of scoring rubrics: reliability, validity and educational consequences. Educ Res Rev 2(2):130–144. https://doi.org/10.1016/j.edurev.2007.05.002

Kohn A (2006) The trouble with rubrics. English J 95(4):12–15. https://www.alfiekohn.org/teaching/rubrics.pdf

Laal M (2011) Lifelong learning: what does it mean? Procedia Soc Behav Sci 28:470–474. https://doi.org/10.1016/j.sbspro.2011.11.090

Malmquist C (2022) Curiosity, motivation, autonomy, and lifelong learning in education and the United States marine corps. Theses, student research, and creative activity: department of teaching, learning and teacher education, p 135. https://digitalcommons.unl.edu/teachlearnstudent/135

Mckeown R (2011) Using rubrics to assess student knowledge related to sustainability: A practitioner's view. J Educ Sustain Dev 5(1):61–74. https://doi.org/10.1177/097340821000500110

Menéndez-Varela JL, Gregori-Giralt E (2018) Rubrics for developing students' professional judgement: a study of sustainable assessment in arts education. Stud Educ Eval 58:70–79. https://doi.org/10.1016/j.stueduc.2018.06.001

Moore J (2005) Barriers and pathways to creating sustainability education programs: policy, rhetoric and reality. Environ Educ Res 11(5):537–555. https://doi.org/10.1080/13504620500169692

Neus E, Stevenson RB, Lasen M, Ferreira J, Davis JM (2017) Approaches to embedding sustainability in teacher education: a synthesis of the literature. Teach Teach Educ 63:405–417. https://doi.org/10.1016/j.tate.2017.01.013

Nicol D, Macfarlane-Dick D (2006) Formative assessment and self-regulated learning: a model and seven principles of good feedback practice. Stud High Educ 31:199–218. https://doi.org/10.1080/03075070600572090

Nicol D (2019) Reconceptualising feedback as an internal not an external process. Italian J Educ Res, Special Issue, pp 71–83. https://ojs.pensamultimedia.it/index.php/sird/article/view/3270

Niemiec CP, Ryan RM (2009) Autonomy, competence, and relatedness in the classroom: applying self-determination theory to educational practice. Theory Res Educ 7(2):133–144. https://doi.org/10.1177/1477878509104318

Nolet V (2009) Preparing sustainability-literate teachers. Teach Coll Rec 111(2):409–442. https://doi.org/10.1177/016146810911100207
Orr DW (2004) Earth in mind: on education, environment, and the human prospect. Earth Island Press, Washington, DC. ISBN, 1559632941, 9781559632942
Orsmond P, Merry S, Reiling K (2002) The use of exemplars and formative feedback when using student derived marking criteria in peer and self-assessment. Assess Eval Higher Educ 27(4):309–2. https://doi.org/10.1080/0260293022000001337
Reddy YM, Andrade H (2010) A review of rubric use in higher education. Assess Eval High Educ 35(4):435–448. https://doi.org/10.1080/02602930902862859
Romanowski MH (2004) Student obsession with grades and achievement. Kappa Delta Pi Record 40(4):149–151. https://doi.org/10.1080/00228958.2004.10516425
Ryan RM, Deci EL (2000) Self-determination theory and the facilitation of intrinsic motivation, social development, and well-being. Am Psychol 55:68–78. https://doi.org/10.1037/0003-066X.55.1.68
Sadler DR (1987) Specifying and promulgating achievement standards. Oxford Rev Educ 13:191–20. https://www.jstor.org/stable/1050133
Tractenberg RE, FitzGerald KT, Collmann J, Evidence of sustainable learning from the mastery rubric for ethical reasoning. Educ Sci 7(1):2. https://doi.org/10.3390/educsci7010002
UNESCO Bangkok (2007) Asia-Pacific guidelines for the development of national ESD indicators. https://bangkok.unesco.org/content/asia-pacific-guidelines-development-national-esd-indicators
Van Etten S, Pressley M, McInerney D, Liem A (2008) College seniors' theory of their academic motivation. J Educ Psychol 100(4):812–828. https://doi.org/10.1037/0022-0663.100.4.812
Villarroel V, Bloxham S, Bruna D, Bruna V, Herrera-Seda C (2018) Authentic assessment: creating a blueprint for course design. Assess Eval High Educ 43(5):840–854. https://doi.org/10.1080/02602938.2017.1412396
Wiggins G (1990) The case for authentic assessment. Pract Assess Res Eval 2(2):28–37. https://doi.org/10.7275/ffb1-mm19

From TBLA to QBLA: Interpreting and Redesigning Business Sustainability Education Through the Business Sustainability Cycle

Tony Bradley

Abstract Assisting students to interpret the meaning and practice of sustainability is a significant challenge in current business education. In recent years, interpreting the *purposes* of sustainable businesses has become a vital aspect; when *purpose* is added to the mix, businesses can look at the full spectrum of their sustainability activities and operations, in a cycle from People, through Purpose, and Planetary to Profitability dimensions, or quadruple bottom-line accounting (QBLA). Consequently, business education for sustainability can be redesigned to reflect this fourfold approach and cyclical model. The chapter considers the development of the *Business Sustainability Cycle* (BSC), which assists students in understanding how enterprises can develop their overall sustainability. It considers the significance of adopting a 'full-spectrum' or 'integral' sustainability approach, in both undergraduate and postgraduate teaching and research, and considers some cases as applied by undergraduate and postgraduate students.

Keywords Business sustainability cycle · Triple and quadruple bottom-line accounting · Ideal types · Jungian quaternity archetype · Undergraduate courses · Postgraduate consultancies

1 Introduction

The teaching of business sustainability, sustainable business and business and the environment has accelerated in recent years, in step with the incidence of references to sustainability on the internet and in the media. Almost thirty years ago McKibben (1996) commented that 'sustainability' had become a 'buzzless buzz word'. The term can be robbed of any real meaning, which is why we need to help and educate students

T. Bradley (✉)
SEARCH (Sustainable Economy Action Research Centre at Hope), Liverpool Hope University Business School, Liverpool, UK
e-mail: bradlet@hope.ac.uk

T. Wall et al. (eds.), *Sustainability in Business Education, Research and Practices*, World Sustainability Series, https://doi.org/10.1007/978-3-031-55996-9_17

in respect of how to understand business sustainability and its varied discussion, given that it has become ubiquitous (Grober and Cunningham 2012).

As presented here, business sustainability can be understood in terms of a 'four Ps' perspective: people, purpose, planet, and profit. This pattern of four perspectives presents an ideal type for understanding 'integral' or 'full-spectrum' sustainability. As discussed below, the familiar three-legged triple bottom-line (TBLA) model of sustainability—focusing on economic, social and environmental dimensions—represents the dominant paradigm for sustainability. But, when purpose is added to the mix it is possible to redraw the model, from lines of accountability and measurement to a cycle of business sustainability.

The objective of this chapter is to explain this model and how it is utilised by turning from four poles of sustainability towards a 'business sustainability cycle' (BSC). Following a discussion of the critical literature, related to TBLA, the business sustainability cycle is introduced and explained. Its application in undergraduate taught courses and through postgraduate consultancies is referenced. The intention is to stimulate the use of the BSC, as a model for deepening our exploration and widening our critical appraisal, as students and investigators, looking at sustainable business.

2 The TBLA Meaning of Sustainability

Seeking an interpretive understanding of sustainability is a highly complex process. But, to communicate its meaning to students some clear definitions are required. A very general definition is: 'the ability of human society to live within the regenerative capacity of the biosphere' (Wackernagel et al. 2002).

This definition has conventionally been understood to meet the demands of Triple Bottom-Line Accounting (TBLA) (Elkington and Rowlands 1999). It constructs an ideal type following the three-legged stool approach of the Brundtland Report, *Our Common Future* (1987), to understand sustainable development. Here businesses are required to account, not only in financial terms (Profit), but also in respect of social needs (People), and the long-term preservation of natural and ecological systems (Planet) (Palmer and Flanagan 2016, Bocker and Meelan 2017, inter alia).

Equally, other three-legged ideal typologies have focused on ESG (environmental, social and governance) reporting (Bose 2020; Arvidsson and Dumay 2022; Darnall et al. 2022) and the 'triangle of sustainability' (Campbell 2016). These originated at the Rio Earth Summit, 1992 (ecology, economy, and social justice). Nevertheless, as Campbell (ibid.) indicates, these three-fold principles have evolved considerably over the past three decades. In particular, the United Nations' Millennium Development Goals, to serve until 2015, were superseded by 17 Sustainable Development Goals, which are to serve until 2030 (Clegg 2015; McCloskey 2015).

Other three-fold ideal types include the three rings of sustainability (Barbier 1987), and "the fellowship of the rings", derived from the Rio Earth Summit (1992) (Cocklin and Moon 2020). Additionally, a more contemporary perspective is reflected in the balanced sustainability triangle, considering inter-generational equity across the three spheres of ecology, society, and economy (Munasinghe 2017). Then, in terms of corporate accountability, the Global Reporting Index (GRI) standards, which is a series of modular accounting measures, also reflects the ESG framework (Sustainability Report 2021).

Purvis et al. (2019) suggest that the origins of the three-fold sustainability model are hard to locate, but, one of the earliest versions was published in a small monograph, at a north of England college, by Spreckley (1981). The brilliance of Spreckley's early work arose from his recognition that 'triple bottom-line accounting' challenged capitalism—and its inherent business model of profit maximisation—at its core. The three-fold ideal type of sustainability needed to incorporate the values of a rapidly changing world.

3 Origins of a Critique

That world, from the early 1970s, was beginning to recognise that market crises indicated more deep-seated social and ecological fractures. A distorted economic system, which placed economic imperatives ahead of those of society and the planet, within the framework of neoclassical, mainstream economics, was unable to either account or compensate for its so-called social and ecological 'externalities' (cf Daly 1977, inter alia). This was the era when neoliberal Thatcherism, Reagonomics, and 'the wolves of Wall Steet' rose to dominate international trade (Marshall and Arestis 1989). Contrastingly, the three rings model idealised interdependent economic, social, and ecological spheres of human agency.

But, arguably, it failed to locate a key meaning of sustainability, which is to address future *purpose*. As such, students of sustainable development, over two generations, were taught a model that reflected the equivalent significance of economic, societal, and natural capital processes. Triple bottom-line accounting presented a flawed interpretive model of sustainability to students. It reflected a technocratic interpretation for business engagement in solving the climate and biodiversity crises, instead of one founded in emerging humanistic values.

This has been recognised by one of the model's prime exponents, John Elkington, who has called for a 'product recall' of this ideal type (Elkington 2018). His main critique is that the model was never meant to be a tool used by accountants, to legitimise businesses using social and natural capital valuation measures, to justify their 'corporate social responsibility'. But that is what had ensued. As he asserts, TBLA had become an instrument for accounting practices that "…move heaven and earth to ensure that they hit their profit targets, the same is very rarely true of their people and planet targets" (ibid n.p.).

Nevertheless, in the two decades since the turn of the millennium, alternative economic perspectives have become increasingly prominent, advocating such approaches as ecological economics (cf Jackson 2009, Gilding 2011). These developed from the questions that Daly and others asked, regarding whether technological development could 'de-couple' growth from reaching (or breaching) environmental limits. Major reports, such as the leading-edge Club of Rome's original *Limits to Growth* (Meadows and Meadows et al. 1972) had indicated the reality of planetary limits, and the dangers that economic 'overshooting' of these ecological and societal boundaries posed, in respect of the planet's 'carrying capacity'.

In turn, this perspective drew on earlier thinking that considered economics within the context of the natural science laws of physics, particularly the second law of thermodynamics. As early as the mid-Victorian era the social philosopher and political economist John Stuart Mill (1806–1873) had argued for stationary economics:

> "If the earth must lose that great portion of its pleasantness which it owes to things that the unlimited increase of wealth and population would extirpate from it, for the mere purpose of enabling it to support a larger, but not a better or a happier population, I sincerely hope, for the sake of posterity that they will be content to be stationary, long before necessity compels it" (in Winch, 2001 p.756).

Others, sought authority in the work of Marx, for attempting to integrate the second law with the labour theory of value (Podolinski 1880; Bellamy Foster and Burkett 2004). Neurath (1882–1945) (in Eubel and Cohen 2004) developed this socialist calculus, challenging the fundamental principle that prices and money valuations can ever reflect real world social and, hence, environmental, values. Frederick Soddy (1877–1956) shared Neurath's view that money valuations were entirely subjective and that real wealth, demonstrated by the use-value of actual things, was continuously being converted, within money markets, into debt.

Perhaps the most important physicist turned economist, reflecting on the implications of the second law of thermodynamics for economics, and establishing the scientific foundations for ecological economics, was Nicholas Georgescu-Roegen (1906–1994) (Martinez-Alier 1987). Critically, he questioned the view, as had Soddy, Neurath, Podolinski and Marx, before him, that money values and prices, in any sense, reflected the long-term value to society of planetary sources (Georgescu-Roegen 1976). But, he did so by connecting economics with the physics principle of entropy, that all matter and energy in the universe is degrading from a state of concentrated high intensity to one of low energy dispersal and chaos. As he commented:

> "Only economists still put the cart before the horse by claiming that the growing turmoil of mankind can be eliminated if prices are right. The truth is that only if our values are right will prices also be so" (Georgescu-Roegen, ibid. xix).

In other words, value is established through values and purposes. And, in sustainability terms, such accurate values need to take the planetary needs of society and nature into full account, if we are to establish meaningful understandings of an intergenerationally, just future. Such a view has led some of the current generation's most influential economists to adopt what had, hitherto, been seen as entirely heterodox positions. For example, as the India-UK economist Dasgupta (2021) puts it:

> "…the macroeconomic growth and development theories that have shaped our beliefs about economic possibilities and our understanding of the progress and regress of nations do not recognise humanity's dependence on Nature" (p 38).

In a similar vein, Helm (2015) argues that policymakers need to deploy a 'non-declining natural capital' rule. This should ensure that natural capital is valued—and sustained—through the decision-making process, in making trade-offs between policy objectives. Overall, Helm is equally sceptical about the capacities of institutions to properly price the value of natural capital, and of environmentalists, who challenge the possibility of generating natural capital substitutes. He argues that our explicit, stated values need to reflect our valuations, and vice versa.

4 Purpose—The Fourth Dimension of QBLA

The significance of this shift in economic thinking is to emphasise the need to incorporate the fourth dimension of *Purpose* into the ideal type of sustainability. As such, the signification model shifts from TBLA to Quadruple Bottom-Line Accounting (QBLA). This has become increasingly fundamental in recent years (Hadders 2017; Elkington 2018).

The demands on business, from consumers, other economic stakeholders, and society, itself -to engage in a debate about the 'meaningfulness' of business—has accelerated the shift to measuring valuations from within a QBLA framework. Businesses need to account for their actions based on ethics, in terms of governance, culture, values, and the commitments they align with, to ensure that life has significance, meaning (Drewell and Larsson 2017) and purpose (Stroehle et al. 2019).

Even so, it is hard enough to measure monetary, social, and natural capital values, without asking how to account for meeting the need for meaning. Yet, that is what society, increasingly, craves, especially in an era where the call is to 'build back better', after a global pandemic. And, when we do insert Purpose into the other three poles of People, Planet and Profit we can identify that sustainability reflects a process of transformation (see Fig. 1).

On this basis it can be argued that sustainability is not a thing-in-itself (sui generis) but, rather, a cyclical process of increasing refinement. Shifting from the model of TBLA to QBLA represents an interpretive modification that can provide students with an alternative window through which to examine business sustainability. Instead of considering sustainability as a way of valuing the present in the light of the future, in accounting terms, it becomes a way of connecting societal, cultural, systemic, and economic ways of understanding business development. The rubric of 4 P's: People, Purpose, Planet and Profit becomes a mnemonic for considering how sustainable a business' development, strategically and operationally, is becoming. This process is conceptualised as the Business Sustainability Cycle (BSC).

We can see that the Business Sustainability (transformation) Cycle parallels other fourfold models of systems theory. Examples are the Shewart (1931/1980) process of economic control (Best and Neuhauser 2006), the Deming Cycle (Scherkenbach 1986, based on the work of S Mizuno at the *Tokyo Institute of Technology*, in 1959), David Kolb's (Sims 1983) experiential learning cycle, *Motorola*'s 6-sigma (Tennant

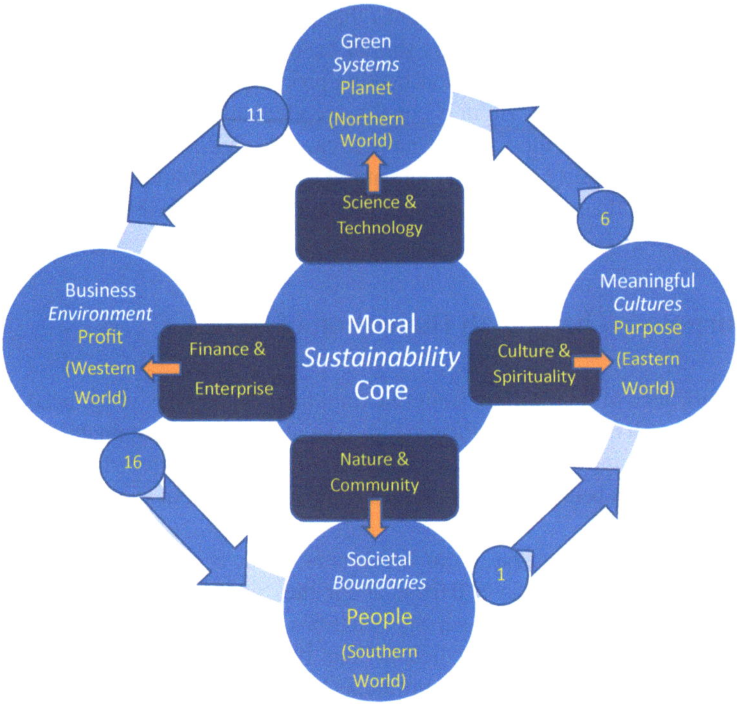

Fig. 1 The business sustainability cycle: transforming business integrally and sustainably: through the ideal type of QBLA. *Source* The author, adapting the integral model of Lessem and Schieffer (2010)

2001) and the generic process of continuous quality improvement (CQI), reflected in the Japanese principle of kaizen (Masaaki 1986).

5 The Jungian GEN*E*tic Basis of the Business Sustainability Cycle

Epistemology, or theories of knowledge, are generated by ontology, or the bedrock nature of models and ideal types. The ontology of the BSC resides in the integral model, which connects the four Ps realities with a dynamic GEN*E*alogical method of transformation, through research-to-innovation (see Lessem and Schieffer 2010, Lessem and Bradley 2019, *inter alia*). The Four Worlds-GEN*E* is based on the work of a collection of scholars, under the auspices of *Trans4m*, formerly based in Geneva. The BSC is a developmental application of this model, which seeks to demonstrate the transformation of businesses at a micro level, based in the model of the GEN*E*, which fits, holographically, within the macro scale of the four worlds, which the 4 P's reflect.

The GEN*E* stands for:

G Grounding business in its community and context (People).
E Emerging its culture through specific values (Purpose).
N Navigating its systems and processes, as "compass bearings", using specific technologies, models, and (scientific) methods, that connect it to the greening of industry and low carbon/ natural capital enhancing futures (Planet).
E *E*ffecting new enterprise forms, through engagement with markets, politics, and societal stake holding (Profit) (ibid.).

The Four Worlds reflect idealised conceptions of the cultural contributions of different global contexts:

> Southern World—focused on relational connections to Nature and Community (e.g. 'ubuntu'— "I am because we are", in sub-Saharan Africa (Mboti 2015)).
>
> Eastern World—focused on reflective meanings drawn from Culture and Spirituality (e.g. 'tao', the middle way connecting 'inner' and 'outer', in Taoism (Coward 1996)).
>
> Northern World—focused on rational systems of Science and Technology (e.g. the scientific method, despite its widespread critique and rejection in much educational theory (Gower 1997)).
>
> Western World—focused on realising enterprise innovation, based in Economics and Politics (e.g. F. A. Hayek's philosophy of the market, as connecting economy, science and politics (Hayek, ed. B. Caldwell 2014)).

The four worlds and the GEN*E* are rooted in C. G. Jung's notion of the quaternity archetype, which, for Jung, was the core archetype for understanding the process of 'individuation' or realisation of the integral self. In popular psychology and human resource management this is most frequently understood in terms of Jung's theory of personality (ideal) types, and derivatives such as the Myers-Briggs Type Indicator (MBTI).

But Jung understood archetypes to be more rooted in material reality than the ideal typologies of Max Weber (1949). He contends that the fundamental four-fold conceptualisation of human psychological processes is rooted in the actual fabric of the material world. Current split-brain theory, evolutionary and molecular biology analysis of the DNA molecule, and embryology, each point to the quaternity as beyond metaphor, being reflective of biological and material existence. Indeed, Anthony Stevens, in his up-dated study of Jung's archetypes, goes so far as to suggest that they are biological entities of the nervous system (1990, p 296).

Jung considered that the production of quaternities—usually circles, divided into four quadrants, such as Tibetan Buddhist mandalas—either in conscious or dream states, was the capacity of the ego to assimilate unconscious material or to seek to avoid psychic disintegration.

> "The quaternity is one of the most widespread archetypes and has also proved to be one of the most useful schemata for representing the arrangement of the functions by which the conscious mind takes its bearings. It is like the crossed threads in the telescope of our

understanding. The cross formed by the points of the quaternity is no less universal and has in addition the highest possible moral and religious significance for Western man. Similarly, the circle, as the symbol of completeness and perfect being, is a widespread expression for heaven, sun, and God; it also expresses the primordial image of man and the soul." (Jung 1983, para. 405).

"From the circle and quaternity motif is derived the symbol of the geometrically formed crystal and the wonder-working stone. From here analogy formation leads on to the city, castle, church, house, and vessel. Another variant is the wheel (rota). The former motif emphasizes the ego's containment in the greater dimension of the self; the latter emphasizes the rotation which also appears as a ritual circumambulation. Psychologically, it denotes concentration on and preoccupation with a centre." (1969, para. 352).

The latter quote indicates something of the way in which the Four World's model is connected to the GEN*E*. The rota—which can also be a spiral through spheres of self, organisation, society, culture—is reflective of the dynamic GEN*E*. This indicates how the cultural process—grounding, emerging, navigating, effecting—becomes shaped into various aspects of economic content. The ego seeks individuation through the GEN*E* process of cycling through the Four Worlds, attempting to integrate principles of Southern, Eastern, Northern and Western worlds into a sustainable business. All derive from the centre; the 'inner moral core'. Equally, all are features of the material world, not simply idealised conceptions.

As such, through the BSC, as an analytical model of the Four Worlds and GEN*E*, a cycle of sustainability is introduced to students. As an ideal type, it functions to offer an interpretive lens to help them understand and conceptualise some of the rapid changes that are taking place, across the globe, in the third decade of the twenty-first century. Elsewhere I have depicted this transformative process using a parallel pattern of 4 P's: personal-place knowledge, pilgrimium, pedagogy and production (Bradley 2019).

Business leaders may adopt a developmental process that takes them on a journey from the place and personal character of their **g**rounding, through an **e**mergent pathway of growth (which can be depicted as pilgrimage, without any religious connotations cf 'the hero's journey' (Campbell 1949/2008), to the adoption of new systems, based in pedagogies of learning, which they use to **n**avigate their way into ***e***ffecting new enterprise and market production.

In most circumstances this remains as an unreflected, innate process. But, occasionally, it surfaces to the level of conscious realisation (see below), whereby the BSC generates an 'aha moment!' of recognition, through which business leaders identify an increasingly sustainable process of development. Our undergraduate and postgraduate students use this as a means for interpreting the sustainability of specific business cases. Thence, our Masters' training consultancy students use the BSC, in 16 steps, as a methodology for organising their consultancies.

6 The 16 Steps of the BSC

Students are presented with the GEN*E* and Four World's ideal types, learning the mnemonic of the 4 Ps, before being introduced to the 16 steps of the BSC. These steps act as a way of helping students to orient where a business is positioned in relation to the cycle of sustainability. The ideal type nature of the cycle is explained.

Students learn that it would be very rare for a business to systematically rotate around the 16 steps in the sequence in which they are presented. Indeed, in my ideal model there are two patterns of rotation. In the first, all the steps of each of the 4 Ps is completed before progressing to the next quadrant of the cycle. In the second, the steps are completed as four rotations of the 4 P's. The latter pattern reveals the four stages that are most definitive of the cycle (stages 1, 6, 11, and 16, see Fig. 1.).

1. People Perspectives

    Step 1: Identifying the business's community grounding.
    Steps 2/5: Assessing the internal business culture/ values.
    Steps 3/9: Connecting the business to its wider network.
    Steps 4/13: Evaluating the developmental/ training needs of the business.

2. Purposive Perspectives

    Steps 5/2: Identifying re-grounding of the business through new values.
    Steps 6/: Assessing the future purpose of the business.
    Steps 7/10: Connecting to local stakeholder values.
    Steps 8/14: Evaluating the business's developmental plan.

3. Planetary Perspectives

    Steps 9/3: Identifying barriers and limits to sustainable business change.
    Steps 10/7: Assessing how to avoid the 'recursive GEN*E*. This is a term we use to refer to the frequent tendency of businesses and organisations to turn towards effects, outcomes, and profitability before they have completed the earlier stages of the cycle, through values transformation and establishing navigational systems and processes.
    Steps 11: Connecting to wider/ global developmental frameworks e.g. UN SDG's.
    Steps 12/15: Evaluating ways of 'closing the loop' for business regeneration.

4. Profitability Perspectives

    Steps 13/4: Identifying (re)sources/of finance for business development.
    Steps 14/8: Assessing levels of investment (human, social, intellectual, financial capital) impact.
    Steps 15/12: Connecting the business ecosystem through a coherent mapping process.
    Step 16: Evaluating the business's regeneration, through sustainability reporting frameworks.

As commented above, the integral 4 Worlds/GEN*E* model can be pictured as a series of Russian dolls (*matryoshka*). Each quaternity nests a further fourfold. Consequently, this 16-steps version can be deployed as the means through which our students evaluate the sustainability dimensions and aspirations of business, either as cases or as consultancy clients.

Aspiration conveys the idea that businesses will tend toward some *ideal* level of sustainable integration. In practice, the 16 steps broadly align with our research aims and objectives. These situate the journeys that the businesses—which our students are evaluating—are on. Inevitably, each is an incomplete journey towards integral sustainability.

7 Operationalising the BSC 1: Evaluating the Sustainability of a Future Business Employer.

All third year (Level 6—higher) undergraduates, studying Business Management at Liverpool Hope Business School, whether as Single or Combined Honours, take a twelve-week course in Business Sustainability. At this level we do not expect students to work through all the 16 steps of the BSC. In lectures they are introduced to the 4 Ps, as pillars, with more attention paid to the People and Planet stages (Stages 1–3, 9 and 11).

From this the students are asked to consider the relative merits of TBLA and QBLA, as typologies for assessing business sustainability, wherein Purpose becomes the basis for an enterprise's sustainability value proposition. The presentation concludes with a case example of the BSC's 16 steps from one of our Masters' consultancy students. But their real work begins in the first Seminar Workshop.

The task for this workshop connects to the reality that they are four months from graduation and seeking to enter the world of work. The activity is:

> "A key aspect of this element is to help you consider your own future in the world of work, in terms of "personal business sustainability". First work on your own, then in groups and, finally, provide some brief feedback.

1. What is your ideal work role and industry/employer after Uni?
2. Name a specific business that you would like to work for?
3. At the start of the element, what do you see as some important sustainability issues relating to this business?

Discuss your personal reflections on the preceding questions in a small group, noting down any points of convergence and divergence across your small group".

This sets the scene for the following 12 weeks. Throughout the course they investigate the sustainability credentials of a named business and industry sector that they will focus their summative assessment on. In subsequent seminars they research the principles and practice of business sustainability in relation to:

- Responsiveness to employees (G), stakeholders (E), suppliers (N) and customers (*E*) in terms of cultural change (People).
- Education and training of staff in sustainability and meeting social needs (People).
- Ethical business standards and values (including so-called 'corporate social responsibility' and ESG) (Purpose).
- Mission and cultural values statements that point beyond the commercial objectives of the business (Purpose).
- Factors enhancing or inhibiting the business' connection to the meaningful economy (Purpose).
- The incorporation of natural capital enhancing processes, standards, and measurement into the business model (Planet).
- Consideration of the 'options for change', in respect of the enterprise's impact on natural capital, from *The Economics of Biodiversity* review (Dasgupta 2021 p.70. (Planet).
- The business' contribution to sustainable finance, ESG investments and the ethical use of funds (Profit).
- Engagement in blended finance and impact investing programmes and responsiveness to the climate-related financial reporting ecosystem (see TCFD 2022) (Profit).
- Assessing their chosen business in relation to my new definition of integral business sustainability, based in the GEN*E* and the 4 P's (see Bradley 2024, forthcoming) (Integral Business Sustainability).

By the end of the third-year undergraduate course students should have an integral framework for assessing a business' stated capacity to function as a sustainable enterprise. Whilst they will have engaged in some critical evaluation, much of their work will have been researching the business' activities, and reporting on them in a descriptive way. A more critical approach is required at first year Masters' level.

8 Operationalising the BSC 2: Critical Analysis of Sustainable Business Model Canvases Amongst Global Corporations.

Our IMBA (International Master's in Business Administration) students are required to undertake an 8 week taught course in International Business Sustainability and Ethics. These students are, also, introduced to the BSC in an alternative 16 steps form, based on the conceptualisation of a four-layered business model canvas (see Fig. 2.). This version of the BSC develops the concept of the original business model canvas (cf Osterwalder and Pigneur 2010). The principle of the canvas is to supersede the need for producing lengthy business plans. Completing the nine building blocks of the canvas, in graphical form, should show, in a single illustration, the inputs, throughputs and outputs of any business.

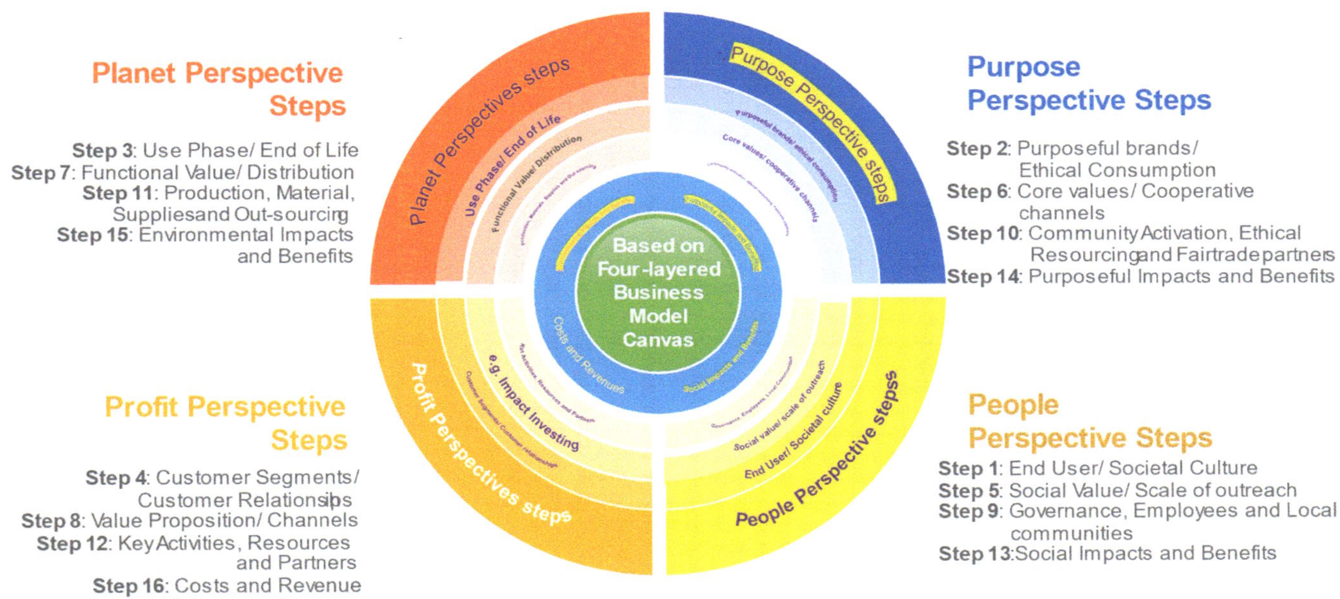

Fig. 2 The BSC based on a four-layered business model canvas. *Source* The author

Reflecting TBLA, the canvas has been reconstituted as a multi-layered form, usually in three layers, to reflect social stakeholder (People) and environmental life-cycle (Planet) dimensions, in addition to the economic (Profit) layer (Joyce and Paquin 2016). In turn, I have added a fourth layer to the canvas, which leads to the BSC model displayed in Fig. 2, to transition from TBLA to QBLA, and an integral understanding of business sustainability. Other versions of sustainability-related canvases have been produced to reflect the need for businesses to introduce increasing circularity into their operations (Daou et al. 2020).

Building on this foundation our taught Masters students are required to produce weekly formative assessments. For each they critically reflect on a global corporate business, to report on its sustainability. The model is simplified so that they report on 6 C's. These are: Context, Community (People), Culture (Purpose), Conservation (Planet), Commerce (Profit) and, critically, Challenges. As such, the 4 P's are re-framed, at Masters level, requiring students to interpret the context of the business and the main sustainability challenges which they identify in relation to its stated strategy and operations.

Student reporting should account for various steps in this second version of the BSC. As such, students learn to critically evaluate the sustainability business models of these global businesses. The assessment states:

> "Formative assessment: Each week students will select a global sustainable business case to write/ produce a critical evaluation on, using the 6 Cs reflective approach, presented in the Course, taking into account the BSC model, in its integral, four-layered, sustainable business model canvas form".

Additionally, the businesses that they report on are selected from the Four Worlds, to help students recognise the global spread of business sustainability. For example, in the first two weeks of reporting, as organised in the preceding academic year 2022–23, they were given the choice to report on the following businesses:

Week 1: Introducing Case Studies in impact investment: Industry focus on Professional Services sector

- Primary Western Case: *Impact Invest, Sweden*
- Secondary Western Case: *LGT Lighthouse, Liechtenstein*
- Secondary Southern Case: *Phatisa, Mauritius*
- Additional Eastern Case: *FMO, Mongolia*
- Additional Northern Case: *EBRD with Tauron Polska Energie, Poland.*

Week 2: Introducing Case Studies in Global Compact business action: Industry focus on business within the UN Global Compact.

- Primary Western Case: *Islandsbanki, Iceland*
- Secondary Western Case: *Curis Network, Cyprus*
- Secondary Southern Case: *Ferreycorp, Peru*
- Additional Eastern Case: *Free-Sal, Lebanon*
- Additional Northern Case: *Adjara Group, Georgia.*

Over the course the cohorts of Masters' students rotate around the Four Worlds. The 2022–23 cohort began with Western cases, related to the Profit pole, because the previous cohort had covered Southern, Eastern and Northern cases, in that order. The selection of global cases, which the students choose from, relates to the cases contained in my, forthcoming, global business sustainability casebook. The focus of this formative work, at taught Masters' level, is on the students developing a clear lens for engaging in detailed critical analysis of a business's sustainability claims and practice.

Once again, the models used to assist this process of critical reflection are ideal types, with the objective of enabling the development of an interpretive pedagogical learning process. As one female student from India commented:

> "Your lectures were informative, engaging, and thought-provoking, and I am grateful for the knowledge and insights that I have gained from them. I found myself fully engrossed in every lecture and learning something new each time. Your passion for sustainability was evident in your lectures, and your ability to explain complex concepts in a clear and concise manner was truly remarkable. You have a unique talent for making the subject come alive and inspiring us to become more conscious of our impact on the environment. Your lectures have broadened my perspective on sustainability. I loved this subject so much that now I am planning my dissertation around it" (J.D. Pers. Comm., 2023).

Whilst it is encouraging to receive such comments, their significance relates less to the quality of the lecturer and more to the ideal types framework within which the topics were contained. It is the capacity of ideal typologies to generate interpretive lenses, which aid students to understand the complexity of business sustainability, almost in an intuitive way—as Jung identified in relation to the quaternity archetype—that is their power as theory of change agents (cf Van Poeck et al. 2017).

9 Operationalising the BSC 3: Business Sustainability Consultancies with Research-Level Master's Students

Our highest-level business sustainability evaluation is reserved for second year Master's students. They are given a choice between completing a research dissertation or a sustainable business consultancy. The vast majority take the dissertation option. About 20% of students in each cohort opt to complete a business consultancy, as they identify it as a more direct route to the labour market. In many cases the consultancy students are the highest achieving students in the cohort.

Each student is placed with a business which is concerned to develop some aspect of their sustainability portfolio, operating in the Liverpool City Region. The businesses are introduced to the 16 steps of the BSC and given the opportunity to express a preference for which sustainability aspects they wish the trainee consultant to focus on. Some businesses recognise the cyclical nature of the process. For others this passes them by.

The consultancies follow a standard five stage process: entry, initial negotiation, main consultancy phase, reporting, exit (cf Schein 2006). Generally, they are intended to run for 4–5 months, but, frequently, the organisational dynamics mean that they are, effectively, three months, which is insufficient time to complete a single rotation of the BSC. We estimate that this should take a minimum of a year. But, the timetabling of our Masters' programmes doesn't allow for that length of training consultancy.

Recent business sectors have included advanced manufacturing, airline, retail, neurological health, hospitality, and community development. The students are carefully supervised as trainee consultants, but a considerable part of their development rests on their abilities to be self-starting. Furthermore, it is only in the post-pandemic period that we have re-oriented the training consultancies towards issues of business sustainability, and the BSC, as the focus of our School has pivoted in this direction.

Table 1. presents secondary data on the depth of engagement with the 16 steps of the BSC during training consultancies in 2022–23. Each of the steps was allocated a score on a scale of 1–5, according to the depth of engagement that each step has reached during the consultancy, for each of seven consultancies.

1 = Business community identified
2 = Business community worked with
3 = Mutual development (client and trainee consultant)
4 = Advanced mutual development
5 = Completed mutual transformation.

Table 1 Depth of engagement with the 16 steps of the BSC during training business sustainability consultancies, 2022–23

	Scores of engagement with BSC steps at 5 levels of intensity					
	Level 1 = 1	Level 2 = 2	Level 3 = 3	Level 4 = 4	Level 5 = 5	Aggregate engagement
Step 1	0	6	3	8	5	22
Step 2	2	0	3	4	10	19
Step 3	1	2	6	8	5	22
Step 4	0	2	9	8	5	24
Step 5	1	2	3	8	5	19
Step 6	1	2	9	8	0	20
Step 7	2	2	9	0	0	11
Step 8	1	6	3	8	0	18
Step 9	1	4	6	0	5	16
Step 10	1	2	0	0	0	3
Step 11	2	0	0	0	0	2
Step 12	2	4	0	0	0	6
Step 13	2	2	0	8	0	12
Step 14	1	6	0	0	5	12
Step 15	2	2	0	0	0	4
Step 16	1	0	0	0	0	1

On this basis complete engagement with a BSC step would equal 35 (7 × 5), using this matrix. As is to be expected most consultancies involved engagement with the first two quadrants, under the People and Purpose dimensions. Even so, there was only a relatively limited ability to engage with valuing the activity of stakeholder networks (step 7). This has led me to address the development of new ways of measuring the value of societal contributions to sustainable development (Bradley 2022), which can be developed, with students, through subsequent training consultancies, using the BSC.

Equally, client businesses engaged their training consultants in identifying barriers to organisational change, in respect of Planetary transitions to cleaner production and 'net zero' carbon emissions. But, more detailed engagement with Planetary dimensions was lacking. This reflected the fact that most businesses, which our students conduct their consultancies within, continue to lack confidence in even inviting anyhelp with their addressing the challenges they face with implementing Planetary sustainability dimensions. Of course, such reticence remains a significant barrier to making the 'green 2030' transitions that society requires.

By contrast, there was a further strengthening of consultancy engagement in relation to the financial management issues of identifying resources for business development and sources of capital investment. Furthermore, the latter step (14) covered the spectrum of six capitals, including human, social, knowledge/intellectual, and natural, as well as manufactured and financial. This reflected the needs of client SME businesses to address questions of finance sustainability dimensions. The longer-term issues of Planetary dimensions are crowded-out by the shorter-term requirements of Profitability. As such, using the BSC, postgraduate students were able to identify the peaks and troughs in the cycle of integral business sustainability, helping them to address these, for themselves, in their own business development.

Following a very successful consultancy, another Indian student, wrote, as I was completing this chapter:

> "Thank you very much for giving us the opportunity to work with you on the consultancy project. It has provided me with much needed grounding and experience. I have learnt lots in these few months and hope to be able to apply my learnings in my career. Through this consultancy, I understood much about sustainability in enterprises and am thoroughly awestruck by how seamlessly the four Ps are intertwined in the working of the commercial world. I'm sure this knowledge will help me build a sustainable enterprise myself" (AKS, Pers. Comm., 2023).

10 Conclusion

The purpose of this chapter has been to indicate the ways in which an ideal typological approach to business sustainability can help students understand, interpret and work with models, in relation to real world cases. The 4 P's, connected to the integral GEN*E* and Four World's typologies, provide students with a sequence of mnemonics which help them to grasp the complex meaning of business sustainability. Equally, using an interpretivist framework helps students to contextualise the more detailed scientific, quantitative, and factual aspects of sustainability within the human, purposive, processual, and commercial aspects of businesses. The aim has been to demonstrate how this approach has been developed within one UK Business School. It is hoped that similar approaches may be replicated and further refined in other contexts. These will help students work with, and in, businesses that are increasingly sustainable. Their education, through the Business Sustainability Cycle, should help them to resolve issues of human, purposeful, natural capital, and financial systems, in more integral ways, for addressing the sustainability crises of the twenty-first century.

References

Ansell C (2015) Pragmatist interpretivism, Chapter 6. In: Bevir M, Rhodes RAW (eds) Routledge handbook of interpretive political science. Routledge

Aronovitch H (2012) Interpreting Weber's ideal-types. Philos Soc Sci 42(3):356–369

Arvidsson S, Dumay J (2022) Corporate ESG reporting quantity, quality and performance: where to now for environmental policy and practice? Bus Strateg Environ 31(3):1091–1110

Barbier EB (1987) The concept of sustainable economic development. Environ Conserv 14(2):101–110. Summer

Bellamy Foster J, Burkett P (2004) Ecological economics and classical marxism: the "Podolinsky Business" reconsidered. Organization Environ 17(1):32–60. March

Best M, Neuhauser D (2006) Walter A Shewhart, 1924, and the Hawthorne factory. BMJ Qual Saf 15(2):142–143

Böcker L, Meelen T (2017) Sharing for people, planet or profit? Analysing motivations for intended sharing economy participation. Environ Innov Soc Trans 23:28–39

Bose S (2020) Evolution of ESG reporting frameworks. Values at work: sustainable investing and ESG reporting, pp 13–33

Bradley T (2024) (Forthcoming) Business Sustainability: and integral perspective. Edward Elgar Publishing, Cheltenham

Bradley T (2019) Towards an integral arts communiversity. unpublished PhD thesis, Da Vinci Institute of Innovation Management. Da Vinci Business School, Johannesburg

Bradley T (2022) Valuing civil society environmental engagement. Chapter 16. In: Cadman T, Sarker T (eds) De Gruyter handbook of sustainable development and finance. De Gruyter, Bonn/Boston, M.A.

Campbell SD (2016) The planner's triangle revisited: sustainability and the evolution of a planning ideal that can't stand still. J Am Plann Assoc 82(4):388–397

Campbell J (1949/2008) The hero with a thousand faces. Princeton University Press, Princeton

Clegg L (2015) Benchmarking and blame games: exploring the contestation of the millennium development goals. Rev Int Stud 41(5):947–967

Cocklin C, Moon K (2020) Essay on environmental policy and sustainability. In: Kobayashi A (ed) International encyclopedia of human geography, 2nd edn. Elsevier, Amsterdam

Coward H (1996) Taoism and Jung: synchronicity and the self. Philosophy East and West 46(4):477–495

Daly H (1977/1991) Steady-state economics, 2nd edn. Island Press, Washington D.C.

Daou A et al (2020) The Ecocanvas as a business model canvas for a circular economy. J Clean Prod 258:120938

Darnall N, Ji H, Iwata K, Arimura TH (2022) Do ESG reporting guidelines and verifications enhance firms' information disclosure? Corp Soc Responsib Environ Manag 29(5):1214–1230

Dasgupta P (2021) The economics of biodiversity: the Dasgupta review. H M Treasury

Drewell M, Larsson B (2017) The rise of the meaningful economy: a megatrend where meaning is a new currency. Foresight Group
Elkington J, Rowlands IH (1999) Cannibals with forks: the triple bottom line of 21st century business. Altern J 25(4):42
Elkington J (2018) 25 years ago I coined the phrase "Triple Bottom Line". Here's why it's time to rethink it. Harvard Business Review. June 25
Farjoun M, Ansell C, Boin A (2015) Perspective—pragmatism in organization studies: meeting the challenges of a dynamic and complex world. Organ Sci 26(6):1787–1804
Gilding P (2011) The great disruption: why the climate crisis will bring on the end of shopping and the birth of a new world. Bloomsbury Press, New York
Gower B (1997) Scientific method: an historical and philosophical introduction. Routledge
Grober U, Cunningham R (2012) Sustainability: a cultural history. Green Books, Cambridge, p 156
Hadders H (2017) The adaptive quadruple bottom line scorecard: measuring organizational sustainability performance. Available online from the Canadian Sustainability Indicators Network: www. csin-rcid. ca/downloads/csin_conf_henk_hadders. pdf. Accessed on 19, p. 2022.
Hayek FA (2014) The collected works of F A Hayek, vol 15, Caldwell B (ed) The market and other orders. The University of Chicago Press Ltd.
Helm D (2015) Natural capital: valuing the planet. Yale University Press, Boston
Hosagrahar J, Soule J, Girard LF, Potts A (2016) Cultural heritage, the UN sustainable development goals, and the new urban agenda. BDC. Bollettino Del Centro Calza Bini, 16(1):37–54
J.D (2023) Pers Comm Reflection on the international business sustainability and ethics course. Liverpool Hope Business School
Jackson T (2009) Prosperity without growth: economics for a finite planet. Earthscan
Joyce A, Paquin RL (2016) The triple layered business model canvas: a tool to design more sustainable business models. J Clean Prod 135:1474–86. 1 November
Jung CG, Adler G, Hull RFC (1969) Aion: researches into the phenomenology of the self, collected works of C. G. Jung, vol 9. Princeton University Press
Jung CG (1983/ 1998) *The Psychology of the Transference.* Routledge.
Kellner D (2003) Toward a critical theory of education. Democracy & Nature 9(1):51–64
Kleine A, Von Hauff M (2009) Sustainability-driven implementation of corporate social responsibility: application of the integrative sustainability triangle. J Bus Ethics 85:517–533
Lessem R, Bradley T (2019) Evolving work: employing self and community. Routledge
Lessem R, Schieffer A (2010) Integral economics: releasing the economic genius of your society. Routledge
Marshall MG, Arestis P (1989) 'Reagonomics' and supply-side economics: a British view. J Econ Issues 23(4):965–975
Martinez-Alier J (1987) Ecological economics: energy. Environment and Society, Blackwell
Masaaki I (1986) Kaizen: the key to Japan's competitive success. McGraw-Hill Ltd.
Mboti N (2015) May the real ubuntu please stand up? J Media Ethics 30(2):125–147. 3 April
McCloskey S (2015) From MDGs to SDGs: we need a critical awakening to succeed. Policy and practices a development education review, p 12
McKibben B (1996) Buzzless buzzword. New York Times. April 10
Mill JS (2001) Principles of political economy, with some of their applications to social philosophy, Books 4 and 5. In: Winch D (ed) Penguin Books
Munasinghe M (2017) Economic, social and environmental impacts and overall sustainability of the tea sector in Sri Lanka. Sustain Prod Consumption 12(1):155–169. https://doi.org/10.1016/j.spc.2017.07.003
Neurath O (2004) in Uebel T, Cohen RS (eds) Economic writings. Open Library
Osterwalder A, Pigneur Y (2010) Business model generation: a handbook for visionaries, game changers, and challengers, vol 1. Wiley, Hoboken
PA (2023) Pers Comm "Thank you for the consultancy experience". Liverpool Hope Business School

Palmer TB, Flanagan DJ (2016) The sustainable company: looking at goals for people, planet and profits. J Bus Strateg 37(6):28–38
Podolinski S (1880) Le Socialisme et l'unité des forces physiques (Socialism and the Unity of Physical Forces). La Revue Socialiste, 8(1):353–365. June
Van Poeck K, Læssøe J, Block T (2017) An exploration of sustainability change agents as facilitators of nonformal learning: mapping a moving and intertwined landscape. Ecol Soc 22(2)
Purvis B, Mao Y, Robinson D (2019) Three pillars of sustainability: in search of conceptual origins. Sustain Sci 14(1):681–695
Sustainability Report (2021) Major update to GRI Standards raises bar for human rights reporting—sustainability-Reports.com. Accessed 7 June 2023
Schein EH (2006) From brainwashing to organizational therapy: a conceptual and empirical journey in search of 'systemic' health and a general model of change dynamics. A drama in five acts. Organization Stud 27(2):287–301
Scherkenbach WW (1986) The Deming route to quality and productivity: road maps and roadblocks. George Washington University Department of Management, Washington D.C.
Sims RR (1983) Kolb's experiential learning theory: a framework for assessing person-job interaction. Acad Manage Rev 8(3):501–508
Soddy F (1934/ 2003) The role of money: what it should be, contrasted with what it has become. Routledge
Spreckley F (1981) Social audit. A management tool for co-operative working. Beechwood College, Leeds
Stapley E, O'Keeffe S, Midgley N (2021) Essentials of ideal-type analysis: a qualitative approach to constructing typologies. American Psychological Association
Stroehle JC, Soonawalla K, Metzner M (2019) How to measure performance in a purposeful company? Analysing the Status Quo. British Academy, Future of the Corporation Working Paper Series
TCFD (2022) Taskforce on climate-related financial disclosures: overview. TCFD
Tennant G (2001) Six sigma: SPC and TQM in manufacturing and services. Gower Publishing, Ltd.
Wackernagel M, Schulz NB, Deumling D, Linares AC, Jenkins M, Kapos V, Monfreda C, Loh J, Myers N, Norgaard R, Randers J (2002) Tracking the ecological overshoot of the human economy. Proc Natl Acad Sci 99(14):9266–9271
Williams M (2002) Generalization in interpretive research. Qual Res Action, pp 125–143

Putting Sustainability Theory into Practice: Developing Networking and Consultancy Opportunities in the Curriculum to Create Real Value Add

Jan Brown, Track Dinning, Tony Wall, and Adam Shore

Abstract This paper presented the findings from an innovative collaborative final year undergraduate module that has been designed to focus on real life sustainability challenges. The module has been designed based on the premise that young people, who are transitioning from higher education into management roles, have the potential to be key players in championing and driving sustainability and sustainable development goals (SDGs). For students to 'be the change' universities have the opportunity to work collaboratively with a wide range of stakeholders to create innovative opportunities to allow students to put theory into practice in safe and supported places and spaces. Using an appreciative inquiry approach (Cooperrider and Srivastva, Pasmore and Woodman (eds), Research in Organization Change and Development, 1, JAI Press, Greenwich, CT, 1987;Cooperrider and Srivastva, Pasmore and Woodman (eds), Research in Organization Change and Development, 25, JAI Press, Greenwich, CT, 2017;) the module was introduced in the academic year 2020/21, developed further in 2021/22 and 2022/23 in the Liverpool John Moores University (LJMU) Business School. The module uses a reformative and capacity building approach (Sterling in Higher education and the challenge of sustainability: contestation, critique, practice, and promise. Kluwer Academic, 2004) with students provided with the opportunity of connecting their theoretical understanding of sustainability with real life unique projects sourced from real life clients. These projects focused on SDGs 7, 8, 9 and 12. This paper presents the results achieved to date from a growing number of projects. Details of an ongoing research project that

J. Brown (✉) · T. Dinning · T. Wall · A. Shore
Liverpool Business School, Liverpool John Moores University, Liverpool, UK
e-mail: J.Brown3@ljmu.ac.uk

T. Dinning
e-mail: T.M.Dinning@ljmu.ac.uk

T. Wall
e-mail: T.Wall@ljmu.ac.uk

A. Shore
e-mail: A.P.Shore@ljmu.ac.uk

T. Wall et al. (eds.), *Sustainability in Business Education, Research and Practices*, World Sustainability Series, https://doi.org/10.1007/978-3-031-55996-9_18

focuses on developing more complex measures of value to add the understanding value from a multi-stakeholder perspective are also presented.

Keywords Student leadership · Anchor institutions · Sustainability · Business consultancy · multiple stakeholders

1 Introduction

There is an urgent need for global citizens, of all ages, to be aware of and play a part in reaching sustainability goals as the 2030 Agenda for Sustainable Development Goals (SDGs) (United Nations 2015[1]) draws ever closer. Although all citizens will be impacted by the failure to achieve these sustainability goals the group of citizens who will be the most impacted and for the longest time period are young people. As 41% of the global population, of 7.7 billion, are aged 24 years or under (Tisdall 2019) young people will not only be impacted but will also have to be part of the communities that lead the way in finding innovative solutions to meet the SDGs.

To equip young people for these challenges, education will be key. The importance of education is highlighted by the 2030 Agenda and is one of the seventeen SDGs. Quality education (SDG 4) that ensures "…inclusive and equitable quality education and promotes lifelong learning opportunities for all" is deemed as vital to global sustainability. To achieve this education will have to span a lifetime and be delivered in many formats, in both formal and informal environments. This will require innovative joined up thinking but has the potential of being both impactful and transformative. To achieve these widespread collaborations will be needed. These collaborations will need to be both local and global and include a wide variety of people at various stages of their life journeys and professional careers. From a higher education perspective leaving these interactions to be stimulated and activated outside of and post university may not provide the optimum solution to meet the SDGs.

In the higher education sector universities play a key role in educating young people to be potential future leaders as they transition into higher education, during their studies and as they transition out of higher education and into management roles. These young leaders have the potential to play key roles in championing SDGs leading up to and well beyond the 2030 deadline as their careers develop. By allowing young people to incubate their understanding, develop networks and gain practical experience of sustainable development issues and actions while at university SDGs can be explored in safe, supported and connected environments. Providing young people with innovative ways to explore sustainability issues in the curriculum, sustainability will be at the heart of everything that they do. As their understanding of complex issues and the actions required to meet the SDGs develops the leadership skills required to meet these challenges have the potential of being developed too.

[1] https://sdgs.un.org/2030agenda.

By acting as anchor institutions (Birch et al. 2013) that bridge these student transitions and by working innovatively, universities can help prepare young people for these leadership roles. Universities have deep roots in local and global networks universities and can work collaboratively with organisations that are committed to SDGs to help students gain real life experiences and understanding of SDGs in practice. These collaborations can not only act as live case studies but can also provide industry networks that can help students translate theory into practice. As Schlipp (1949) identifies, making the connection between a body of knowledge and concrete real-life problems is an important skill that students need to learn. Creating opportunities for students to do so in the curriculum offers both academics and practitioners some exciting and innovative opportunities.

2 Sustainability, SDGs and Anchor Institutions

In an ambitious attempt to address sustainability issues globally, the United Nations, in 2015, adopted a historic decision and committed members to a 2030 Agenda for sustainability and the achievement of specific SDGs (United Nations 2015). In the 2030 Agenda for Sustainable Development, the global and connected nature of the actions needed to tackle sustainability issues were clearly identified. Stating "never before have world leaders pledged common action and endeavour across such a broad and universal policy agenda… devoting ourselves collectively to the pursuit of global development and of "win–win" cooperation" signatory countries identified that a connected approach would be needed to create win–win outcomes (United Nations 2015[2]). This connected and collaborative approach is challenging as transformations towards more sustainable lifestyles will require global citizens of all ages to be involved (Maller 2023). However, to meet the sustainable development goals action will need to be transformative with global goals being implemented at a local level. At a local level the role of geographic place and place-based institutions are increasingly becoming drivers of modern urban development and change (Birch et al. 2013). In these places, anchor institutions defined as "…anchors of place, for by their practices, they "root" or otherwise "moor" the people of the urban place" (Birch et al. 2013, p. 8), have the potential to be key drivers of change relating to sustainability issues (Ehlenz 2018). One such anchor institution is the university and its students.

2.1 Student Leadership

As anchor institutions, universities have the opportunity to introduce and shape undergraduate students' views, understanding and actions relating to sustainability. To

[2] https://sdgs.un.org/2030agenda.

achieve this the ethos of sustainability should be a core value to all actors involved. However, as Maller (2023, p. 259) notes "transformations towards more sustainable lifestyles, including the generation of values that promote stewardship and care of nature is not easy to achieve". By recognising that "...schools and individual teachers within schools are a major influence, alongside the family, the media and the peer group, on the developing values of children and young people, and thus of society at large" (Halstead and Taylor 1966, p. 3) universities can offer opportunities for sustainability issues to be explored in safe, supportive and collaborative ways. By designing transformational opportunities in the curriculum, such as the business consultancy approach, universities can contribute to a more sustainable conscious society in a supportive and critically analytical environment (Wamsler 2020).

Providing opportunities for students to be at the heart of these investigations' students can explore and develop their own personal values and leadership styles from a transcendental values perspective. Those are values that focus on helping others and recognising the connectedness to natural, social or spiritual systems (Cardona 2000; Sosik 2005). By framing sustainability from a transcendental values perspective students can explore the complexity and connectedness of these issues and goals from a 'bigger picture perspective' and can creatively work within their university studies to develop their understanding, ideas and potential solutions that could be of great value, not only during their studies but valuable as they transition out of university and evolve into the next generation of business leaders. This echoes Dyer and Dyer's (2017) view that higher education plays a critical role in moving society towards sustainability with students being at the forefront and leading the way. Over the last decade, young people globally have become increasingly vocal about the need to directly address sustainability issues with strong youth leaders such as Greta Thunberg leading the way (Thew et al. 2020). As youth leadership grows to demand action on sustainability issues universities can also demonstrate their own values and leadership by developing curricula that addresses sustainability issues too. As Burns (2003, p. 211) states "leaders embrace values; values grip leaders. The stronger the value systems, the more strongly leaders can be empowered and the more deeply leaders can empower".

For example, evidence highlights that student leadership of sustainability initiatives or activities can provide a very powerful and mobilising form of motivation to stimulate engagement (see Ferdiansyah et al. 2022; Lee et al. 2023). Specifically, this student leadership perspective generated a new sense of agency in relation to sustainability and learning (Ferdiansyah et al. 2022) and helped to directly "overcome barriers such as lack of knowledge and feelings of powerlessness" and that "community-building, flexibility and a sense of ownership were key strengths [which] contributed toward student leaders' feelings of hopefulness, self-confidence and inspiration" (Lee et al. 2023, p. 553). This was especially the case when developing and delivering curricula in partnership with organisations beyond the higher education institution (Ordaz et al. 2021).

2.2 Developing a Sustainable Community of Practice

The need for the development of sustainable management practices is noted by the United Nations (2015), who state, "we recognise that social and economic development depends on the sustainable management of our planet's natural resources." These management practices require the inclusion of a wide range of actors who are in all stages of the career journey, from senior executive level through to novice, entry-level. At university, students can be classed as novice or entry level members with students developing their conceptual understanding of management practices theoretically through academic research and publications and applying this theory to a wide range of practical applications including case studies, simulations and projects. For entry-level actors to develop, novices learn and understand more than just technical competencies and entry-level practical skills (Wenger 1998). By socialising undergraduate students into a community of practice with exposure to actors at all stages of their career journey and with experience of sustainability, students have the opportunity of putting the theory taught at university into real life practice outside of the university. This allows students to develop through "the experience of socially situated signifying practices" with "students…exposed not only to what can be seen to be done but also to that which is hidden through tacit understanding, transmission of meaning, contextualisation of tools and techniques, all of which render the experience more meaningful" (Hanney 2020, p. 164).

By using a reformative and capacity building approach a sustainability community of practice can be built into the existing system with the focus being on education for sustainability and learning for change (Sterling 2004). This reformative approach utilises systems thinking that argues that "valid knowledge and meaningful understanding comes from building up whole pictures of phenomenon, not by breaking them into parts" (Flood 2010, p. 133). SDGs can be used to focus theoretical understanding around specific goals and projects can be recruited from real life practitioners that focus on the chosen goals. In doing so projects sourced from real life clients can be used to demonstrate the value and importance of this theory in practice to the novice, the student, while at the same time allowing them to develop contacts that will allow them to enter and develop in a community of practice that has the potential to support them through their own career journey as they transition out of university. For this case, study SDG 7—affordable and clean energy, SDG 8—decent work and economic growth, SDG 9—industry, innovation and infrastructure and SDG 12—responsible consumption and production were chosen as the focus for the projects as these specific SDGs aligned to the Liverpool City Region Combined Authority's (2020) local industrial strategy.

2.3 *Multi-Stakeholder Approach to Sustainability*

This case study is based on an innovative module that has been developed at LJMU, UK. LJMU prides itself on being both local and global and being "rooted in the Liverpool City Region but with a global presence" and has a clear collaborative civic agenda of "growing and supporting our community…to work in partnership on an inclusive and positive future for all of those around us" (LJMU 2023a[3]). As part of LJMU, the Liverpool Business School (LBS) has a stated mission of seeking "…to be an engaged and inclusive business school at the heart of the Liverpool City Region with an economic and social impact beyond borders. We are driven by our commitment to innovation and creativity, which permeates to all aspects of our teaching and learning, research and enterprise." (LJMU 2023b[4]). One element of innovation is the commitment to work alongside the Liverpool City Region Authority Board to contribute towards its stated intention of growing the Liverpool City Region economy (Liverpool City Region Combined Authority 2023[5]). The core of that strategy is sustainability.

In line with the local industrial strategy published by the Liverpool City Region Combined Authority in 2020,[6] a strategic decision was taken by LBS to increase the number of projects focusing on sustainability in the academic year 2020/21. To achieve this a dedicated Business Clinic was established (LJMU 2023c[7])to liaise with organisations wanting to collaborate with LJMU. Business Clinic project officers undertook a market audit of the key business networks in the Liverpool City Region including Chambers of Commerce, business events etc. at which sustainability was being discussed and attended with a stated intent of recruiting organisations interested in sustainability projects.

Using an appreciative inquiry approach (Cooperrider and Srivastva 1987, 2017) an innovative collaborative module was developed in a large final year undergraduate business programme in LBS, in the academic year 2020/21 and developed further in 2021/22 and 2022/23. Titled Business Consultancy the module was developed based on the organisational needs established through multiple meetings and dialogues with organisational leaders within the Liverpool City Region. Two of those stated needs are projects that focus on sustainability issues and projects from organisations that are focusing on sustainability and organisational support. Live client projects and organisational clients were recruited via a dedicated Business clinic that is operational throughout the year (LJMU 2023c) with dedicated project officers working with organisations to produce appropriate project briefs, see Fig. 1 for an example of a sustainability project brief. At the start of the module, students reviewed all projects

---

[3] https://www.ljmu.ac.uk/about-us/our-vision-and-values.

[4] https://www.ljmu.ac.uk/about-us/faculties/faculty-of-business-and-law/liverpool-business-school.

[5] https://www.liverpoolcityregion-ca.gov.uk/growing-our-economy/.

[6] https://www.liverpoolcityregion-ca.gov.uk/wp-content/uploads/LCRCA_LIS_AUG19.pdf.

[7] https://www.ljmu.ac.uk/about-us/faculties/faculty-of-business-and-law/liverpool-business-school/liverpool-business-clinic.

recruited for the academic year and a structured bidding process was established that meant student teams could bid for the project that they would most like to work on. Based on a published set of criteria the Business Clinic matched each student team with unique projects in week 2 of the module. Student teams worked on the individual and unique project throughout the 12-week guided module (Fig. 2).

During the module students communicated with their client at least three times, in weeks 2, 6 and 10, and were supported weekly with a dedicated professional business consultant, who was a specialist in the area the project focused on. Business consultants are used to help students put theory into practice, develop the student team's business networks and practice their business sense making. Business consultant meetings took place onsite at LJMU within the module's scheduled timetable.

As part of the summative assessment for the module student teams were required to send a final project report, based on the aims and objectives stated in the project brief, to their client at the end of the module and ask for feedback from the client. This report was also submitted for academic assessment and had a 60% summative assessment weighting. A reflective report that included a reflection on the student team's experiences and satisfaction when working on the module and a reflection on the client's feedback on the project and the final report was also required for summative academic assessment. This reflection had a 40% summative assessment weighting.

3 Methodology

The research used both quantitative and qualitative methods. Since academic year 2018/2019 the number of unique business consultancy projects have been recorded. These projects were categorised to track the number and growth in different types of projects. One recorded category is that of projects that focused on sustainability. Sustainability was defined as projects whose client was an organisation who produced and/or marketed sustainable products and services and projects that required actions based on sustainability issues. The total number of projects was recorded each year and the percentage of sustainability projects identified each academic year.

As part of the business consultancy module's summative assessment students and clients were required to reflect on the value of the project qualitatively at the end of the module. A content analysis, defined as "…a systematic, replicable technique for compressing many words of text into fewer content categories based on explicit rules of coding" (Stemler 2001, p. 1), was undertaken on the student team and client feedback. The sustainability projects from academic years 2021/22 & 2022/23 were identified and analysed using the emergent coding technique. Codes from both the student teams and the clients were reviewed and common themes were identified.

Name of business/organisation/department: Eco [redacted]

Address: [redacted]

Type:	NGO		Charitable Organisation		Ltd Co.	x	Partnership		Sole Trader		Other	

Number of people employed (if business). Three (one person permanent and two contract workers)

Please give a brief description of what your business/organisation/department does:

Eco [redacted] is a brand-new startup venture that puts people and planet first. The organisations vision is built on three key foundations:

1. To produce technical (waterproof/water-resistant) sports apparel without detriment to the environment.
2. To ensure that the manufacture is rooted in ethical principles of fairness and opportunity.
3. To be a pioneering brand that offers an alternative to fast fashion and unnecessarily damaging clothing production.

The milestones of Eco [redacted] thus far include:

- Completely designed a full range of eco outdoor water-resistant clothing for the everyday user, using UK designers and developers from the freelance market.
- Sourced a range of eco award winning materials and comments for use in the range. The materials have been selected as they provide environmental benefit over existing in market solutions.
- Secured a deal with an ethical UK manufacture to produce the range. The manufacture gives the opportunity women leaving prison and people from a disadvantaged background to undergo training as clothing machinists. Those trainees and graduates are currently producing the prototypes and will manufacture the full range moving forward.
- Produced the first prototypes in August & September 2021 with the second prototypes being made in February 2022.

Please give a brief description of the problem/project/task that you want our students to engage in:
If this is significantly large or falls within two different areas of work we may break the project into two.

<u>Marketing / Social Media</u>

To develop a marketing and social media strategy that gets Eco Pursuit out there!

Outcomes

- Create a full concept and marketing plan.
- Create a social media strategy.
- Present and talk through the plan with the organisation, explaining how the strategy can be successfully implemented across different platforms and communication channel.
- Produce a summary of critical performance metrics.
- Create content and rollout in line with the plan.
- Measure the success of the campaign with measures to be defined during the marketing plan.

Are there any deadline to the project?

No deadlines

Fig. 1 Example of a project brief based on sustainability

What are the outcomes of the project? Do you have specific evaluation methods for this project? E.g. number of increased sales, website hits, increased footfall, etc.

Success measures to be defined during the marketing plan however some suggestions include:

- Website hits
- Website and content interaction
- Pre orders for sales
- Capture of potential customer feedback

How often are you able to meet or speak with the students?
(Please state what restrictions you think there may be)

Please tick:	Weekly	x	Fortnightly		Monthly		Quarterly		Other (please state)	

Do you require a non-disclosure agreement? E.g. If students are working with confidential information			

Name of principle contact: [redacted]		Email: [redacted]
		Phone: [redacted]

Location of project
No day-to-day travel required however, if the student would like to visit the manufacturers or designers this can be arranged. The manufacturer is based in London and the designers are just outside of Manchester.

Please indicate what your expected budget is for the project and where you expect these costs to occur (e.g. travel costs, phone calls or other general admin if applicable):

Lower-level marketing budget due to the start-up nature of the business. The actual budget can be defined/suggested as part of the concept plan.

All travel and expenses to the manufactures and designers will be paid for.

Are you available to be present during the final presentations? (April/May)			Yes	

Risk Assessment and insurance:
Note: these will only be required should the project included the students working at your premises. The Business Clinic Officer will advise.

Risk assessment attached?		*Date of assessment:*	
Insurance certificate attached?		*Date of expiry:*	

Declaration statement
I declare that I have full responsibility for the decision making process within the company at this level and that all the information provided is true to the best of my knowledge.

Signature [redacted]

Fig. 2 Example of a project brief based on sustainability—continuation

4 Results

Universities can act as anchor institutions that bring multiple stakeholders together intentionally to drive actions that meet contribute to the sustainability agenda and SDGs locally and globally. In these institutions, business schools have the opportunity of leading this initiative. By creating innovative modules and programmes with the specific aim of achieving this business schools can create real life value for organisations both operationally and strategically while at the same time incubating future business leaders.

4.1 Data Collection and Analysis

The total number of unique projects has grown year-on-year since the academic year 2018/19. Starting with a relatively small number of unique projects, 11, in the academic year 2018/19 to a significantly larger number of projects, 94, in the academic year 2022/23, see Table 1. In 2018/19 82% of the projects had sustainability themes. However, as the demand for unique projects grew in the next two academic years, the percentage of sustainability projects decreased. This was due to a wide range of organisations seeking support from the LBS Business Clinic as awareness, and success, of the module offer grew within the general business population in the Liverpool City Region. Since the 2020 publication of the Liverpool City Region Local Industrial Strategy and the LBS's commitment to supporting this strategy, a refocus on project recruitment was made in that year. This resulted in an increase in the percentage of sustainability projects to approximately 50% (58% in 2021/22 and 46% in 2022/23) of the total number of projects being worked on in the module. It is expected that this ratio will be maintained in future years.

To review the success of the sustainability projects from a student team and a client perspective an analysis of the feedback and reflection on the project was undertaken, see Table 2. Student teams and clients were asked to review their satisfaction with

Table 1 Number of unique projects with a sustainability focus

Academic year	Total no. of unique projects	No. of SDG/ sustainability projects	% of sustainability projects (%)
2018/2019	11	9	82
2019/2020	32	12	38
2020/2021	62	21	34
2021/2022	53	31	58
2022/2023	94	43	46
2023/24	Est.120	Est. 60	50
Est.2028/2029	Est.180	Est. 90	

Table 2 Percentage satisfaction rate for sustainability projects

Academic year	No. of SDG/sustainability projects	Student team satisfaction (%)	Client satisfaction (%)
2018/2019	9	100	100
2019/2020	12	100	100
2020/2021	21	100	100
2021/2022	31	100	100
2022/2023	43	100	100

the project using a 5-point Likert scale (0 = not satisfied, 5 = satisfied). The team and client satisfaction results were analysed separately. The results showed that these projects achieved a 100% satisfaction rate from both student teams and clients.

4.2 Client Satisfaction

Exploring client satisfaction feedback two themes emerged. The first theme related to the sustainability projects helping clients launch or promote sustainable products and/or services.

Value add was created by producing marketing plans and websites:

"Team XXX Consultancy has added real value to my organization by providing me with a new website and providing me with a marketing plan that I can follow in the forthcoming year." Client working with Student Team 1 (2021/22).

New ideas for marketing communications:

"The fresh set(s) of eyes really cut through the challenges as a business owner of being too close to a project by offering some great angles on things we can do and indeed have and will work towards implementing…in particular the ideas around the use of TikTok as a marketing medium have directly resulted in an improved social media presence and additional subscribers to the service." Client working with Student Team 4 (2022/23).

Ideas for a forthcoming event:

"We recognize your creativity and innovative skills when brainstorming ideas and we are excited to use your ideas within our event." Client working with Student Team 6.

The second theme related to how sustainable projects helped the organisation become more sustainable.

Ideas relating to reducing waste:

"Effective strategies for reducing plastic waste onsite." Client working with Student Team 2 (2022/23 project).

Contributing to a larger organisational strategy:

"This was an interesting report in which some innovative ideas were presented. These ideas could contribute to XXX's strategy of becoming carbon neutral by 2030

and encourage the reduced use of paper... Your ideas were interesting, and I will pass them onto the XXX sustainability team for further consideration. Thank you for all your efforts." Client working with Student Team 3 (2022/23).

Applying certain concepts or models:

"Thank you for the recommendation to try Carrolls CSR Pyramid to learn more about the triple bottom line (Profit, People, Planet), and for the detailed analysis of incremental score increases that would be possible and how to achieve them...this has been a helpful first step towards B-Corp certification." Client working with Student Team 5 (2022/23).

The value of the business consultancy projects to the clients was clearly shown in the two emergent themes. The value of student teams helping clients get their sustainability ideas offers into action was clearly shown in the first theme. In the second theme student teams helped clients think strategically about sustainability issues and allowed student teams to show how their academic understanding of sustainability can be applied in practice.

4.3 Student Team Satisfaction

From a student team satisfaction perspective, four main themes emerged.

Understanding sustainability in business practice:

"...as a team, we are both very satisfied with how this project was carried out...we identified Eco XXX as a business we wanted to work with due to the growing importance of CSR and Sustainability in business..." Student Team 1 (2021/22).

Linking theory to practice:

"*The consultancy module has yielded huge personal growth in understanding business processes and how the external environment can affect business decisions. We have been able to link theoretical models and analyse industry sources to understand how then can benefit the team in framing our ideas.*" Student Team 4 (2022/23).

Practicing research and communication skills:

"We conducted a thorough analysis of current paper usage in the organisation,

developed a comprehensive set of ideas for reducing paper consumption, and Presented our findings and recommendations to XXX." Student Team 3 (2022/23).

"*This project has drastically improved all of the team's confidence with networking and relationships as we had to obtain sponsors and guest speakers for the festival. All the team also feel that they have more confidence in their qualifications and skills after this project as we had industry experience with a real business client and participated in the operations of the business therefore preparing us for life after university in the world of work.*" Student Team 6 (2022/23).

Working collaboratively:

"*The team worked collaboratively and efficiently towards the project's goals, resulting in high team satisfaction. Team members felt valued, engaged, and motivated throughout the project, contributing to its overall success and satisfaction for both the client and team members involved*" Student Team 2 (2022/23).

"*A happy team is a productive one! …keeping track of the various achievements in the space of 12 weeks has helped keeping the morale of the team up giving us the necessary confidence that we would meet the deadlines for the project.*" Student Team 5 (2022/23).

From a student perspective, the business consultancy projects allowed students to add value by providing opportunities in the curriculum to apply relevant theory into practice by working on real life, unique projects that created real value add. These opportunities not only allowed students to develop their knowledge but also to develop their business and leadership skills. By working professionally with both a live client and a professional business consultant over an extended time period students developed their business networks and practiced their business skills in safe and supportive environments.

Using a strategy of intentionally recruiting clients and projects that have an explicit focus on the SDGs students have the opportunity to frame their understanding of sustainability issues from a real-life business perspective. By providing students with the opportunity to work with professionals who have active leadership roles and are facing real life sustainability challenges students can engage in live business dialogues with practitioners and work towards providing actionable solutions to the challenges defined in the project brief. At the end of the module students were expected to reflect on the experiences gained during the module and reflect for future action. This reflection helped students think about their own leadership skills and what they think will be required as they transition out of university.

5 Conclusion

The opportunity for universities, and their business schools, to act as anchor institutions which have an intentional aim of driving the sustainability agenda can be achieved by working collaboratively with multiple stakeholders. By creating innovative modules and programmes that meet multiple stakeholders needs value can be added to both students and organisations.

The case study and the results presented in this paper highlight one way that LJMU is driving the sustainability agenda from a student and an organisational perspective. The satisfaction rates for the sustainability projects on the module to date from both a client and a student perspective demonstrate that this curriculum innovation has been successful, to date. However, to demonstrate true leadership and remain agile it is acknowledged that, as with all collaborative initiatives, constant feedback and adjustments will be needed in future years to maintain these satisfaction levels and to meet all stakeholders' needs.

The different types of value created by the projects to the clients and the students also offer an interesting insight into what has been created to date and this insight can be used to help staff at LBS plan and develop future iterations of this module and others in the future. By intentionally focusing on sustainability value add can be provided over the longer term and supported by a stream of research developed to support and conceptualise this innovation.

By putting students at the heart of this innovation and allowing them to develop their understanding of theory and practice in a safe, supportive and collaborative environment students can develop their leadership skills with sustainability being at the heart of these skills. Equipping students to 'be the change' while at university can help develop values and actions from a 'bigger picture perspective'. As part of the bigger picture complexity is embraced and actions creatively developed by young leaders to ensure that their futures are as sustainable as possible.

References

Birch E, Perry DC, Taylor HL (2013) Universities as anchor institutions. J High Educ Outreach Engagem 17(3):7–15

Burns JM (2003) Transforming leadership. Atlantic Monthly Press, New York

Cardona P (2000) Transcendental leadership. The Leadership and Organization Dev J 21(4):201–206

Cooperrider DL, Srivastva S (1987) Appreciative inquiry in organizational life. In: Pasmore W, Woodman R (eds) Research in organization change and development, 1. JAI Press, Greenwich, CT, pp 129–169

Cooperrider DL, Srivastva S (2017) The gift of new eyes: personal reflections after 30 years of appreciative inquiry in organisational life. In: Pasmore W, Woodman R (eds) Research in organization change and development, 25. JAI Press, Greenwich, CT, pp 81–142

Dyer G, Dyer M (2017) Strategic leadership for sustainability by higher education: the American college & university presidents' climate commitment. J Clean Prod 140(1):111–116

Ehlenz MM (2018) Defining anchor institution strategies: comparing theory to practice. Plan Theory Pract 19(1):74–92

Ferdiansyah S, Winarno A, Ardhita Z (2022) Service learning in Indonesia: developing undergraduate students' leadership during COVID-19 pandemic. Higher Educ, Skills and Work-Based Learn 12(5):884–899

Flood R (2010) The relationship of "Systems Thinking" to action research. In: Reason P, Bradbury H (eds) Handbook of action research—participative practice and enquiry. Sage Publications, London

Halstead JM, Taylor MJ (eds) (1966) Values in education and education in values. The Falmer Press, London, UK

Hanney R (2020) Making projects real in a higher education context. In: Morley DA, Jamil MG (eds) Applied pedagogies I higher education: real world learning and innovation across the curriculum. Palgrave MacMillan, Switzerland

Lee B, Liu K, Warnock TS, Kim MO, Skett S (2023) Students leading students: a qualitative study exploring a student-led model for engagement with the sustainable development goals. Int J Sustain High Educ 24(3):535–552

Liverpool City Region Combined Authority (2020) Local industrial strategy. Liverpool City region combined authority. Retrieved 12 June 2023, from https://www.liverpoolcityregion-ca.gov.uk/wp-content/uploads/LCRCA_LIS_AUG19.pdf

Liverpool City Region Combined Authority (2023) Growing the Liverpool city region economy. Retrieved 12 June 2023, from https://www.liverpoolcityregion-ca.gov.uk/growing-our-economy/

Liverpool John Moores University (2023a) Our mission and values. Retrieved 12 June 2023, from https://www.ljmu.ac.uk/about-us/our-vision-and-values

Liverpool John Moores University (2023b) Liverpool business school: an innovative and nurturing business school. Retrieved 12 June 023, from https://www.ljmu.ac.uk/about-us/faculties/faculty-of-business-and-law/liverpool-business-school

Liverpool John Moores University (2023c) Liverpool business clinic. Retrieved 12 June 2023, from https://www.ljmu.ac.uk/about-us/faculties/faculty-of-business-and-law/liverpool-business-school/liverpool-business-clinic

Maller C (2023) Turning things around: a discussion of values, practices, and action in the context of social-ecological change. People & Nature 5:258–270

Ordaz K, Tan K, Skett S, Herremans IM (2021) Developing leadership qualities in environmental sustainability through university co-curricular activities. Int J Sustain High Educ 22(7):1609–1629

Schlipp PA (ed) (1949) The library of living philosophers, vol VII. Open Court, Evanston, Illinois

Sosik JJ (2005) The role of personal values in the charismatic leadership of corporate managers: a model and preliminary field study. Leadersh Q 16:221–244

Stemler S (2001) An overview of content analysis. Pract Assess Res Eval 7(17):17

Sterling S (2004) Higher education, sustainability and the role of systemic learning. In: Corcoran PB, Wals AEJ (eds) Higher education and the challenge of sustainability: contestation, critique, practice, and promise. Kluwer Academic

Thew H, Middlemiss L, Paavola J (2020) "Youth is not a political position": exploring justice claims-making in the UN climate change negotiations. Glob Environ Chang 61:102036

Tisdall S (2019) About 41% of the global population are under 24. And they're angry... The Guardian. 26th Oct. Retrieved from https://www.theguardian.com/world/2019/oct/26/young-people-predisposed-shake-up-established-order-protest#:~:text=It%20is%20as%20if%20the,is%20aged%2024%20or%20under

United Nations (2015) Transforming our world: the 2030 agenda for sustainable development. Retrieved 12 June 2023, from https://sdgs.un.org/2030agenda

Wamsler C (2020) Education for sustainability: fostering a more conscious society and transformation towards sustainability. Int J Sustain High Educ 21(1):112–130

Wenger E (1998) Communities of practice. Learning, meaning and identity. Cambridge University Press, New York

Digital Pedagogies for Sustainable Futures: The Role of Gamification to Enhance Student Learning

Sue Cronshaw and Michael Drummond

Abstract This chapter explores whether gaming can provide a clear route to undergraduate students engaging with and understanding sustainability issues in business and management. A qualitative evaluation survey of gamification in the classroom was undertaken by 106 students across 3 years of study to identify what specific elements of gamification was successful in engaging and motivating them and how it can translate in sustainability education. By understanding student perspectives on game based learning (GBL), this research aimed to explore the use of this digital pedagogy to identify the potential of gamification for increased student motivation, engagement and learning. The findings demonstrated that students mostly enjoyed using gamification in the classroom but with a sizable number undecided on its use. Student's alluded that they would benefit more from the experience should technical support be available at a tutor level which was inconsistent across the 3 years of study.

Keywords Gamification · Sustainability · Student engagement · Digital pedagogies

1 Introduction

There has been a growing initiative within the higher education sector, exploring how institutions can develop an empirical and conceptual understanding of what they can do to support the realisation of SDGs (Chankseliani and McCowan 2021). As part of this process, there has been a recognition that the commitment to change must extend further, considering the education of future leaders who can understand the complexities of achieving the SDGs. This is supported by the growing

S. Cronshaw (✉) · M. Drummond
Liverpool Business School, Liverpool John Moores University, Liverpool, UK
e-mail: S.Cronshaw@ljmu.ac.uk

M. Drummond
e-mail: M.G.Drummond@ljmu.ac.uk

T. Wall et al. (eds.), *Sustainability in Business Education, Research and Practices*, World Sustainability Series, https://doi.org/10.1007/978-3-031-55996-9_19

need of companies' hiring sustainability literate graduates (Lozano et al. 2015). Having a unified approach is key, as Chankseliani and McCowan (2021) argue, it is important to assess the concord between teaching, community engagement and university activity to ensure alignment. This is echoed by Avelar et al (2019) who attests to HE institutions considering sustainability across all facets of the institution to change behaviour and attitudes, and Lozano (2012) who states that stronger alignment between ethical values and the activity within higher educational institution is fundamental for student futures in business. Research by Hernandez-Lopez et al. (2020) identifies the impact of such an approach, concluding that a lecturer's ethical leadership and the University's ethical climate directly impact a student's understanding of ethical business. In recent years and in line with the UN's PRME initiative, there is an ongoing focus in business schools on 'greening the curriculum' (Donath et al 2020) with a view to preparing students for sustainable business environments. Similarly, there has been an emergent development of digital pedagogies in education for sustainable development (Neriz Jara et al. 2020). Here, there are many digital platforms and mechanisms currently utilised in HE such as gamification, Virtual/Augmented/Mixed Realities, web design and podcasts. This research aimed to explore the use of these digital pedagogies for the purposes of identifying the potential of gamification for a deeper understanding of the UN SDGs.

This chapter begins with an introduction to digital pedagogies in the context of gamification followed by the method chosen to obtain the data. The chapter continues with the findings from the data covering both positive and negative experiences. A discussion is then derived from the findings before conclusion of the chapter.

2 Digital Pedagogies

In academia, a widely used definition of gamification, and one adopted for this study, comes from Deterding et al. (2011) who sees gamification being the application of gaming mechanics in a non-gaming context. More holistically, gamification can be seen as a value creating process for an end users experience (Huotari and Hamari 2012). Whilst gamification in education is on the increase, there is still some apprehension when using it as a method of learning. On the face of it, the apprehension may be for several reasons, from dismissing gamification as nothing more than 'points and badges' (Robertson 2010) to a nervousness from staff utilising gamification due to lack of digital confidence, teaching students often coined as 'digitally native' (Kirschner and De Bruyckere 2017: 136).

Gamification does have its own variances of terminology with the concept sometimes being referred to as 'game-based learning, serious games and learning by design' (Hung 2017: 57). However, irrespective of the variety of the definitions that exist, it can be said that the focus of gamification is to provide a lasting and memorable experience for an end user. With gamification therefore being subjective in its definition, application, and experience of the end user (Huotari and Hamari 2012), it does allow for a variable application of gaming elements to be used.

When considering this approach, there is the challenge of identifying the levels of student engagement the game process stimulates and encourages. Engagement can be defined as the 'behavioural intensity and emotional quality of a person's active involvement during a task' (Reeve et al. 2004: 143). However, there is a need for a clearer definition of the term, considering the level of student involvement (Axelson and Flick 2010), largely due to the variability and subjectivity of how engagement can be impacted in students. Fredricks et al. (2004) review the idea of engagement and conclude that it is a 'meta construct' (p. 60) that includes not only 'behavioural' (participation, good behaviour, effort) but also adds 'emotional' (interest, positive emotions), and 'cognitive' (psychological involvement in learning, self-regulation) dimensions. For the purposes of this study, student engagement followed this definition due to the appreciation for, not only behaviour, but also the emotion and cognition of students when it comes to interacting with gamification. The level of engagement is also reflected in the motivation of the student to participate. Gatti et al. (2019) highlight the importance of prior motivation to learn for successful integration of digital games. They argue a student's interest and understanding will increase if they begin with the motivation to attend. This is echoed by Hamari et al (2014) who argue that applied gamification and its success is very context-dependent, because of the different intrinsic and extrinsic motivators associated with the different activities. The common use of extrinsic rewards such as badges and points can undermine intrinsic motivations which can arguably destabilise the student's engagement in the game (Knaving and Bjork 2013).

It has been argued that education for SDGs (ESDGs) should avoid a formal teaching style and aim for a more transformative learning approach (Rieckmann et al. 2017), with the use of more sophisticated teaching techniques to develop students' awareness and engagement (Cottafava et al. 2019). This approach can allow students to test their knowledge whilst also creating new knowledge through the application of theory in different contexts. According to Gatti et al. (2019) experiential learning methods can be a positive approach for teaching sustainability, generating cognitive and affective outcomes that in turn may help to develop skills in critical thinking. As suggested by Molderez and Fonseca (2018) education on sustainability requires a 'dynamic and a holistic' way of teaching, considering what Annan-Diab and Molinari (2017) refer to as interdisciplinary education, reflecting a range of perspectives and understanding of sustainable development. To facilitate this, Universities should ensure adequate resources are in place to support the intention of experimenting with new pedagogical approaches. Active and open learning can be more suitable to engage students (Braßler and Block 2017) therefore the mechanisms of gamification lend itself to the pedagogical requirements of education for sustainable development.

3 Methodology

The methodology used online qualitative questionnaires, distributed to students studying business and management in one business school, to assess the role of alternative pedagogical mechanisms, i.e. digital games, in their learning. They were all registered for modules that utilised digital gaming mechanisms for the duration of the semester. The aim of the approach was to not only identify if they felt motivated and engaged by the games, but to explore the use of games to deepen their understanding of topics. 106 students across 3 years of study participated after using a digital gaming mechanism across a 12 week module. Responses were exported from JISC Online Surveys into a Microsoft Excel file. Using Pivot Tables, the respondent data was visualised to summarise the findings. Ethical approval was granted for obtaining data from students across 3 years of study, through the university's approval process.

Thematic analysis was deemed the most appropriate method for the qualitative data, due to the inherently interpretive nature of qualitative research (Billups 2021). This approach allowed for themes and sub themes to be identified, alongside a priori themes. The a priori themes of motivation, student engagement and depth of understanding were categorised based on the researchers experience of using games within modules and in acknowledging that these themes had emerged based on the rationale for using such digital pedagogies.

4 Findings

Focusing on two open ended questions that asked what was liked and disliked about using gamification in the classroom, four key themes, adapted from Thebes et al. (2014) Cluster of Master Mechanics and Dynamics in Gaming, were used to thematically analyse the responses alongside fifteen sub-themes.

To separate the negative and positive aspects from the broad themes was dependent on the specificity of the comment made by the participant. A negative connotation was considered as a comment that alluded to frustration or a barrier to an otherwise successful experience i.e. "*difficult*" or "*couldn't*". Otherwise, a positive connotation was considered to be a comment that suggested the participant enjoyed or benefitted from the experience i.e. "*allowed*" or "*opportunity*".

Theme: challenges	Attempt at engaging a student through the systems use of challenges
Sub themes:	
Accomplishment	There is a sense of achieving something
Theme: social influences	Considers the use of social factors within the process
Sub themes:	
Competition	The competitive aspect of the system i.e., points scoring, leader boards

(continued)

(continued)

Group work	The interactivity between users of the system
Theme: system design	Looking at the system functionality and the mechanisms outside of the system that support the experience i.e., academic content
Sub themes:	
Application Functionality	How the system operates at a functional level i.e., entering data, saving progress
Simulation	How the system replicates a real-life experience
User interface	The look and feel of the system
Content	The content both in and out of the system that is in place to provide educational content
Lack of autonomy	The way in which as student felt in control of their system use
Support and guidance	The information within the system and outside of the system for aid and assistance in using the system
Not met expectation	A student's expectation of the system was not what they anticipated it to be
Academic Procedure	The processes and procedures in use that align with the system either in support or for added subject context
Theme: user specifics	Individual feelings towards using the system
Sub themes:	
Engagement	How a student felt more compelled to interact with the system
Motivation	How a student wanted to continue using the system
Demotivation	How a student didn't want to continue using the system
Complex/confusion	How a student felt when interacting with the system

Positive Experiences

The findings of this study provide an insight into both the positive and negative variables for students using games as part of their modules in business and management. Students determined there were many positive aspects to using gamification as a tool for learning, highlighting both 'Social Influences' and 'System Design' as prominent aspects. Some students suggested that it was the social element that not only aided their learning but also provided an opportunity to broaden their knowledge as "*It allowed me to interact with others which helped my understanding of the topic*" and provided an "*…opportunity of growth*". Taking Bukowitz and Williams' (1999) definition of "knowledge sharing as an activity trough which knowledge is exchanged among people", it appears that the use of a digital tool has only amplified this knowledge sharing in this research.

This provides further clarity in other studies considering digital use in knowledge sharing with Eid and Al-Jabri (2016) noting the positive enjoyment between students when collaborating with online discussion. This does suggest an intrinsic motivation towards the sub-theme of 'Group Work', but this may be a behaviour or view that the individual student had going into education rather than the tool influencing the motivation. Nevertheless, it is evident that group work can have a positive role to play for engaging students in education through gamification, as supported by Smiderle et al. (2020) who argues that the use of gamification in an educational context, can result in increased student engagement. This is echoed by de la Torre and Berbegal-Mirabent (2020) who claim the use of gamification can increase student engagement, both in the activity and the subject, in some cases improving short-term knowledge acquisition (Chen et al. 2019) and supporting student learning and motivation (Naik 2017).

When asked about the role of the game in promoting understanding, most students either fully agreed or agreed that gamification helped improve their understanding of the topic. However, the size of students who were undecided cannot be ignored. This could be due to the expectations of the gamification use, as their responses suggest uncertainty about whether the software was being used to learn something new or to apply current learning. Selim's (2007) study into eLearning critical success factors determined that students should be made aware of what the difference is when learning by "*construction*" and by "*absorption*" when it comes to gaining value in the classroom. Evidently, managing the expectation of the student would perhaps determine their understanding on the topic.

What is more explicit in the open-ended question responses is the extrinsic motivation students have towards the use of gamification. The 'Social Influences' themes and 'Competition' sub-theme, again, have strong correlation with the comments made that would lend themselves well towards a labelling of extrinsic. Students favoured the "*competitiveness of the program trying to improve to beat other people*" and "*trying to win*". Whilst these appear self-motivated, there was also a united motivation amongst students when in groups by "*trying to get high in the leaderboard with peers*" and students using the tool "*discussing tactics to win*". Landers and Landers (2015) argue that leaderboards have the potential to increase the amount of time a student engages with a task. This is supported by Butler (2013) who states that success relative to other players does affect the player's perception of fun, meaning that the position on a leaderboard does not negate the players perception of positivity towards the game. These findings support past research that suggests gaming experiences amongst students in learning is often met positively (Burguillo 2010).

It appears that students favour group work when there is a dedicated mechanism or tool to help facilitate interaction gamification compared to a more traditional academic view of group work, for example, presentations. However, that is not to suggest students prefer individual work, although they do if it is for self-graded work (Walker and Shore 2015). Therefore, it would suggest that it should an academic objective be to work collaboratively and maximise engagement, then utilising a tool for this would be beneficial assuming no grade is impacted.

Negative Experiences

The use of gamification was not considered a wholly positive learning experience, as students did reflect on the negative aspects of gamification as a tool for learning, identifying 'System Design' as the most problematic alongside 'User Specifics'. Whilst it could be argued that there is some commonality across the two themes,

'System Design' was coded when there was an explicit view about the system, either in the tool itself, or the supporting processes in place. Irrespective of the differentiation, both themes note the importance of the user interface, or aesthetics, when it comes to engaging with digital software. This is consistent with existing perspectives (O'Brien and Toms 2010).

Overbeeke et al. (2003) state that a user's expectation with regards to software application is that they should not only be functional but also engaging. This research provides further evidence of this with many students being disappointed in the way the tool worked, linking to the sub-themes 'Application Functionality' and 'User Interface'. A lot of the frustration around functionality was "*that all players can't save at same time*" which proved to be problematic in group work. This was followed by "*difficult to access in the beginning*". This had an impact on a student's engagement in using the tool but also, more holistically, perhaps in the educational content itself.

Given the negativity surrounding the system design and challenges in access, it is unsurprising to see the sub-theme 'Support and Guidance' being clearly shown. Should a student be unsure how something works or having difficulty then they would seek guidance. However, the guidance was found to have a *"lack of explanation on some parts of the game and how they affect things"*, which could suggest in-game guidance. However, some comments note that the tool "*wasn't explained well in the beginning*" which could suggest there was a lack of introduction from academic staff or even a lack of confidence in the tool from staff. The former suggestion shares commonality with studies on the lack of digital confidence amongst academic staff (Kirschner and De Bruyckere 2017). If a student is finding using the software difficult and cannot find any support and guidance to help with their issue, then they are less likely to maintain focus. The support of an academic staff member using gamification in the classroom is deemed key for it to be a success amongst students with Selim's (2007, p. 16) study concluding that "*Instructors should be able to handle the technological tools used in e-learning-based courses, such as e-mail, e-discussion and website maintenance*". This study suggests that this was lacking.

Linking to 'Support and Guidance' another negative sub-theme was 'Academic Procedure' or how the tool was designed for use in the classroom. Students didn't appreciate that the tool "*determines whether I pass or fail the module as I feel like it is not a good representation of my knowledge on the subject*". Complementing this view, a study by Kwon and Ozpolat (2021) identified that integration with gamification in education as part of assessment had a significant and negative impact on students' grades. It could be suggested then that if engagement and motivation is to be maximised in the classroom then linking it to assignments may not necessarily be the best course of action.

5 Discussion

Based on the findings of the questions about engagement and motivation, it is striking how similar the responses are. Whilst most are positive about gamification use, many are undecided (*Undecided* was a specific option available for selection on the likert scale within the survey). The themes identified do lend themselves to explanations of these findings such as, 'System Design' and 'Social Influences'. For the latter theme, students predominantly enjoyed using gamification for the competitive element against peers and for the interactivity for group work. However, it could be argued that the former theme provides an explanation for their undecided view on engagement or motivation to use gamification. The sub-themes, notably 'User Interface' and 'Support and Guidance' provide further insight into why the students are 'undecided' in their views. There is an apparent reluctance to fully endorse the role of the game due to students feeling they are not entirely supported in the process. It is important therefore that students are given the right guidance and clear instruction.

Should the tutor want to maximise engagement and motivation for using gamification then clear direction and support is required from staff leading the sessions. It is important to recognise the role that lecturers play in facilitating the learning environment and supporting the students before, during and after the game. Competencies of leaders and staff is argued by Mindt and Rieckmann (2017) as an important factor in the success of implementing sustainability in education. There are also aspects of system mechanics, namely, engineering, design, and management that require adequate explanation (Lin et al 2012). Lessons should be clear in the aims and objectives of the use of gamification software; is it to demonstrate current understanding? or is it to learn something new? The objective set should determine what mechanics are more favourable to be used. This analysis suggests that if learning something new, students should have more autonomy and guidance on their progress whereas, if seeking to apply existing learning, then the competitive elements in gamification can be deemed more successful.

In addition, the user interface of the software appears to be key in how students take in information and whether they want to engage with the software more. Whilst group work and competition are celebrated as a success for using gamification in the classroom, it does appear that it should only be used in certain circumstances. From this study, to maximise engagement and motivation, a student's grade should not be determined by their participation of group input, as students feel that they have less autonomy or control on their desired outcome. However, if group work is to be a factor in grades, then the system needs to be designed to cope with multiple interactions amongst students at any given time, bringing the 'System Design' into the spotlight for being key to positive interactivity in collaboration.

6 Conclusion

Business schools are recognising the need for a commitment to responsible management education but the mechanisms by which the issues of sustainability are delivered must deviate from the traditional pedagogies to encourage students to engage with the topic at a deeper level with a more industry focused approach. Students need to understand the implications of SDGs for both business and society, which requires them to have both the knowledge and the motivation to learn (Kopnina 2020). This research has demonstrated the impact of gamification on student motivation and engagement, with 61% highlighting the increase in their motivation for learning during the process of the game and 70% emphasising improved engagement. As highlighted by Hamari et al. (2014) gamification does produce positive effects and benefits and can be implemented to be ideal learning environments (Khaleel et al. 2015).

It is argued that to understand the complex reality of social issues, it is beneficial to consider a combination of both micro and macro level perspectives (Boeren 2016). This combined approach lends itself to the integrated nature of gamification, considering individual standpoints alongside the industry outlook. Sustainability is a highly complex topic, involving multiple stakeholders and agencies which can be a challenge for educators aiming for student engagement and understanding in the subject. Engaging with digital pedagogies provides students with the opportunity to be autonomous, share their views and perceptions about the key issues, giving them a voice and a place to action and strategize ways forward, whilst supporting their understanding.

References

Annan-Diab F, Molinari C (2017) Interdisciplinarity: practical approach to advancing education for sustainability and for the sustainable development goals. Int J Manage Educ 15(2): 73–83

Avelar ABA, da Silva-Oliveira KD, da Silva Pereira R (2019) Education for advancing the implementation of the sustainable development goals: a systematic approach. The Int J Manage Educ 17(3). https://doi.org/10.1016/j.ijme.2019.100322

Axelson RD, Flick A (2010) Defining student engagement, change: the magazine of higher learning, 43(1): 38–43

Billups F (2021) Qualitative data collection tools: design, development, and applications. SAGE Publications, Inc., https://doi.org/10.4135/9781071878699

Boeren E (2016) Lifelong learning participation in a changing policy context: an interdisciplinary theory. Palgrave-Macmillan, London

Braßler M, Block M (2017) Interdisciplinary teamwork on sustainable development—the top ten strategies based on experience of student initiated projects. In: Handbook of theory and practice of sustainable development in higher education. Springer, New York, NY, pp 65–77

Braun V, Clarke V (2022) Thematic analysis. A practical guide. SAGE

Bukowitz WR, Williams RL (1999) The knowledge management field book. Financial Times (FT) Press, Prentice Hall, Upper Saddle River, N.J.

Burguillo JC (2010) Using game theory and competition-based learning to stimulate student motivation and performance. Comput Educ 55(2):566–575
Butler C (2013) The effect of leaderboard ranking on players' perception of gaming fun. In: Ozok AA, Zaphiris P (eds) Online communities and social computing. OCSC 2013. Lecture Notes in Computer Science, vol 8029. Springer, Berlin, Heidelberg. https://doi.org/10.1007/978-3-642-39371-6_15
Chankseliani M, McCowan T (2021) Higher education and the sustainable development goals. High Educ 81(1):1–8. https://doi.org/10.1007/s10734-020-00652-w
Chen SW, Yang CH, Huang KS, Fu SL (2019) Digital games for learning energy conservation: a study of impacts on motivation, attention, and learning outcomes. Innov Educ Teach Int 56(1):66–76. https://doi.org/10.1080/14703297.2017.1348960
Cottafava D, Cavaglià G, Corazza L (2019) Education of sustainable development goals through students' active engagement: a transformative learning experience. Sustain Account, Manage Policy J 10(3):521–544. https://doi.org/10.1108/SAMPJ-05-2018-0152
de la Torre R, Berbegal-Mirabent J (2020) Using game-based principles to empower students in non-STEM academic programmes. Innov Educ Teach Int 57(5):511–520. https://doi.org/10.1080/14703297.2020.1727352
Deci E, Ryan R (1985) Intrinsic motivation and self-determination in human behaviour. Plenum Press,New York, NY
Deterding S, Sicart M, Nacke L, O'Hara K, Dixon D (2011) Gamification: toward a definition. In: Proceedings of the CHI 2011 gamification workshop. Vancouver, British Columbia, Canada
Donath L, Mircea G, Rozman T (2020) E-learning platforms as leverage for education for sustainable development. Eur J Sustain Dev 9(2):1–19
Fredricks JA, Blumenfeld PC, Paris AH (2004) School engagement: potential of the concept, state of the evidence. Rev Educ Res 74:9–109
Gatti L, Ulrich M, Seele P (2019) Education for sustainable development through business simulation games: an exploratory study of sustainability gamification and its effects on students' learning outcomes. J Clean Prod 207:667–678
Hamari J, Koivisto J, Sarsa H (2014) Does gamification work? A literature review of empirical studies on gamification. In: Proceedings of the 47th Hawaii international conference on system sciences. HICSS
Hernandez-Lopez LE, Álamo-Vera FR, Ballesteros-Rodríguez JL, De Saa-Pérez P (2020) Socialization of business students in ethical issues: the role of individuals' attitude and institutional factors. The Int J Manage Educ 18(1):1–12. https://doi.org/10.1016/j.ijme.2020.100363
Hung ACY (2017) A critique and defence of gamification. J Interact Online Learn 15(1):57–72
Huotari K, Hamari J (2012) Defining gamification: a service marketing perspective. In: Proceedings of the 16th international academic MindTrek conference. Academic MindTrek Conference, New York, NY, pp 17–22
Khaleel FL, Ashaari NS, Tengku Wook TSM, Ismail A (2015) User-enjoyable learning environment based on gamification elements, IEEE, pp 221–226. https://doi.org/10.1109/I4CT.2015.7219570
Kirschner PA, De Bruyckere P (2017) The myths of the digital native and the multitasker. Teach Teach Educ 67:135–142
Knaving K, Björk S (2013). Designing for fun and play: exploring possibilities in design for gamification. ACM Press, pp 131–134. https://doi.org/10.1145/2583008.2583032
Kopnina H (2020) Education for sustainable development goals (ESDG): what is wrong with ESDGs, and what can we do better? Educ Sci 10(10):261. https://doi.org/10.3390/educsci10100261
Kwon HY, Özpolat K (2021) The dark side of narrow gamification: negative impact of assessment gamification on student perceptions and content knowledge. INFORMS Trans Educ 21(2):67–81
Landers RN, Landers AK (2015) An empirical test of the theory of gamified learning: the effect of leaderboards on time-on-task and academic performance. Simul Gaming 45(6):769–785

Lin RJ, Chang H, Kai-quan Z, Chou SC (2012) Technologies and innovation matrix for behaviour-motivation-embedded web 2.0 services. In: 8th international conference on computing technology and information management (ICCM). IEEE, 2012
Lozano JF (2012) Educating responsible managers. the role of university ethos. J Acad Ethics 10:213–226. https://doi.org/10.1007/s10805-012-9166-3
Lozano R, Ceulemans K, Seatter CS (2015) Teaching organisational change management for sustainability: designing and delivering a course at the university of Leeds to better prepare future sustainability change agents. J Clean Prod 106:205–215
Mindt L, Rieckmann M (2017) Developing competencies for sustainability-driven entrepreneurship in higher education: a literature review of teaching and learning methods. Teoría De La Educación. Revista Interuniversitaria 29(1):129
Molderez I, Fonseca E (2018) The efficacy of real-world experiences and service learning for fostering competences for sustainable development in higher education. J Clean Prod 172:4397–4410. https://doi.org/10.1016/j.jclepro.2017.04.062
Muñoz-Merino PJ, Molina MF, Muñoz-Organero M, Delgado Kloos C (2014) Motivation and emotions in competition systems for education: an empirical study. IEEE Trans Educ 57(3):182–187
Mustafa IM, Al-Jabri IM (2016) Social networking, knowledge sharing, and student learning: the case of university students. Comput Educ 99:14–27
Naik N (2017) The use of GBL to teach mathematics in higher education. Innov Educ Teach Int 54(3):238–246. https://doi.org/10.1080/14703297.2015.1108857
Neriz Jara L, Núñez A, Fuentes-Caceres VA, Ramis F, Jerez O (2020) Simulation-based training as a teaching and learning tool for management education. Innov Educ Teach Int 57(6):701–713. https://doi.org/10.1080/14703297.2019.1631874
O'Brien HL, Toms EG (2010) The development and evaluation of a survey to measure user engagement. J Am Soc Inform Sci Technol 61(1):50–69
Overbeeke K, Djajadiningrat T, Hummels C, Wensveen S, Frens J (2003) Let's make things engaging. In: Blythe MA, Monk AF, Overbeeke K, Wright PC (eds) Funology: from usability to enjoyment. KluwerAcademic Press, Amsterdam, pp 7–17
Reeve J, Jang H, Carrell D, Jeon S, Barch J (2004) Enhancing students' engagement by increasing teachers' autonomy support. Motiv Emot 28:147–169
Rieckmann M, Mindt L, Gardiner S, Leicht A, Heiss J (2017) Education for sustainable development goals—learning objectives
Robertson M (2010) Can't play, won't play. http://www.hideandseek.net/2010/10/06/cant-play-wont-play/. Accessed 04 Oct 2022
Selim HM (2007) E-learning critical success factors: an exploratory investigation of student perceptions. Int J Technol Mark 2(2):157–182
Smiderle R, Rigo SJ, Marques LB (2020) The impact of gamification on students' learning, engagement and behaviour based on their personality traits. Smart Learn Environ **7**(3). https://doi.org/10.1186/s40561-019-0098-x
Thebes S, Lins S, Basten D (2014) Gamifying information systems, a synthesis of gamification mechanics and dynamics. In: European conference on information systems (ECIS). Tel-Aviv, Israel
Walker CL, Shore BM (2015) Myth busting: do high-performance students prefer working alone? Gifted and Talented Int 30(1–2):85–105

Developing Entrepreneurial Sustainability Thinking and Action at 'The World Café'

Jan Brown, Track Dinning, and Tony Wall

Abstract This paper uses a case study approach to explore how collaborative initiatives can be developed and used to stimulate inspiring conversations that lead to innovative actions that contribute to meeting the 2030 sustainability goals. As the challenging 2030 targets get closer the need to germinate and incubate innovative ideas and actions becomes a key responsibility for a wide range of stakeholders. As anchor institutions in their communities universities, globally, have the opportunity to play a key role in stimulating ideas and championing actions. This case study focuses on one initiative developed by the Liverpool Business School (LBS) in Liverpool John Moores University, UK. Using an appreciative inquiry approach (Cooperrider and Srivastva in Research in organization change and development. JAI Press, Greenwich, CT, pp. 129–169, 1987; Cooperrider and Srivastva in Research in organization change and development. JAI Press, Greenwich, CT, pp. 81–142, 2017) the LBS 'World Café' (Brown and Isaacs 2005) was developed and launched in 2022. This café aims to develop promising and impactful ideas, practices or provocations that spark co-created learning, innovation and action. Based on multi-stakeholder dialogues, wide-ranging debates are facilitated to germinate and incubate innovative ideas in safe and supportive spaces. In this paper, the themes of ideas and actions developed are explored in detail based on a content analysis of the programming to date. Reflections from co-creators and key learnings, both expected and unexpected, are provided. Details for future sessions and multi-stakeholder developments are presented and commitments for action identified.

Keywords Sustainability · Co-created action · Dialogue · Innovation · Incubation · Safe spaces

---

J. Brown (✉) · T. Dinning · T. Wall
Liverpool Business School, Liverpool John Moores University, Liverpool, UK
e-mail: J.Brown3@ljmu.ac.uk

T. Dinning
e-mail: T.M.Dinning@ljmu.ac.uk

T. Wall
e-mail: T.Wall@ljmu.ac.uk

T. Wall et al. (eds.), *Sustainability in Business Education, Research and Practices*, World Sustainability Series, https://doi.org/10.1007/978-3-031-55996-9_20

1 Introduction

Business schools play a key role in stimulating sustainable thinking, innovation and action in their own community and for a wide range of stakeholder groups. With access to a wide range of academic and industry sources and the ability to connect a diversity of stakeholders in innovative ways, business schools can intentionally champion ideas and actions that can contribute to global change. One such set of global actions is the sustainable development goals (SDGs) identified by the United Nations (2015)[1] in the 2030 Agenda for Sustainable Development.

To achieve such actions business schools need to encourage and promote practices that can directly lead to change. As Mihov, chair of the Principles of Responsible Management Education (PRME), notes, "…higher education and the business school ecosystem of schools, businesses and communities have an opportunity to equip the current and next generation of leaders with the skills to advance a broader role of business as a force for good" (PRME 2021, p. 1). By drawing on and connecting this diverse ecosystem business schools can create safe spaces in which everyone feels comfortable in expressing themselves and participating fully without fear of denial of experience (Brown and Issacs 2005). In these safe spaces, innovative practices can be used to explore a wide range of relevant issues, challenges, and opportunities intentionally with the express purpose of creating actions for change.

There are multiple ways of achieving innovative practice. This paper focuses on one such practice, the World Café, and demonstrates how using this practice can create innovative action. Developed in 1995 by Juanita Brown and David Issacs and based on the concepts of appreciative inquiry, bohemian and civic dialogue, collective wisdom and intelligence a community of practice is intentionally created in which safe spaces are created to stimulate authentic conversation and dialogue. These safe spaces encourage diversity and conversational leadership with the view of facilitating global citizenship and social innovation (World Café 2023a).[2]

It is not uncommon for café type events to be used to engage academic staff in conversation about past and future projects. Burke and Sheldon (2013) suggest that encouraging the conversation can in fact bring about innovation in the workplace. The focus of the café style event used in this case study is to foster dialogue between staff to promote shared intelligence and future planning and actions (Brown and Isaacs 2005). Based on seven design principles, see Table 1, the World Café model provides business schools with an easy-to-follow process in which a diversity of ideas can be connected for action.

---

[1] https://sdgs.un.org/2030agenda.

[2] https://theworldcafe.com/key-concepts-resources/glossary/.

Table 1 World café design principles

	Design principle
1	Clarify the context
2	Create a hospitable space
3	Explore questions that matter
4	Encourage everyone's contribution
5	Connect diverse perspectives
6	Listen together for patterns and insights
7	Share collective discoveries

Source:Adapted from World Café (2023b)[3]

2 SDGs, PRME and Initiatives

As places in which knowledge is co-created and actions formulated to drive change universities have a key role in actioning the 2030 sustainability agenda and the SDGs (United Nations 2015).[4] These universities' business schools have the opportunity to recognise this and develop innovative programmes and initiatives to stimulate ideas and actions internally and across multiple stakeholders. The importance and central role of business leadership needed to achieve the SDGs is recognised by the United Nations-supported initiative, PRME (principles of responsible management education), which states that "business and management schools, as well as management and leadership development institutions, are among the most influential actors in the world, as they shape the skills and mindset of future leaders" (PRME 2023).[5]

As a signatory of the PRME the Liverpool Business School (LBS) at Liverpool John Moores University (LJMU), UK, is working in line with PRME's mission "...to transform management education and develop the responsible decision-makers of tomorrow to advance sustainable development" (PRME 2023). To achieve this transformation LBS encourages innovative initiatives to stimulate debate, connections and ideas that can be incubated, and new connections and actions created. One such initiative is the LBS World Café series. Devised to kick-start impactful conversations and actions the first of the series was scheduled to coincide with the LBS launch of their United Nations Principles of Responsible Management Education commitment and the new LBS Impact Incubator in 2022.

[3] https://theworldcafe.com/key-concepts-resources/design-principles/.

[4] https://sdgs.un.org/2030agenda.

[5] https://wawww.unprme.org/about.

2.1 Anchor Institutions

Based on the notion that universities are anchor institutions, defined as "enduring organisations that are rooted in their localities" (AITF 2023),[6] that have significant physical infrastructure in that place; LBS can use this sticky capital, that is fixed investments in place such as dedicated physical spaces (Ehlenz 2018), to host innovative events, outside of the mainstream curriculum, that can kick-start actions that can lead to change (Fulbright-Anderson et al. 2001). Drawing on the university's ability to bridge and connect a wide variety of stakeholder groups in innovative ways, participants at such events can be varied to suit the intended outcomes. Attendees within the university and beyond can be brought together to stimulate debate and action in safe, accessible and supportive environments and spaces to discuss important topics such as sustainability.

By envisioning themselves as anchor institutions business schools can remain agile and meet the needs of multiple stakeholders. In a rapidly changing world in which mobility is increasing globally, and with the rapid rise in digital and virtual ecologies technologies, universities need to ensure that they remain connected internally and responsive so that they meet the changing needs of both local and global environments (Kendal and Raymond 2019). The LBS World Café series has been created to seed and develop as an evolving series of events that can embody principles based on sustainability issues.

2.2 Co-created Learning

With the stated mission of "...inspiring conversations around promising and impactful ideas, practices, or provocations, to spark our own learning and innovation" (LJMU 2022)[7] the LBS World Café was devised as a co-created learning space and set of activities that allows all participants to get actively involved in a supportive environment. Based on a community of practice approach (Lave and Wenger 1991; Wenger 1998) knowledge was viewed as being co-created through the interaction and construction of multiple perspectives and/or artefacts (Paavola et al. 2004). Using Leonard and Swap's (1999) five-stage process for sparking creativity (preparation; innovation opportunity; divergence; incubation and convergence) the series was devised to welcome and enable LBS colleagues, wider LJMU colleagues and a range of external stakeholders interested in sustainability to attend, debate and collaborate on ideas that were of mutual interest. At the first LBS World Café participants were encouraged to submit ideas for debate and action from their own perspective and in the format, they felt, was most appropriate. No limitations on content or format were made however engagement through the incubator workshops identified key areas to be explored and impact frameworks were developed accordingly.

---

[6] https://www.margainc.com/initiatives/aitf.

[7] https://www.ljmu.ac.uk/about-us/events/2022/6/14/world-cafe-2022.

By creating environments and opportunities in which learning is viewed as a social process in which all participants have the opportunity to spark ideas creatively, the LBS World Café series aims to create new opportunities for dialogue, inquiry and debate in as supportive an environment as possible (Mahat and Dollinger 2019). Providing not only the space but sufficient time in which participants can focus exclusively on the 'ideas into action' process, the traditional academic timetable and schedule is disrupted to allow new opportunities and actions to emerge (Leonard and Swap 1999). In these intentional environments learning can be co-created and incubated to use a divergent to convergent approach to stimulate new action by the end of the event. In creating a series of such events participants can get used to processes and learning remains dynamic and develops in line with a wide range of stakeholder needs, some of which are complex and rapidly changing (Prahalad and Ramaswamy 2004).

Defining co-creation as "a participative process when people and organisations together generate and develop meaning" (Ind and Coates 2013 p. 86) co-creation can be stimulated for a variety of purposes. As Sheth and Uslay (2007) identify, co-creation occurs on a spectrum and includes co-conception, co-design, co-production and co-maintenance. The LBS World Café series has been designed not only to stimulate co-learning but also to encourage other forms of co-creation that could enable future activities and actions. Using co-creation which spans the design process (Sanders and Stappers 2008), co-design was built into the LBS World Café series so that actions are committed during the event, based on new ideas and connections the participants had developed during the day. This rapid ideas into action approach allowed the 'fuzzy' front-end ideas to be processed quickly into less ambiguous actions and sees "co-design as a form of co-creation [that] comes from sustained effort in design thinking to create 'user-centric' approaches" (Mahat and Dollinger 2019, p. 247).

2.3 *Incubators*

As the 2030 timescale for the SDGs draws closer the United Nations (2019, p. 2) admits that "it is abundantly clear that a much deeper, faster and more ambitious response is needed to unleash the social and economic transformation needed to achieve our 2030 goals." Innovative approaches and practices will be needed for these responses to be enacted. Universities and their business schools are in ideal positions to help stimulate and facilitate this. One way to stimulate these new approaches and practices is through incubation spaces, events, and programmes. In business, incubators are not a new phenomenon and were developed in 1959 in New York, USA. In today's business environment incubators are common practice with Hansberg and Korreck (2020, p. 152) identifying that "incubators have become an integral part of the modern entrepreneurial ecosystem, supporting the growth of new ventures based on a broad range of measures". In universities, incubators are also becoming increasingly common practice to stimulate entrepreneurial ideas and actions and bridge the academic and industry divide (Jamil et al. 2015).

By creating events that incubate innovative ideas and actions business schools can facilitate opportunities that create deeper, faster and potentially more ambitious responses to a wide range of complex sustainability challenges. These responses will vary depending on the aims of the incubator and the issues focused on. Outputs can also vary and can have implications for both practice and theory. However, all incubators aim to ignite new ideas by purposively creating flash points in which innovative co-creations are developed. By using a bounded time period participants keep an intense focus on key issues with the ideas progressing to create explosive action.

The LBS World Café series aims to deepen responses to sustainability challenges by providing safe spaces for a wide range of stakeholders to meet, discuss and action ideas that include cross-disciplinary and cross-industry responses. The explicit intention of the LBS World Café series is "… to spark our own learning and innovation" (LJMU 2022)[8] with the focus on 'spark', that is to cause new ideas to be formed within a short timescale, encourage a community of practice that starts to get used to the concept of forming and acting on ideas quickly. As the series develops and the outputs created from the series are highlighted and celebrated more ambitious responses to sustainability will be encouraged based on reflections from participants co-learning and co-design experiences and future needs.

2.4 *The Coffeehouse Concept and World Café*

Using the 'coffeehouse' and 'café culture' concepts which creates sites "…of pleasurable leisure linked to both the birth of consumer culture and the less frivolous public sphere" (Karabab and Ger 2011, p. 737) events can be created that disrupt participants everyday experiences in the hope of sparking ideas and creativity in supportive and pleasant environments. Drawing on the deep and global roots of the coffeehouse that founded the seventeenth-century public sphere (Karabab and Ger 2011) the LBS World Café series' ambition is to recreate a public sphere that focuses on sustainability issues and puts ideas into action. Defined by Habermas (1992) as "…discursive sphere where people from different parts of the society get together and engage in debates about matters of mutual interest, thus forming public opinion and, if possible, reaching a common judgment about a debate" the public sphere provides a wide range of stakeholders with a place to co-learn and co-create with the stated intention of action (Karabab and Ger 2011, p. 746).

Based on the world café approach designed by Brown and Issacs (2005) in the early 1990s that uses "an organisational or social design process [that]…offers a practical way to enhance the human capacity for collaborative thought…using dialogue

[8] https://www.ljmu.ac.uk/about-us/events/2022/6/14/world-cafe-2022.

methodologies" (Schieffer et al. 2004) the LBS World Café series provides collaborative opportunities for participants to engage in innovative co-learning that sparks action. By using dialogue methodologies, the igniting creativity process (Leonard and Swap 1999) can be followed allowing all participants to be actively engaged. The importance and central role of the relationship developer in the facilitation of the café process is highlighted by Gulati et al. (2008). Defined by Brown and Isaacs (2005) as the convenor or 'barista' in industry one of the main aims of this role is to ensure that the participants are comfortable, and included and that the conversation keeps flowing. Drawing on their work with Starbucks they note the importance and value of relational capital by having "...an appreciation of the role the baristas, or coffee brewers, played in creating a comfortable, stable, and entertaining environment for the company's customers" (Gulati et al. 2008). Using baristas to facilitate the LBS World Café sessions and incubate conversations and actions between sessions emphasised the importance of keeping participants motivated and content.

3 Methodology

Qualitative research was used to explore the experiences and outputs from the first event of the LBS World Café series (Eriksson and Kovalainen 2008). Using an appreciative inquiry approach (Cooperrider and Srivastva 1987, 2017) to help boost innovation that leads to action the current best practices, strategic planning, organisational culture and initiatives in LBS were analysed and an opportunity to develop a series of supportive innovative events that focused on sustainability identified. These events were named the LBS World Café series and created to use "inquiry to create positive, optimistic visions of the future to inspire and awaken 'the best in people' (Cooperrider and Whitney 1999; Kessler 2013)" (Moore 2019).

The first event was held in June 2022 between 10 am and 3 pm. The date and time were chosen to allow as many colleagues as possible to participate. Colleagues were asked to contribute to 'best time' and 'best day' decisions as part of the scoping of current practice feedback exercise. Colleagues were encouraged to attend the full-day event and asked to bring all their ideas to the café. The event was held in the executive suite rooms of LBS and the spaces in between, which are based on the fifth floor of the building. The space was bright and airy and offered panoramic views of the city centre of Liverpool. These spaces were chosen to reflect the open and outward-looking concept behind the LBS World Café series and in all the spaces participants were provided with numerous ways to record inputs and actions, for example, flip chart paper, blank and with provocations, to encourage contributions. The opening and closing sessions were held in the same room, Café 521 (denoting room number 21 on the fifth floor). Three morning and afternoon sessions were concurrently in rooms next to each other. Between sessions participants were encouraged to visit the rooms for the sessions that they had not attended, to review the activities and outputs created from that session and contribute to the discussion by using the recording methods used in that session. Participants were also encouraged to review the provocations in the flip charts in the in-between spaces and add any contributions they saw fit.

At the beginning of the event, the event's organisers set the context for the day. In the spirit of collaborative inquiry during the event, the organisers and baristas spoke to as many participants as possible to elicit their views on the sessions and encouraged them to contribute in whatever ways the participant felt comfortable. As the World Café (2023a)[9] highlights the practice of collaborative inquiry "emphasizes that dialogue contributes to collective thought and learning by encouraging the group to attend collectively, to learn and watch for and experience its own tacit (previously undiscussed) process in action. Once noted and discussed, new ways of thinking can occur." At the end of the event, all activities and actions noted during the event were consolidated. These actions then formed part of future meetings that were already in place in LBS and outputs were recorded.

Leonard and Swap's (1999) five-stage process for sparking creativity process was used to structure the analysis with each stage of the event. An exploratory approach was used as this was the first in the series of LBS World Café and no prior data had been collected on this series of events. Multiple data sources were used to analyse the event. These sources included individual proposals; feedback from the event's organisers, baristas and participants; photographs taken during the event and the lists of actions summarised at the end of the event and identified during post-LBS World Café incubator events.

4 Results

The creation of intentional events, such as a world café series, provides business schools with the opportunity to stimulate sustainability ideas and actions in a supportive environment. By providing sufficient time for sparks to be created and actions to be formed intentional events provide a wide range of stakeholders both in and outside the university to proactively work for change.

4.1 Data Collection and Analysis

The first event of the LBS World Café series was analysed using Leonard and Swap's (1999) five-stage process to spark creativity. These stages allow for an analysis of each stage of the ideation and action process with the outcomes from the event identified. The stages of the process are preparation, innovation opportunity, divergence, incubation, and convergence.

[9] https://theworldcafe.com/key-concepts-resources/glossary/.

4.2 *Stage 1: Preparation of the LBS World Café and the First Event*

To build on the work already being undertaken by LBS in relation to sustainability and to coincide with the launch of LBS's commitment to PRME and the LBS Impact Incubator the LBS World Café series was launched in 2022, see Fig. 1.

Scheduled outside of the main teaching period to ensure inclusivity and allow as many colleagues and stakeholders as possible to attend the event was held on 14th June 2022. Sufficient time was planned to allow the five-stage sparking of the creativity process to take place (Leonard and Swap 1999) by the end of the first event and the advantages of the process of incubation being used by the LBS World Café series were clearly seen by the participants. The event was promoted widely to LJMU staff and key stakeholders with an easy-to-register link provided.

LBS colleagues and key stakeholders were invited to submit a proposal for the event using a standard template. Key criteria required by each submission and included on the template were take-away; outline of L&T café conversation; evaluation or evidence of impact so far; theory/evidence/referencing the activity/provocation on evaluation so far (if any); planned next steps (if any) and questions for conversation. See Table 2 for a summary of proposals submitted directly relating to sustainability.

To highlight the participatory nature of the event and desire to create an inclusive community of practice through the LBS World Café series the proposal template intentionally included words such as 'conversation', 'sharing', 'provocation' and 'next steps'. Proposals were not limited in scope but by linking the event to the

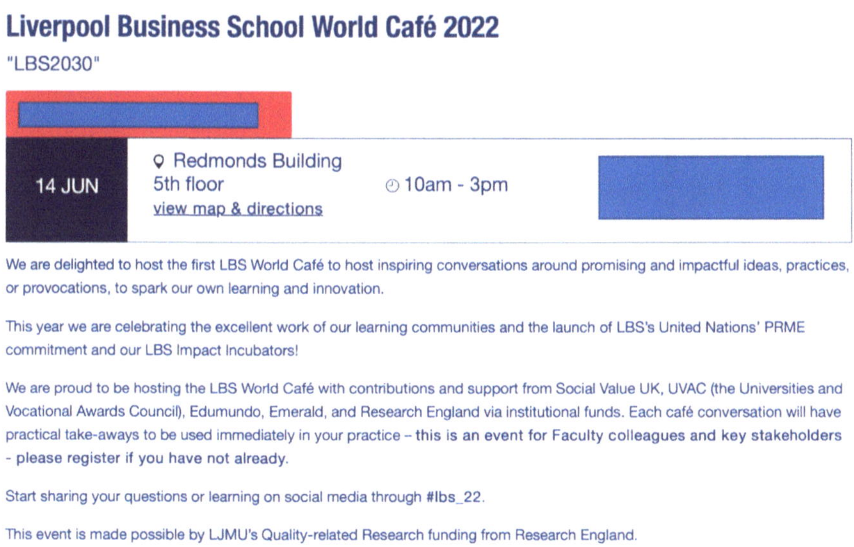

Fig. 1 Launch of the LBS world café series and promotion of the first event

Table 2 Summary of LBS world café proposal submissions specifically relating to sustainability

Proposal	Take away	Outline	Evaluation of impact	Theory/ provocation	Next steps	Questions
Sustainability	Building collaborations	Aim to develop a social system to handle problems to devise a solution	ROMA, LREN, and REAP can reach a much larger audience than full articles GTRSB—LJMU and NRN 'ripple' effect of impact PRME	Harrison et al. (2020); Collins et al. (2019, 2020)	PRME and you	Join PRME Develop or co-develop a short piece, for instance, CABS short blog/ Innovations in Practice/ PRISM/AACSB
	Carbon literacy	Climate, carbon literacy and the curriculum	Improved employability skills and certificates to evidence CPD Influence employers/family/ peers through individual or group actions Train-the-trainer opportunities	Students shared experiences of training on social media Case study presented at PRME 2021 conference Spawned requests for Carbon Literacy training from staff	Pre-empt government plans by proactively introducing Carbon Literacy training for staff and students in 2022/23 and 2023/24	What actions could we recommend SMEs to take on their journey towards Net Zero? How could we measure the impact of Carbon Literacy training? Could we work with other Liverpool institutions to make Liverpool a Carbon Literate city? How are we approaching the Green Jobs agenda?

(continued)

Table 2 (continued)

Proposal	Take away	Outline	Evaluation of impact	Theory/ provocation	Next steps	Questions
Sustainability	Aligning curricula to the UN SDG commitments (using experiential simulations)	Developing Socially conscious Mindsets An experiential approach for students to consider the UN SDGs Aligning curricula to the SDG commitments	These new developments rolled out for Sept/ Oct 22 Edumundo simulations currently in use at 300 universities 80 000 students take part in Edumundo simulations each year	UN Sustainable Development Goals PRME	Roll out the sustainability simulations On-board new lecturers Discussion around future developments/collaborations eg. Social enterprise simulation	Would this new approach to experiential learning help you align your curricula? What else would you like to see for e.g. 23/ 24? Would you be interested in partnering/ collaborating to create something completely new?

launch of LBS becoming a signatory of PRME and the LBS Impact Incubators a clear sustainability thread was established.

4.3 *Stage 2: Maximising Innovation Opportunities*

Once all proposals had been received the organising team clustered the proposals into themes that had the potential to maximise debate and stimulate innovative ideas and actions. For the first event, six themes or clusters were created with equal time allocated to each theme.

After an initial welcome to the LBS World Series, the first event was framed using an introduction and closing session. Both sessions explicitly used reflective practice techniques action (Gibbs 1998) and the incubation process was explained to set the scene for action. In the introduction session, see Fig. 2, five speakers introduced their visions and expectations for 2030 explicitly relating to the United Nations (2015)[10] 2030 Agenda for Sustainable Development. Speakers were chosen to gain a multi-stakeholder perspective of 2030 and an innovative approach to how data and actions from the event would be captured during the day, bibliopoetry, was introduced by the bibliopoetry practitioner Lisa Rossetti.

Three themes were introduced concurrently. In the morning timeslot, the themes of the sessions were personal tutoring, sketching and writing your teaching and learning; and sustainability and PRME were introduced, see Fig. 3. Each session was facilitated by a barista whose role was to encourage dialogue, debate and potential actions between presenters and between presenters and the audience following the igniting creativity and action process (Leonard and Swap 1999). In the personal tutoring and sustainability sessions each of the three conversationalists presented their ideas with divergent ideas encouraged and incubated. At the end of each of the sessions, the barista converged the ideas and identified specific actions that came out of the session. Feedback from each barista was given to the event's organisers at the end of the sessions.

Participants were provided with individual copies of each submitted proposal, see Fig. 4, and a collaborative inquiry approach was used to encourage active participation (Brown and Isaacs 2005).

In the afternoon timeslot, three further themes were introduced, see Fig. 5. These themes were mentoring, engaging students and alumni and simulations and games. In each session, the same process for igniting creativity and action (Leonard and Swap 1999) was used as the morning sessions.

Active communities of practice were encouraged in each session, see Fig. 6, with lists of actions encouraged during and at the end of each session.

The event was closed by reflecting on the activities undertaken during the event in the form of an ethnographic poem. Ideas and actions incubated in and between

---

[10] https://sdgs.un.org/2030agenda.

10-10.15 - Café 521

Welcome refreshments and conversations (tea, coffee, water, fruit, KitKats)

10.15-11.15 - Café 523

"My visions for 2030"
Sheldon Chadwick
LJMU Student and CEO of Showmen's Mental Health Awareness Charity (7 mins)

"A 2030 vision for difference in education: From Marvin Gaye to..."
Adam Vasco
Every Day Heroes Lead LJMU Student and staff (7 mins)

"How do we know we're making a difference?"
Catherine Manning Social Value UK (7 mins)

"HE skills 2030"
Mandy Crawford Lee (CEO, UVAC) (7 mins)

"LBS 2030"
Adam Shore Director of School of Business & Management (7 mins)

"Anansi and pot of wisdom"
Lisa Rossetti, international bibliopoetry practitioner and LJMU graduate

Q&A

Fig. 2 Welcome and introduction session used to frame the event

sessions were identified and plans going forward were identified by the Director of LBS, Adam Shore (Fig. 7).

4.4 Stage 3: The Role of the Baristas in the Divergence Process

To incubate and stimulate the creativity process the role of the barista was created for the event. Based on the barista principle introduced by Starbucks (Gulati et al. 2008) each barista was requested a comfortable, stable, and entertaining environment for the participants. In order to encourage divergent thinking the barista identified key points from each presenter, encouraged debate and dialogue with and between participants, and presenters, and at the end of the session encouraged a list of actions to be drawn up that participants could follow up on after the event, see Fig. 8. This speedy ideas into action format was new to many participants and the use of a barista to stimulate this process in a comfortable and supportive manner was key to the success of the event.

11.30-12.40

Café 518 - Personal Tutoring
Your barista is Lucy

- Gwenda Mynott
- Lucy McGrath
- Andy Doyle

Café 519 - Sketching out & Writing up your T&L
Your baristas are Tony and Laura

- Traffic light technique for sketching ideas
- Rapid writing a practitioner piece ready for publication
- Warming up for a T&L writing retreat 29th/30th June!

Café 520 - Sustainability and PRME
Your barista is Helen

- Building, collaborating and sustaining local partnerships - Helen Collins and Tricia Joliffe
- Carbon literacy training - Ann Hindley
- Sustainability gaming - Leon Lloyd and Milo Hendriks

Fig. 3 Morning session themes

Fig. 4 Active participation taking place during one of the morning sessions. *Source* LJMU used with permission

1.30-2.40

Café 518 - Mentoring
Your barista is Jan

- Jan Brown
- Debbie Callaghan
- Amanda Mason

Café 519 - Engaging students and Alumni
Your baristas are Katie and Andy

- Andy Doyle
- Katie Hyslop

Café 520 - Simulation and games
Your barista is Clay

- Gaming for learning - Leon Lloyd and Milo Hendriks
- Gaming - Peter Barton and Clay Gransden
- Coins - Polly Wilson

Fig. 5 Afternoon session themes

Fig. 6 Creating a community of practice in the afternoon session of Café 520. *Source* LJMU used with permission

2.40-3pm Cafe 521

- Celebration with international bibliopoetry practitioner Lisa Rossetti narrating an ethnographic poem from the day with refreshments (prosecco, tea/coffee, water)
- Closing comments from Adam Shore Director of School of Business & Management

Fig. 7 Closing session

Fig. 8 The baristas sparking debate in the afternoon session of Café 519. *Source* LJMU used with permission

4.5 Stage 4: Incubating and Igniting Sparks During and Between Sessions

At the end of each session, each barista was tasked with briefing the organising team about the details of the sessions and the actions produced. Baristas were also encouraged to keep stimulating the conversations started in the sessions during the breaks and lunch periods. This continued incubation of ideas into action by both the organising team and the baristas meant that the intention to create an active community of practice was continued throughout the event. This ensured that any flash points that had not been sparked during the sessions were ignited and any disruptive ideas generated between sessions developed. Examples of feedback from participants included comments about the energy created during the session with participant A commenting that the event was “incredibly exciting”. The value of

Table 3 Examples of actions and outputs resulting from the first world café event

Actions and outputs resulting from the world café	
1	New Digital Transformation Consortium involving the Open University, Jisc, ALT, the Association of Vietnamese Universities and Colleges
2	A new project to explore staff and student belonging
3	A new project to evidence the impacts of business clinics and clinical practice
4	A new national project to collaborate on examining the societal impacts of MBA projects in collaboration with the UK's Quality Assurance Agency for Higher Education, and a dozen universities
5	Led to global symposium outlined and discussed elsewhere in this book
6	Clearer promotion pathways highlighting the role and importance of impact in and outside the university (including associate and full professorship opportunities)
7	European Symposium conference
8	A series of collaborative blogs published in the Chartered Association of Business Schools blog section
9	Increase in sign-up for carbon literacy training

the event was commented on by Participant B who states that it "makes a massive difference to the work we do here". The event sparked discussions that gave "a clear direction to our research and impact work" (Participant C).

4.6 *Stage 5: Innovative Ways of Converging Ideas into Action*

To develop a supportive community of practice the ethnographic poem developed during the event was used as an innovative way of reflecting on progress made during the event. A list of sparks created during and between sessions was highlighted and commitments to action developed during the event were noted. These actions included ideas for projects, research and publications; signing up for carbon literacy training; sharing tutoring mentoring experience and ideas for development and setting up appointments to understand the use of simulations and games further, see Table 3 for examples of actions and outputs resulting from the World Café. These examples were created as a direct result of the igniting creativity process (Leonard and Swap 1999) that was used during the event.

5 Conclusion

The creative use of events such as the World Café can allow business schools to effectively incubate innovative ideas and actions regularly. Basing such events on sustainability issues business schools can lead the change as envisaged by PRME (2021).

Framing these events around a meta issue such as sustainability other issues, in this case, issues of personal tutoring, student and alumni engagement and mentoring, can all be contextualised with sustainability being at the heart of all actions produced.

By creating safe spaces in which authentic action-driven conversations are incubated over time communities of practice can be formed that have the potential of driving the sustainability agenda. Using civic dialogue to encourage a diversity of opinions from a wide range of stakeholders allows conversational leadership to emerge as seen by the outputs created from the LBS World Café event analysed in this paper. Based on a structured process that sparks creativity and develops ideas into actions in one specific event actions can be created.

Creating a series of events allows ideas and actions can be developed over time. As cumulative intelligence and wisdom are built and participants become familiar with the World Café approach provocations can be developed that explore sustainability in more complex and detailed ways. Longitudinally, actions from the events can be analysed and the values created measured. Actions can then be adapted and developed to suit the community of practice's needs and the inclusion of a wide range of stakeholder groups allows for the convening abilities of business schools to be communicated to the wider civic realm. The utilisation of this convening power in turn allows business schools to remain agile and adaptable to an increasingly complex but connected business environment (Kendal and Raymond 2019).

References

Anchor Institutions Task Force (2023) What is the Anchor Institution Task Force? Marga Inc. https://www.margainc.com/initiatives/aitf. Accessed 16 June 2023

Brown J, Issacs D (2005) The world café: shaping our futures through conversations that matter. Berrett-Koehler Publishers, California

Burke C, Sheldon K (2013) Encouraging workplace innovation using the 'world café' model. Nurs Manage 7(7):9–14

Cooperrider DL, Srivastva S (1987) Appreciative inquiry in organizational life. In: Pasmore W, Woodman R (eds) Research in organization change and development. JAI Press, Greenwich, CT, vol 1, pp 129–169

Cooperrider DL, Srivastva S (2017) The gift of new eyes: personal reflections after 30 years of appreciative inquiry in organisational life. In: Pasmore W, Woodman R (eds)

Cooperrider DL, Whitney D (1999) Appreciative inquiry: a positive revolution in change. In: Holman P, Devane T (eds) The change handbook: group methods for shaping the future. San Fransisco, USA: Berett-Koehler Publishers Inc

Collins H, Harrison PA (2019) Supporting Liverpool's Roma community: an illumination via Bourdieu's theory of capital. Innovations Pract13:10–15

Collins H, Harrison PA, Palace M (2020) Out of the shadows: a young woman's journey from hiding to celebrating her identity. Qual Rep 25:4310–4325

Ehlenz MM (2018) Defining anchor institution strategies: comparing theory to practice. Plan Theory Pract 19(1):74–92

Eriksson P, Kovalainen A (2008) Qualitative methods in business research. Sage Publications, London

Farnsworth V, Kleanthous I, Etienne Wenger-Trayner E (2016) Communities of practice as a social theory of learning: a conversation with Etienne Wenger. Br J Educ Stud 64(2):139–160

Fulbright-Anderson K, Auspos P, Anderson A (2001) Community involvement in partnerships with educational institutions, medical centers, and utility companies. Aspen Institute, USA
Gibbs G (1998) Learning by doing: a guide to teaching and learning methods. Oxford Polytechnic Press, Oxford
Gulati R, Huffman S, Neilson G (2008) The barista principle – Starbucks and the rise of relational capital. PWC Publications, Strategy & Business
Habermas J (1992) The structural transformation of the public sphere. Polity Press, Cambridge
Hansberg JP, Korreck S (2020) Business incubators and accelerators: a co-citation analysis-based, systematic literature review. J Technol Transfer 45:151–176
Harrison PA, Collins H, Bahor A (2020) We don't have the same opportunities as others: shining Bourdieu's lens on UK Roma migrants' Precarious (Workers') habitus Work, Employ Soc 36:217–234
Ind N, Coates N (2013) The meanings of co-creation. Eur Bus Rev 25(1):86–95
Jamil F, Ismail KA, Mahmood N (2015) University incubators: a gateway to an entrepreneurial society. J Econ Sustain Dev 6(6):153–161
Kendal D, Raymond CM (2019) Understanding pathways to shifting people's values over time in the context of social-ecological systems. Sustain Sci 14:1333–1342
Karabab E, Ger G (2011) Early modern Ottoman coffeehouse culture and the formation of the consumer subject. J Consumer Res 37:737–760
Kessler EH (ed) (2013) Encyclopedia of management theory. USA: Sage Publications
Lave J, Wenger E (1991) Situated learning: legitimate peripheral participation. Cambridge University Press, Cambridge
Leonard DA, Swap WC (1999) When sparks fly: igniting creativity in groups. Harvard Business School Press, Boston
Liverpool John Moores (2022) Liverpool business school world café 2022. https://www.ljmu.ac.uk/about-us/events/2022/6/14/world-cafe-2022. Accessed 12 June 2023
Mahat M, Dollinger M (2019) Mind the gap: co-created learning spaces in higher education. In: Fisher K (ed) The translational design of university learning environments: an evidence-based approach. Sense Publishers, Rotterdam, pp 221–238
Moore C (2019) What is appreciative inquiry? (definitions, examples, & model). https://positivepsychology.com/appreciative-inquiry/#model. Accessed 12 June 2023
Paavola S, Lipponen L, Hallarainen K (2004) Models of innovative knowledge communities and three metaphors of learning. Rev Educ Res 74(4):557–576
Prahalad CK, Ramaswamy V (2004) The future of competition: co-creating unique value with customers. Harvard Business Press, Boston
Principles of Responsible Management (2021) Annual Report 2020/21. United Nations Global Compact
Principles of Responsible Management (2023) What is PRME? https://www.unprme.org/about. Accessed 16 June 2023
Research in organization change and development. JAI Press, Greenwich, CT, vol 25, pp 81–142
Sanders EBN, Stappers PJ (2008) Co-creation and the new landscapes of design. Co-design 4(1):5–18
Schieffer A, Issacs D, Gyllenpalm B (2004) The world café. World Business Academy—Transformation, 18(8):18–25
Sheth J, Uslay C (2007) Implications of the revised definition of marketing: from exchange to value creation. J Public Policy Mark 26(2):302–307
United Nations (2019) The sustainable development goals report 2019. United Nations, New York
United Nations (2015) Transforming our world: the 2030 Agenda for sustainable development. https://sdgs.un.org/2030agenda Accessed 12 June 2023

Wenger E (1998) Communities of practice: learning, meaning and identity. Cambridge University Press, Cambridge
World Café (2023a) Glossary. https://theworldcafe.com/key-concepts-resources/glossary/. Accessed 12 June 2023
World Café (2023b) Design principles. https://theworldcafe.com/key-concepts-resources/design-principles/. Accessed 16 June 2023

MIX
Papier aus verantwortungsvollen Quellen
Paper from responsible sources
FSC® C105338

If you have any concerns about our products,
you can contact us on
ProductSafety@springernature.com

In case Publisher is established outside the EU,
the EU authorized representative is:
Springer Nature Customer Service Center GmbH
Europaplatz 3, 69115 Heidelberg, Germany

Printed by Libri Plureos GmbH
in Hamburg, Germany